Analysis *for* Public Decisions

Analysis *for* Public Decisions

E.S. Quade

ELSEVIER · NEW YORK
NEW YORK · OXFORD

Elsevier North Holland, Inc.
52 Vanderbilt Avenue, New York, N.Y. 10017

Sole Distributors outside the United States and Canada:
Elsevier Science Publishers B.V.
P.O. Box 211, 1000 AE Amsterdam, The Netherlands

Sixth Printing, 1981

Library of Congress Cataloging in Publication Data

Quade, Edward S.
Analysis for public decisions

Includes bibliographies.
1. Policy sciences. I. Title
H61.Q16 309.2'12 74-19545
ISBN 0-444-00153-0

Manufactured in the United States of America

CONTENTS

PREFACE

Decision-makers in the public service, and in the private sector as well, are seeking better ways to uncover and select goals that are in the public interest, better ways to design and choose alternatives to achieve those goals, and better ways to see that the alternatives selected are implemented properly.

The need is acute. We see signs everywhere of ineffective programs, wasted money, and unsolved problems. There are complaints that the solutions being proposed range from poor, at best, to counterproductive, at worst. One possibly minor contributor to this state of affairs is the fact that we are not making full use of our best methods to achieve understanding, to discover our goals, or to solve our problems.

Hope may lie in the greater use of analytic methods for public policy decision-making, which is the thesis of this book. Although it is true that such methods are widely used, they are often used inappropriately, frequently to advocate a predetermined position rather than to discover what the position should be, and even more frequently with excessive emphasis on quantification. The analysis of public policy is, moreover, largely a new field, and practitioners of new disciplines tend to regard them as panaceas. Therefore, too often, such approaches as systems analysis and program budgeting have been oversold along with computers and purely quantitative techniques.

As a survey of the nature, aims, limitations, and the help one can expect from policy analysis and related aids to decision-making, this book attempts to suggest alternatives to traditional methods of decision with emphasis on the public policy area. The book is written largely for the inexperienced among those who would use analytic help with public decisions, but to some extent it is also intended for those who would provide it.

If one were to set out to train policy analysts, one would see that they were taught microeconomics, decision theory, organization theory, linear programming, probability and statistics, and so forth. A thorough working knowledge of these disciplines may extend the capabilities of the practitioner and the policy-maker greatly. Nevertheless, I have not included these topics. There are many fine textbooks that do, and the topics themselves have their most fruitful application elsewhere than in public policy problems.

This book does not attempt to lay down cookbook rules for analysts or their clients which, if followed, would result in the "right" policy analyses. In concept, it might be possible to provide a set of clear-cut decision rules for likely contingencies in at least some specific problem areas, and do it with a level of detail such that analysts could carry out relevant and competent analyses, simply by following those rules. In reality, however, we find this completely impossible and believe it will never become possible. No public policy question can be answered by analysis alone, divorced from political considerations and judgment and intuition play a large role.

Working through this book will not make anyone an accomplished policy analyst. One would need to master a great deal of additional material. He should, however, obtain an appreciation of the limitations and capabilities of analysis applied to public decisions and some insights into aspects not found in the usual text.

The distinctive features here are coverage of a broad spectrum of analytic aids to decision-making; an attempt to organize materials and allocate attention to them according to the needs of the users rather than according to the interests of analysts, or according to some abstract logic; a demonstration that analysis is not limited to questions that can be quantified; and an attempt to provide information—both practical and theoretical—to questions that are important to those who need analytical help and to those who are responsible for sponsoring, evaluating, and implementing the analysis of others.

The Rand Corporation sponsored this book as part of its research program on the methodology of analysis, allowing me to rearrange my duties in order to do the necessary study and writing.

Pacific Palisades, California E. S. Quade

ACKNOWLEDGMENTS

My work reflects the efforts of many colleagues, so many I cannot name them all. These are people who worked on actual policy analyses, developing concepts and techniques and writing about their experience. With very few exceptions, the ideas presented here originated with my Rand colleagues and often I do not know the exact source. In addition, I am indebted to a number of colleagues and excolleagues who have read sections of the manuscript and whose comments led me to make many improvements: Kathleen Archibald, Gene Fisher, R. D. Specht, and Frank Trinkl. I also made extensive use of the Rand writings of Norman Dalkey, Olaf Helmer, Charles Hitch, James Schlesinger, Arthur Swersey, and Peter Szanton.

Chapter 1

ANALYSIS AND PUBLIC DECISIONS

The world today, and most of its subdivisions, face a great many complex problems—some are very new, others we have made and are still making constant efforts to solve. In the United States solutions have often been announced, but when they have been tried out, they are disappointing at best, and frequently seem to leave the situation worse off than if nothing had been done.

There may be good reasons for our failures. For one thing, the capability of changing the state of society by a few deliberate decisions, except in minor ways, simply may not exist. For another, the causes of many of society's most urgent problems are inadequately understood and, until they are, the solutions proposed may well turn out to be not only inadequate but possibly in the wrong direction. Some problems are insolvable; a few cannot be solved for lack of a way of deciding what constitutes a solution. At other times, but seldom, the trouble might also lie with the intentions of public officials and others who command the required resources; personal gain or prejudice may direct their actions. And even in some instances when this appears to be the case, the basis again may be lack of understanding; the decision-makers may not really know what their goals should be. The difficulty here does not seem to lie so much in discovering what the interests of society and its people are, but what, in fact, is to their interest. There are still other possibilities, however. For one, we may not be making use of our best methods to achieve understanding, to discover our goals, or to solve our problems.

The Unsatisfactory State of Public Policy-making

Dissatisfaction with the results of the decision-making processes in use by government is apparent. Complaints about ineffective programs and wasted money are increasing. Growth in industry and population has not turned out to be the unqualified blessing so many governments sought. These governments are now subject to pressures from sources that are different from those of the past. One pressure is simply the growing load on the environment. Another pressure is the withdrawal of tacit acceptance of authority, which is the source of its legitimacy and most of its power. Today rumors of a policy change affecting a particular neighborhood or a minority group will normally find those affected alert, articulate, and determined to participate in the making—or blocking—of any decision that concerns them. And the concerns are not only those of the minority; the female majority is beginning to demand fundamental changes.

Education is typical of the areas in which our efforts at improvement have had little success. In the United States, for instance, voters are increasingly rejecting the proposals of elected officials and bureaucrats. Gorham (1971) cites school bond issues as an example. He notes that from 1963 to 1970 the percentage of school bond issues approved by the voters fell from 72 percent to 53 percent. Moreover, this practice grew steadily worse in spite of the efforts of school officials who became more cautious and put far fewer issues on the ballot in later years—only 1216 in 1970 as compared with 2048 in 1963.

However, difficulties are not unique to education. The United States is also faced with the critical issues of poverty and unemployment. The federal minimum wage law, established under the Fair Labor Standards Act of 1938, was designed to help with this problem—to eliminate low wages without eliminating jobs. But how has it worked in practice?

Even advocates of the law admit that increases in the minimum wage have had adverse effects upon employment. Those who retain their jobs and receive higher pay are better off, but their gains turn out to be at the expense of others who lose or cannot get jobs, particularly, blacks, females, teenagers, unskilled workers, and people living in economically depressed areas.[1] This was certainly not the intent of Congress.

Even technological problems present difficulties:

> . . . As a familiar example in the environmental field we can consider the problem of controlling Southern California's photochemical smog. The contribution of automobiles to this problem was recognized more than twenty years ago, when my colleague Haagen-Smith showed how unburned hydrocarbons and oxides of nitrogen were both required for the photochemical reactions which produce smog. It was decided very early to try and control hydrocarbons as being easier. It was also decided at the same time by the other authorities that health problems might be associated with the existing emission levels of carbon monoxide, which has no known connection with smog. Regulations resulted curbing both carbon monoxide and unburned hydrocarbons. The natural response of the automobile industry was to raise the flame temperatures in engines to maintain the so-called high performance of cars, thereby increasing the emission of the oxides of nitrogen and largely cancelling the benefit of the regulation of hydrocarbons. If we include the increase in car population, we find that smog has not been reduced at all. (Gell-Mann, 1971)

In addition to providing everyone with a decent wage and eliminating smog, there are other problems we have declared our intent to solve, have often stated we had solutions for, and have sometimes enacted as laws. Our efforts, however, have not been outstandingly successful; often there has been no noticeable improvement and sometimes, as was implied earlier, we seem to have made the situation deteriorate more rapidly than it would have without intervention. Moreover, the beneficiaries of a program are not always those intended; for

---

[1] For further discussion, see Kau and Kau (1973) and Miller and Williams (1972).

example, legislation to help the family farmer has proved to be a bonanza for the large corporate farms.

One government official (Carlson, 1969) summed up the situation this way:

> An increasing number of critics are . . . beginning to say that the federal government has developed two defects that are central to its existence: (a) it does not know how to tell whether many of the things it does are worth doing at all; and (b) whenever it does decide that something is worth doing, it does not know how to create and carry out a program capable of achieving the results it seeks.

The situation is not unique to the United States (Ritchie, 1971):

> Like many other advanced nations Canada finds itself in this age of scientific sophistication and technological wonders still making critical decisions on the basis of good will, intuition, and hope. Effective democracy implies the opportunity to make informed choices. It is rarely that the electorate has such a happy privilege, nor do policy makers themselves find the opportunity much more frequent. The common pattern is for information to be inadequate, alternatives to be unidentified or unappraised. More often than not, we appear to drift into solutions, or expedients. forced by circumstances to take action even before having obtained any clear picture of the complexities of the problem. A pessimist might say that we react to crisis by sowing the seeds of new crises as yet unforeseen.

A number of reasons cited for this failure of intent to match outcome follow.

Some Possible Reasons

A public policy decision is a decision made by society for itself (the election of a President, for instance) or for society by its elected representatives—decisions taken by individuals or groups that have material effects on individuals other than those involved in making the decision.

In the past, when events moved more slowly, the corrective effects of experience played a very much larger role than they do today. Through trial and error and political give and take, it was possible for public officials to develop policies that took into account the objectives, estimates, and values of everybody in society, or at least everybody who had influence. This is no longer the case; technology and events move so rapidly that natural trial and error—give-and-take processes can become too catastrophe-prone for comfort before the process approaches completion—not only war, but population pressure, resource shortages, and environmental deterioration are in this category.

One difficulty is that the organizations and bureaucracies the public decision-maker must work with—his own and those that interface with it—are often beset with red tape, poor communication, low morale, inadequate staff, incomplete records, and pressures from special interest groups with ready-made solutions. Government organizations do not have the same reputation for attracting or for making efficient use of individual skills and competence that is usually attributed to industry. Also, the legislative process itself is not a model of efficiency.

One possible reason for poor results is that despite the many new and more complex problems now facing government, the methods used by public authorities for seeking their solutions have hardly changed since the beginning of government. Or, if that statement seems too strong, at least the rate at which improved methods are being adopted is far less than the rate at which public policy issues are increasing in complexity.

An important reason is stated by Schultze (1970):

> . . . government programs rarely have an automatic regulator that tells us when an activity has ceased to be productive or could be made more efficient, or should be displaced by another activity. In private business, society relies upon profits and competition to furnish the needed incentives and discipline and to provide a feedback on the quality of decisions. The system is imperfect, but basically sound in the private sector—it is virtually nonexistent in the government sector. In government, we must find another tool for making the choices which resource scarcity forces upon us.

Today, with worldwide industrial growth bringing an end to natural resources as well, the problem is compounded. There is no automatic mechanism to constantly monitor what goes on in government as the market does for business in areas where competition is relatively free; as Schultze implies, therefore, we must seek substitutes. These are, at best, partial. Elections are one alternative, but the damage is usually done by the time the incompetents and the rascals are thrown out; there are also press criticism, scholarly critics, and an occasional public scandal to be contended with. Analysis is another means, and it can serve for other problems as well.

The appropriate analysis I call "policy analysis." Although decision-makers have sought help from analysis from time immemorial (with mixed results), it is my belief based largely on faith that a much wider and more intelligent use of policy analysis as it can be achieved today can make a great contribution to better public decisions.

Policy Analysis

In a broad sense policy analysis may be defined as any type of analysis that generates and presents information in such a way as to improve the basis for policy-makers to exercise their judgment. Operations research, systems, cost-benefit, and cost-effectiveness analyses are all of this type, although they all tend to slight certain aspects such as the political and organizational problems associated with decision-making and its implementation.

In policy analysis, the word analysis is used in its most general sense; it implies the use of intuition and judgment and encompasses not only the examination of policy by decomposition into its components but also the design and synthesis of new alternatives. The activities involved may range from research to illuminate or provide insight into an anticipated issue or problem to evaluation of a completed program. Some policy analyses are informal, involving nothing more than hard and careful thinking whereas others require extensive data gathering and elaborate calculation employing sophisticated mathematical processes.

Although the term policy analysis covers an enormous variety of studies, analyses, and research, we talk mainly about the more formal (of the sort contracted for by a government agency). An example would be a study by a private company but supported by the National Institute of Law Enforcement and Criminal Justice, which involves several man-years of effort and designed to aid in developing policy and statutory guidelines for the operation of private police (Kakalik and Wildhorn, 1971). The product of such a study is likely to be a set of conclusions (or sometimes recommendations), possibly presented to the sponsor in an oral briefing but always supported by a written report. Almost all we have to say about policy analysis on this scale applies equally to the more informal analysis or to analysis done by a single expert or consultant.

Policy analysis seeks to improve decision-making in a particular situation. The improvement of the public policy-making process, e.g., by education or legislative reform, is beyond our scope, although it might be the subject of a policy analysis.

Because the term "policy analysis" does not imply anything about quality, rigor, or comprehensiveness, but merely the purpose and the context of the work, it may seem desirable to adopt a new name for the type of analysis that incorporates the improvements we call for. But we have too many qualifying terms already and another would simply add confusion. Hence, we continue to use the term policy analysis.[2] The same term is applied both to the process (or activity) and to the product of that activity. Thus policy analysis produces policy analyses. Generally we are talking about the process, not the product. It is important, however, to distinguish the process from the product.

We should also point out that in this book, we are not talking about the full range of activity that can be put under the policy analysis umbrella. We are discussing analysis for public policy decisions. We say little or nothing about how to do research on social problems, or how to collect and analyze data, or how to design social indicators, all of which can be considered as aspects of policy analysis.

Current State of Analysis in the Public Sector

The belief that policy analysis or related approaches such as systems analysis, or cost-benefit analysis, or operations research have the capability of being very helpful with public policy-making is by no means new. It has been fairly widespread for ten years or so. In fact, a few years ago, far more often than at

[2] As used here, it is also completely different from the same term as used in the behavioral study of policy-making, where the term refers to analysis of the contents and genesis of actual policies. See, for instance, Froman, 1968.

present, we heard that to meet the challenges of our contemporary society—from the international balance of trade to the inadequacies of the United States postal system—we need only turn to the systems analysis techniques developed for military research and development.

For instance, Senator Hubert H. Humphrey, said in 1964:

> We have vast needs in education, in transportation, in communications, in weather control, air and water pollution, in medical facilities and technology, housing and many, many more areas. The national scope of many of these problems, the degree to which they cut across many political jurisdictions, the complexity of disciplines which must be coordinated to solve them—all this precludes ordinary private development.
>
> These problems are admirably suited to the same kind of 'systems-analysis' approach that have paid off so well in defense. (Press release, 1964)

Again, four years later (Humphrey, 1968):

> . . . the techniques that are going to put a man on the Moon are going to be exactly the techniques that we are going to need to clean up our cities: the management techniques that are involved, the coordination of government and business, of scientist and engineer . . . the systems analysis that we have used in our space and aeronautics program—that is the approach that the modern city of America is going to need if it's going to become a livable social institution. So maybe we're pioneering in space only to save ourselves on Earth. As a matter of fact, maybe the nation that puts a man on the Moon is the nation that will put man on his feet first right here on Earth. I think so.

In spite of these optimistic words, however, it is now clear that systems analysis, as it has been practiced, say, in the aerospace industry or for the U.S. Department of Defense, has not been, nor is it likely to be, a spectacular success with educational reform or welfare or even a city police department. For that matter, there have been few, if any, spectacular successes in defense. This is not to say that aerospace technology, operations research, and systems analysis cannot help a great deal with specific, well-defined subproblems and subsystems of our federal, state, and local governments. There is ample evidence that they can but the larger, fuzzy, and ill-defined problem situations that become public issues are another matter. Even the simpler government activities, e.g., something like the postal service, can give us trouble. The difficulties seldom lie with analysis in the strict sense, however. We may be able to design an efficient system on paper, but as yet we have no algorithms for finding ways to overcome the resistance offered by tradition, legal restrictions, and a host of privileged interests that inhibit radical or even morphological change.

The contrast between early expectations and outcome is nowhere more apparent than in one of the four Aerospace Studies, as they have come to be known, initiated in late 1964 by the State of California (Bickner, 1972):

> The original hopes related to the last of the four efforts, the Integrated Transportation Study, were expressed by Governor Brown in these words: "We will ask the systems engineers to study ways to provide a complete transportation network within the State, efficiently coupled into land, sea, and air transportation from out of state.

We will ask them to identify the major patterns of movement of people, merchandise, materials, and food within the State. We will ask them to describe the transportation system which the State will need 30 to 50 years from now to provide efficient movement. And, finally, we will ask them to tell us how much such a transportation system will cost; who should pay for it; who should run it."

The proposal from the winning contractor, North American Aviation, did not express any such intentions; it merely promised the design of a study[3] to outline the content and specifications for a systematic approach toward resolving the basic transportation problem. Even that generated considerable criticism as did the other studies in that set.[4]

From a superficial point of view, there were, and still are, questions about how successful analytic techniques were in defense. There are people who, with little justification, attribute cost overruns and the disasters of the Vietnam war if not to poor analysis, at least to the way analysis has been used. Hoos (1972, p. 59) remarks: "The successes of systems analysis as a method for achieving peace and security seem to be purely fictitious." But it seems far more likely that the decisions to embark on and continue the Vietnam adventure came not from analysis but from the accepted wisdom, which is to say, from mythology. It is clear that the people who used the techniques of systems analysis in defense found them useful and when they have moved into the domestic area they carried the techniques with them.

In spite of the obvious shortcomings of analytic methods and their use, many public officials have been trying, persistently, to get more analytic help, particularly scientific and technological help. The quest for help with social policy problems has not been as persistent but that help has been sought also. Congress, has, in fact, enacted considerable legislation to encourage the use of analysis in federal and state governments. In introducing the government-wide Planning–Programming–Budgeting System (PPBS) in 1965, President Johnson took a step that would require the use of more analysis. Indeed, it is sometimes argued that all the other aspects of the system existed merely to facilitate and support better analysis. But, in spite of these and other efforts, the desired analysis, like PPB itself, turned out to be hard to get going. Policy analysis of reasonably high quality is still a scarce commodity even in many parts of the federal government. There are numerous reasons why this is the case.

Why Analysis Has Trouble

According to some critics, analysis has been tried and has not helped very much. In a sense this may be true, but in my view, it was not a very proper try.

---

[3] In fairness to the analysis, I should mention that the scale of the California Aerospace Studies ($50,000 each) made possible in each case little more than the writing of a long proposal for subsequent research or development.

[4] See, for example, Hoos (1972). Professor Hoos finds much to criticize about system analysis, often with justification. But for a criticism of her criticism see the review by Pollock (1972).

What was tried was often the wrong sort of analysis, by people who did not really understand its many limitations, in an environment beset with politics and special interests. Here analysis was viewed largely as a means to justify a decision already made or to advocate a particular position in policy debate. Moreover, too much was expected from the analysis (and often still is).

Public policy problems tend to be far more messy and ill-defined than are military and industrial problems, for which systems analysis and operations research–the precursors of policy analysis–got their start. The latter depended to a much larger extent upon engineering and technology. Such problems tend to be easier to formulate and quantitative methods can be of more help in finding a solution. To solve a social problem, one must find a way to induce social change–to persuade many people to behave differently than they have behaved in the past. That is, one may have to convince people to have fewer children, or not to drive after drinking, or to hire minority people in their companies. By contrast, resolution of a technological problem involves decisions by many fewer individuals.

It takes far more than the discovery of a solution on paper to eliminate any major public policy problem. There are additional tasks–communicating the solution to all involved and convincing them that it is a valid solution, laying out the jurisdictional boundaries, and marshalling the necessary resources. Responsibility for financing, implementing and managing the programs that result is divided among federal state, and local governments. People have to be motivated or persuaded to behave in accordance with the analytic solution; doing so is a frustrating business. This can even happen when change is in their own best interest. It is seldom obvious to many people that they should forego immediate personal gain or pleasure in favor of long-term social or even personal gain. Moreover, the discovery of a more efficient resource allocation always implies a reallocation. This means that when analysis is most successful in performing its function, it is almost certain to be criticized.

The problems of public policy are thus likely to be "wicked" problems.[5] That is, they may have no definite formulation and no stopping rule to tell the problem-solver when he has a solution. Moreover, a proposed solution is not true or false but good or bad. There may be neither an immediate nor even an ultimate test of a solution–the set of potential solutions is not enumerable; every such problem is essentially unique and is a symptom of another problem. Even for the comparatively simple problems involved in locating a freeway or changing the school curricula there is not a solution but a resolution–and that must rely on political judgment. For instance, a freeway route may be different depending upon whether the criterion is to minimize construction cost, travel time, scenic vistas destroyed, or number of people displaced, etc. The problem has to be "structured" before the traditional methods can apply and in structuring it may no longer remain the same problem.

Analysis alone is clearly not sufficient to provide a solution to questions such as

[5] In the sense of Webber and Rittel (1973); "squishy" to Strauch (1973).

- How much of the county budget should be allocated to welfare and what portion of that to birth control clinics in the several cities?
- Are metropolitan transportation needs better served by a rapid transit system or by more and higher performance freeways?
- Is there some legislative action that might end the increase in juvenile delinquency?

Often such problems cut across established political and administrative boundaries, presenting jurisdictional issues to overcome, as may occur, for instance, when the indirect costs to the public resulting from a project dwarf the direct costs and benefits.[6] The difficulties of many of these problems lie more in deciding what ought to be done than in how to do it. The allocation of resources for efficiency may be secondary and the important questions are ones of equity; "Who benefits?" and "Who pays?" may require far more attention than the question of which policy generates the greater net of benefits over costs.

Finally, too much has been expected from analysis. Sometimes the claims by analysts have been so excessive that no decision-maker could take them seriously. Also, even when the study has included several pages citing omitted factors, approximations, lack of data, assumptions of linearity, and so on, the user has often tended to overlook such caveats in view of the welter of data, tables, and computations likely to be present. If the solution obtained from the model is reasonably in accord with the client's belief, the analysis may become almost gospel. Later, if enough of the uncertainties turn out differently than they were assumed by the analysts, analysis itself gets a bad name.

Governments, federal, state, and local, were set up years ago to handle a set of problems that differed greatly from many of those faced today. They have adapted, of course, but not as rapidly as the problems have developed. One lag has been in the acquisition of an analytic capability. Lack of such a capability may help to explain the shallowness of so many congressional hearings and why the results of some federal programs, such as those in housing or health care for indigents, came as a surprise to those who supported or even originated the legislation. This is not because the impacts were always so difficult to anticipate but that until recently, even at the federal level, few departments had an analytic capability of any consequence.

At the state and city government level the lag has been much greater. With the exception of a few of the larger states and a smaller number of cities, in most areas these jurisdictions do not have either the research traditions or the skills to tackle very many of their problems through analysis.

---

[6] The direct costs of a statewide waste management system projected for California in 1990, as estimated in the Waste Management Study carried out by Aerojet General, was $\frac{1}{2}$ to $1\frac{1}{4}$ billion dollars annually, depending on the levels of pollution tolerated. The estimated indirect cost to the public resulting from the high level of pollution associated with the lower cost figure was set at $2\frac{1}{2}$ billion dollars annually (Bickner, 1972).

What Can Be Expected From Public Policy Analysis?

Policy analysis is not a perfected discipline. We are finding that to be really helpful with broad public problems, it must be practiced and used a great deal differently today than systems analysis, operations research, and cost-benefit analysis were practiced and used a few years ago. Changes in method and attitude arising from the inability of the more quantitative and conventional methodologies to handle the political and social aspects of public problems are needed. Equally important is the need for analysts to pay a great deal more attention to aspects of the problem that in past applications have usually been considered to lie outside analysis proper.

- Winning the cooperation and assistance of the people, including the bureaucracy, currently affected by and dealing with the problem
- Helping the client formulate his decisions or recommendations so that they are not only accepted by his superiors, but at lower echelons as well, by the target group, and even by collateral interests
- Seeing that solutions are designed so that they can be implemented without being vitiated

Success with these requires inclusion within the analysis of an attempt to understand how the policy being investigated will be constrained by the institutions and individuals affected by the adoption of that policy (and the spillovers that go with it).

Can public policy analysis be expected to find a solution to any of the world's major problems? Perhaps, but it is not to be expected, and even if it did the problem would still exist until a policy-maker recognized the solution and took a series of actions to implement it. Thus far it has been more valuable as a means for investigating problems than for solving them, doing more to reveal the complexities of public problems than how to make the choices easier. But this alone can be of tremendous help.

That the problems of society are difficult has been known for a long time; for example, to Edmund Burke in 1791:

> An ignorant man, who is not fool enough to meddle with his clock, is, however, sufficiently confident to think he can safely take to pieces and put together, at his pleasure, a moral machine of another guise, importance and complexity, composed of far other wheels and springs and balances and counteracting and cooperating powers. Men little think how immorally they act in rashly meddling with what they do not understand.

We have not been and never will be able to make policy analysis a purely rational, coldly objective, scientific aid to decision-making that will neatly lay bare the solution to every problem to which it is applied. There are always considerations that cannot be handled quantitatively, maybe not analytically, or even systematically, and there may be problems with no solution. In the end, politics and intuitive judgment must rule.

Nevertheless policy analysis can be a splendid tool to help in the making of public decisions, but there must be decision-makers who appreciate its limitations and who know what to expect from it. Unfortunately, many managers do not yet appreciate how policy analysis can help them; others do not seek analytic help with their decisions unless there is a crisis that threatens their leadership, for the introduction of an analytic capability can change the existing system that brought them to success.

Considering what can be done to increase understanding, the claims for what one can expect from public policy analysis should be rather modest. It can frequently reduce the complexity of problems to manageable proportions (manageable by judgment that is) by identifying and clarifying those elements about which information exists or can be found. By making information available and laying bare hidden assumptions and value preferences public policy analysis can widen the area of informed judgment. It can counter the purely subjective approach on the part of advocates of a program by forcing them to defend their line of argument and talk about the specifics of the situation rather than merely express their personal opinion with statements of noble purpose, thereby raising the quality of public discussion.

Policy analysis is valuable because it can help a decision-maker by providing information through research and analysis, by isolating and clarifying the issues, by revealing inconsistencies in aims and efforts, by generating new alternatives, and by suggesting ways to translate ideas into feasible and realizable policies. Its major contribution may be to yield insights, particularly with regard to dominance and sensitivity of the parameters. It is no more than an adjunct, although a powerful one, to the judgment, intuition, and experience of decision-makers.

References

Bickner, Robert E., Science at the service of government: California tries to exploit an unnatural resource. *Policy sciences,* 1972, **3**, (2) 183-199.

Carlson, Jack W., Can we do anything right? *Washington monthly,* 1969, **1**, (11) 75.

Dror, Yehezkel, *Public policymaking reexamined,* Chandler Publishing Co., San Francisco, 1968.

Froman, Jr., Lewis A., Public policy. *International encyclopedia of the social science,* 1968, **13**, 204-208.

Gell-Man, Murray, How scientists can really help. *Physics today,* **24** (5) 1971, 23-25.

Gorham, William, Ignorance is blissless for government, speech August 16, 1971, Washington, D.C., appearing in *Improving management for more effective government, 50th anniversary lectures, 1921-1971,* General Accounting Office, 1972.

Hoos, Ida R., *Systems analysis in public policy: A critique.* University of California Press, Berkeley, 1972.

Humphrey, Hubert H. (Vice-President), speech at the Smithsonian Institution, Washington, D.C., May 7, 1968.

Humphrey, Senator Hubert H., Press Release dated April 12, 1964.

Kakalik, J. S., & Wildhorn, S. *Private police in the United States: Findings and recommendations.* R-869-DOJ, The Rand Corporation, Santa Monica, California, 1971.

Kau, James B. & Kau, Mary L. Social policy implications of the minimum wage law, *Policy Science,* 1973, **4**, 21-27.

Miller, R. L. & Williams, R. M., *The economics of national issues.* Canfield Press, San Francisco, 1972.

Pollock, Stephen M., Rational methods. *Science,* 1972, **178**, 739.

Ritchie, Ronald S., *An institute for research on public policy information,* Canada, Ottawa, 1971.

Schultze, Charles L., Director, Bureau of the Budget, Statement in *Planning, Programming, Budgeting, Inquiry of the Subcommittee on National Security and International Operations,* Washington, D.C., 1970, 172-173.

Strauch, Ralph E., *A critical assessment of quantitative methodology as a policy analysis tool,* P-5282, The Rand Corporation, Santa Monica, California, 1974.

Webber, M. M., & H. W. J. Rittel, Dilemmas in a general theory of planning, *Policy sciences* 1973, **4**, (2) 155-169.

Chapter 2

WHAT SORT OF PROBLEM? WHAT SORT OF ANALYSIS?

Introductory Remarks

In the previous chapter we expressed the belief that analysis, properly designed and used, could help public decision-makers make better choices. But for what sort of problem and with what sort of analysis?

Analysis can help with almost every public policy decision, from the most routine to the most profound, e.g., from a choice among ways to improve record-keeping in an employment office to critical choices the outcome of which may determine whether we live or die. The extent of help that can be provided will vary just as widely, depending on such factors as the problem itself, the context, the decision-maker, the time available for analysis, and the state of information.

There are contrasting views as to the role policy analysis can play in public decision-making. For people who believe that rational analysis and systematic planning can be done, and done adequately if not perfectly, policy analysis is seen as an important tool. For others, who see much more merit in the traditional bargaining or "muddling through" approach to policy formulation, the role is far more limited. In this latter view, analysis is seen largely as a device to help a decision-maker by contributing to his bargaining position. We.will say a little, but very little, later in this chapter about analysis in this latter role. Our belief is that the first view holds more promise for the improvement of public policy-making.

The purpose of policy analysis is to help (or sometimes influence) a decision-maker to make a better decision in a particular problem situation than he might otherwise have made without the analysis. This is not the same thing as attempting to provide him with a complete description of what should be done in every conceivable contingency that might ensue. The outcome to be expected from the analysis is almost never a clear recommendation for choice. Except in rare instances, when one alternative turns out to be so clearly superior to all others in almost every aspect, the analyst does not even try to offer more than a suggestion as to what the choice should be—there is too much uncertainty and too many differing views of equity and of values for that. His goal should be much more modest, often no more than to increase the amount and quality of the information the policy-makers have on which to base their decision.

The policy analyst cannot take a traditional academic approach and be effective. That is, his choice of hypotheses and research design must not be determined by his disciplinary interest but by the demands of the client's problem. Moreover, his interest does not necessarily stop with the discovery of what appears to be the best course of action, isolated from the political and organizational factors that may influence its acceptance and implementation.

Problems Where Analysis Can Help

The diversity and range of public problems to which analysis can be helpfully applied is extremely great, ranging from low-level "efficiency" problems—How often should the stock of primary school textbooks be inventoried?—to policy questions at the highest level—How much should the dollar be devalued? Also, evaluations of an ongoing state program to rehabilitate delinquent teenagers and of a Highway Department proposal to install a system of traffic control lights on the state's freeways might both be commissioned with improved operations in view but they would involve far different methods and disciplines. On the other hand, these evaluations might be similar in that they are both requested by a decision-maker who has a budget cut in mind and involve much the same method. This sort of thing makes it almost impossible to find a neat scheme for classification.

In view of the above, we will discuss the type of problem where analysis can help under the following five headings: operational efficiency, resource allocation, program evaluation, planning and budgeting, and "strategic" choice. The headings do not by any means represent non-overlapping categories of problems. Program evaluation, for instance, might be carried out to determine a better resource allocation; that, in turn, might hinge on whether or not a way can be found to make the operation of the program being evaluated more efficient. Moreover, the same analytic tasks, say an experiment or the design of a new program, might be carried out to improve efficiency or to make a high-level strategic choice easier.

IMPROVEMENT IN EFFICIENCY OF OPERATIONS

The work of the New York City–Rand Institute with the Fire Department is an example of analysis in this category.[1]

New York City spends something like $200 million a year on fire protection and the costs incurred in insurance overhead and fire damage are even larger. For instance, by examining past records, researchers have worked out schemes to identify the relative probabilities of structural fires, trash fires, false alarms, and other kind of incidents likely to occur at quite specific times and places. These results were then used to enable the Fire Department to position men and

[1] For a fuller description of this work, see New York City–Rand Institute 1972, pp. 499–506.

equipment more effectively. Another study devised new procedures for making the dispatching of fire-fighting equipment both faster and more reliable. Similar investigations led the Department to introduce tactical control units which saved the city the cost of adding the equivalent of 15 full-time companies–in dollars, something like 9 million per year.

Other problems in this category might involve such tasks as designing garbage collection routes, forecasting the demand for timber products, deciding whether computers to handle inventories should be rented or purchased, or determining how many of each type of approved textbook should be distributed to a given school.

Actually a good many public policy questions–typical examples are those involved in helping the New York City Fire Department provide an improved level of protection within its budget–may require little more on the part of the analyst than careful data collection and the skillful application of the fairly standard and largely mathematical procedures that form the basis of operations research.[2] Such questions are typically asked in an attempt to increase efficiency in a situation where it is clear what "efficiency" means. The situation can be simulated by a computer program, or often, with minor modifications, can be modeled by a well-known technique such as linear programming or queueing theory and the analysis reduced to the application of a well-understood mathematical procedure. An "optimum" solution is then obtained by means of a systematic computational routine. The queueing model, for example, may be adapted to many operations of freeways, airports, service facilities, maintenance shops, and so on.

Peter Szanton (1972), former head of the New York City–Rand Institute, spoke of policy analysis in this category as follows:

> Our most common objective has been . . . to improve the ability of city departments to provide improved levels of service within constrained budgets. The gains come from the redeployment of resources in patterns more sensitive to changing demands; from information systems which bring together information on interdependent functions; from methods for more accurately assessing the consequences of alternative procedures; and–more rarely–from the introduction of new technology. This is work readily performed by analysts whose training is in engineering, or operations research, or economics. It is the work to which quantitative analytic tools are best adapted and for which limitations of those techniques are least important. And it is worth doing. In a city which commits some $6 billion annually to the provision or services, efficiency gains even of 1/10th of 1% can pay for a major research institute many times over. Equally important, such gains increase disproportionately the sums available in succeeding budgets for innovative and discretionary uses. By fairly conservative calculation, the implemented results of Institute studies are now saving the City some $20 million annually.

Although analysis probably has its biggest payoff when applied to problems under this heading, we devote little direct attention to such "efficiency" problems

[2] Operations research and related disciplines are defined and their aims and characteristics discussed subsequently in this chapter; see Forms of Policy Analysis, page 21.

or to the disciplines of operations research and management science. Problems of this type in the public sector usually have their counterparts in the private sector and the methods of approach are essentially the same and well documented in numerous excellent textbooks on operations research; see, for example (Wagner, 1969; Morse, 1967; Ackoff & Sasieni, 1968). As a consequence we place our emphasis on problems under the next four headings.

RESOURCE ALLOCATION

In the public sector, resource allocation is an ever-present problem. The allocation of public money starts at the top with the federal or state or city budget and is then reallocated again and again at lower levels, presumably in ways that are intended to be equitable and efficient and to most enhance the public welfare.

Analysts are frequently called upon for help with these problems by providing cost and effectiveness information concerning alternative programs. An example would be the examination of the Motor Vehicle and Passenger Injury Prevention Program carried out for HEW (1966) a number of years ago. Eight alternative programs, including the following were considered:

1. *Driver Licensing:* To establish a medical screening program for licensing which would exclude drivers with certain health problems

   . . .

4. *Seat Belt Use:* To encourage people to use seat belts

   . . .

7. *Driver Skill Improvement:* To establish a nationwide driver training program to improve driver skills by providing experience and guidance in handling automobiles in hazardous situations

Such factors as costs and deaths averted were estimated for these eight programs, indicating, for instance, that the cost per death averted for the proposed Driver Skill Improvement Program would be over 1000 times that for the proposed Seat Belt Use Program. These results were then used to help in establishing funding priorities for fiscal 1968.

Another study of the sort that might be considered here would be one to help the Department of Transportation decide on the most promising place to put R & D support for passenger transportation. Still another might involve an investigation of the income tax, the sales tax, and the property tax as competing sources of additional revenue. Into this category also we might add the design of a rating system to allocate federal funds to water and sewer projects or of any procedure for making trade-offs between programs.

In most governmental divisions funds have been allocated traditionally among programs based on such considerations as: last year's level plus (sometimes, but very rarely minus) an increment based on the over-all budget increase (almost always), a sense of priorities, workload indications, and tactical political considerations. Ideally, for any total use of resources, one would desire an allocation of funds among programs such that the last dollar used in each program would yield equal benefits, that is, no change of funds from one program to another would result in greater aggregate benefits. We are almost always unable to advise how to do this but it defines the direction for good allocation.

It is the measurement and distribution of benefits[3] that loom as the major obstacles in public resource allocation. For example, we are not currently able to determine the economic contribution of slum clearance, if any, to the reduction of crime or fires in the surrounding area, let alone its contribution to the feelings of security and other aspects of city life. Questions of equity are also troublesome because taking funds from one program and adding them to another implies taking money from one set of people and giving it to another.

The disciplines used to tackle the problems in this category go by various names but the approaches and methods are basically the same. Systems analysis is coming to be a most commonly used name for policy analysis of this type but the terms cost-benefit analysis, cost-effectiveness analysis, and operations research are also applied.

PROGRAM EVALUATION

Good public administration requires that effective programs, both ongoing and proposed, be identified and that ineffective ones be terminated or improved. A study in which the effectiveness of ongoing programs are measured and the component strategies and policies that cause them to behave as they do are identified is called an evaluation.[4]

Because the outcome of an evaluation is usually either a change in the resources allocated to a program or a suggestion for improvement in its operation, it might seem that we do not really need this category. However, an investigation to determine whether additional funds should be allocated to a job-training program for delinquent teenagers is likely to require widely different skills and techniques than one to determine whether to allocate funds to buy an additional computer for the Bureau of Motor Vehicles.

One aspect of evaluation lies in the attempt to design an experimental or demonstration program so that the results can be used by policy-makers before implementation of a full-scale program. An experiment was used to investigate the use of educational vouchers, for instance before a full-scale program was started. The design of an experimental program so that the results can be used to give a good idea of whether or not a full-scale program would succeed is becoming fairly common. Of course, experiments are sometimes used to help evaluate ongoing programs as well as proposed programs.

---

[3] Costs, as we shall see later, really involve exactly the same problems.

[4] The title "Evaluation" tends to be associated with studies that attempt to determine the worth of programs in which people are a major element. Investigations of adult education or vocational rehabilitation or of income transfer programs such as Social Security are called evaluations. Sometimes completed programs are evaluated but rarely proposed programs, other terms being used for studies involving the latter.

The analytic techniques used in evaluation (except for experimental design) are basically those used for resource allocation or to increase efficiency. If the objective is to decide whether or not to terminate the program, cost-benefit analysis is usual. Differences in approach exist between the analysis of future programs and the evaluation of ongoing ones, largely because of special problems that arise with the people involved in the latter.

Evaluation of public programs is so important a government activity that it has been treated as the subject of a separate chapter (see Chapter 15).

PLANNING AND BUDGETING

Almost every governmental unit, particularly if it has resources to disburse, does a certain amount of formal planning. This requires determining objectives or goals and then specifying, or attempting to specify, the best means for reaching those goals or objectives. Analysis is, or can be, helpful in all phases of planning. This includes forecasting the contingencies for which one plans, searching for better alternatives, estimating the research requirements, and so on.

Budgeting—the allocation of resources to tasks or to other agencies—also involves analysis if the allocation is to be in any sense an optimal one. It also may require a program, a specific blueprint for a definite course of action, something to implement a plan. This, in turn, involves more analysis to see that actions are ordered properly in time, that resources are available when required, and so forth.

The Federal Planning Programming–Budgeting System (PPBS), installed in the various agencies of the United States Government in 1965 with a great deal of fanfare, did much to increase the demand for analysis and analysts by government. Indeed, one reason for its essential demise as a federal activity by 1971 was the lack of the required analytic capability. The important characteristics of PPBS—emphasis on objectives and on the presentation of a number of options to achieve those objectives—forced a dependence on analysis, for these options had to be discovered, compared, and ranked for effectiveness.

The objective of a PPB system or of the activities of planning, programming, and budgeting as they might be carried out in the absence of a specific system is to improve the quality of resource allocation decision-making. To achieve this objective, decisions have to be made about which programs to initiate, on what scale, and which programs to expand or reduce, and by how much. This implies decisions about combinations of people, equipment, facilities, timing, and technologies.

Analyses are needed for such tasks as: (1) fairly routine evaluations of ongoing or proposed programs or projects with a view to changing the resource allocation or to improving operations with the same allocation; (2) comparisons of the costs and benefits of proposed programs; (3) the investigation of special issues or problems not associated with proposed or established programs but which someone inside or outside the government brings to notice; and (4) detailed preparation of new programs.

"STRATEGIC" CHOICE

The first step in handling any problem is to determine what we want to do. How to go about it and how much money we should allocate to the effort must come later. For example, before we can sensibly allocate funds to fight drug addiction we need to decide whether we should take a "medical" or a "law enforcement" approach. But it is hard to decide what one should attempt without some fairly good idea of feasibility and cost. Analysis can make a large contribution to such questions.

Choice-of-objective problems, however, are only one area in which analysis can help solve the mélange of questions that lie in this category, which we have chosen to call "strategic" choice. Such questions as how to increase (or decrease or control) citizen participation in local government, or what strategy to use in introducing a management information system in a government agency, or how to restructure the government itself belong here. A question of this type would be whether the use of low-sulfur coal should be required in order to reduce air pollution in, say, New York City; there is probably no way except through analysis to learn whether or not this would lead to an increase in strip mining in Montana or elsewhere and inevitably to environmental damage that might be an even greater problem than that the requirement sought to correct.

Even at the very highest level, most policy choices can be assisted by analysis—although analysis at this level is seldom performed by professional analysts.

In his 1960 Godkin Lectures, Sir Charles Snow (1961) remarked that a "bizarre" feature of advanced industrial societies was that the "cardinal choices"—decisions that determine whether we live or die—had to be made, in secret, "by men who cannot have a first-hand knowledge of what those choices depend upon or what their results may be."

Snow's cardinal choices—the decision to develop the fission bomb and to use it against Japan, or, more currently, the decision to accept "parity" instead of "superiority" *vis-à-vis* the Soviet Union—can be assisted by analysis. Such choices do not depend directly on any of the narrowly technical traditional disciplines of natural sciences or engineering—not on physics, electronics, or economics, or on knowledge found in textbooks, but on many factors from a multitude of sources. There are no experts on these factors in the sense that there are experts in navigation or thermodynamics. Any knowledge that exists must come as the result of study applied to the particular question and its context. It may thus not be first-hand knowledge. Given time such questions are open to study and, if the policy choices can be brought within the understanding of analysts, they can be brought within the understanding of those who must make the legal decision. The task of analysis is to bring together knowledge from all the various disciplines that can help and to present at least some of the risks and implications of the possible choices.

As we remarked at the beginning of the section, the classification is not very firm and the categories are not clearly separated. Strategic choice, for instance, may depend on whether an increase in the efficiency of some operation can be found; a budget allocation may depend on a determination of how well we expect to carry out a program.

General Characteristics of Analysis for Policy Decisions

Problems are varied, hence the appropriate analysis must also be varied. Policy analysis may, of course, be asked to help with only part of a decision-maker's problem, say, to investigate the impacts, or even merely one impact, that might follow a suggested decision. If, however, a study is to analyze all aspects of a decision problem, it must include the following:

1. *An investigation of what it is the decision-maker seeks to accomplish:* For an analysis to be commissioned or contemplated, someone must have a problem, that is, be dissatisfied with some aspects of the state of affairs and want to make a decision to alter it without being clear as to how to do this. He will have his own ideas as to what should be done; at the start, however, his objectives may not be very well thought out and may be so vague as to be impractical. Also he may not be aware of some of their implications, which may include things he ought not to want done. These implications should be investigated and operational goals determined and communicated clearly by the analyst.
2. *A search for alternative ways of achieving the objectives:* There should be particular emphasis on the design and invention of new possibilities.
3. *A full comparison of the alternatives (in terms of their impacts):* The study should consider the distributional and spillover effects, and, to the extent feasible, the political and organizational problems associated with acceptance and implementation.
4. *A consideration of all significant aspects of the problem:* How far to go here, as elsewhere, is a matter of judgment.
5. *An iterative approach:* This implies that, if none of the alternatives originally considered can achieve the goals, further alternatives must be sought and, if that fails, the goals must be reexamined and possibly lowered. Alternatively, if it is possible to do significantly better, more ambitious goals should be considered. Analysis of one sort or another has been used to provide advice on public questions for a long time and the fundamentals of what has to be done have been long recognized. What is novel, if anything, about the sort of public policy analysis we are discussing is mostly a matter of emphasis and attitude. The emphasis is that of rational analysis: on the clarification of objectives; on the search for alternatives, including their design and invention; on the attempt to look at the problem as a whole and at the whole problem, including spillovers and distributional impacts; on explicitness; on the recognition of uncertainty; on iteration; and, particularly, on the use of quantitative procedures insofar as this can be done without distortion. Other novel aspects lie in attitude: toward using models and computations as much to supply perspective and to focus judgment as to furnish answers; in the acceptance of quasi-quantitative or even purely

intuitive methods rather than omit significant considerations; and in the attempt to take political and organizational feasibility into account.

What Policy Analysis Is Not

1. Policy analysis is not an exact science nor can it become one. The goals are different; science is concerned primarily with the pursuit of truth, and it seeks to understand and predict. Policy analysis seeks to help a decision-maker make a better choice than he otherwise would have made. It is thus concerned with the more effective manipulation of the real world—even if this may have to be accomplished without full understanding of the underlying phenomena.

Policy analysis attempts to use the methods of science and it strives for the same traditions: results obtained by processes that another analyst can duplicate to obtain the same result; all calculations, assumptions, data, and judgments make explicit and subject to checking, criticism, and disagreement; objectivity, its propositions independent of personalities, reputations or vested interests. Moreover, the methods of policy analysis are not fully scientific. We must, in fact, do things sometimes we think are right but cannot completely justify or even check in the output of our work.

2. Policy analysis is not a panacea for the defects in public decisions. Analysis, no matter how perfectly done, cannot ensure that public policy decisions will all be in the public interest; it cannot, for example, substitute for an honest man or compensate for one with an axe to grind.

3. Policy analysis is not a tool for advocacy on the part of the analyst. If so used, it can and usually does cause the analysis to become distorted. Ideally, policy analysis is unbiased, designed to consider the entire problem, and to give all factors and all sides of the question their proper weight. If the analyst becomes an advocate, his advocacy can too easily affect what the analysis purports to show.

Forms of Policy Analysis

Other names—operations research, cost-benefit analysis, cost-effectiveness analysis, and systems analysis—are often used for studies that we term policy analysis. In this section, we describe these four important types of analysis, tell how each is related to the others, and illustrate the type of problem they are most associated with. The techniques or processes of analysis they employ—simulation, gaming, mathematical modeling, and so on—are discussed in later chapters.

The distinctions among operations research, cost-effectiveness analysis, cost-benefit analysis, and systems analysis are, I must admit, rather arbitrary and arise largely from their origin. Depending on the background of the speaker and the context, a particular study might be characterized as any one of the four types of analysis. As disciplines, they have a great deal in common. They all depend heavily on economic theory and draw from the same stockpile of tools–linear programming, queueing theory, computer simulation, operational gaming, and decision theory, to name a few.

One belief, once very widely held, is that to improve public administration in the United States and elsewhere one need only take an economic approach. It holds that economic theory, particularly microeconomics and welfare economics, plus quantitative decision theory applied in the public interest, can bring about whatever reform is needed. The four approaches characterized in this section all have economics as a core.[5] Our view, however, is that, for the analysis of public policy decisions, methods from other disciplines, the political and social sciences, for instance, have a role to play. Also, techniques such as scenario writing, operational gaming, and the direct use of expert judgment are needed. These techniques, which do not have an economic origin and are at most quasi-quantitative, have come in use in system analysis[6] and operations research to supplement quantitative and economic methods.

A brief characterization of the relationships among operations research, systems analysis, and policy analysis (but possibly a misleading one because the nature and aims of operations research have changed so much over the past twenty years) may be described as follows. Operations research seeks (or sought) to help do something better; systems analysis also seeks to do that and, in addition, seeks to see the right thing done cheaply as well as better; policy analysis seeks to do all that systems analysis strives for and, moreover, to see that it is done with equity. Thus systems analysis may be thought of as encompassing operations research (as originally conceived) plus economic considerations and inquiry into goals and their interaction with means; policy analysis may be thought of as encompassing systems analysis but with an additional concern for the distributional impacts of policy. In addition, policy analysis places more emphasis on implementation and political and organization considerations.

OPERATIONS RESEARCH

Although it did not develop until after World War II, operations research is today the most used and the most useful of the four disciplines.

Operations research, or as the British say, operational research, may be defined narrowly as it was by the British Operational Research Society [1962]:

> Operational research is the attack of modern science on complex problems arising in the direction and management of large systems of men, machines, materials and money in industry, business, government and defense. Its distinctive approach is to develop a scientific model of the system, incorporating measurements of factors such

---

[5] Originally this was not true of operations research and even today depends less on economic theory than do the other approaches.

[6] At least into systems analysis as defined by Quade and Boucher (1968).

> The problems with which economics deals may be classified as maximization problems, completely subject to control through mensuration, and choice-of-objectives problems, which are not. In maximization problems one is attempting to optimize a clearly definable objective function; the solution is embodied in the available data, if they are used correctly. These are the problems for which operations research was initially designed, although one should not underestimate the skill required to select the objective or achieve its maximization.
>
> In choice-of-objective problems, on the other hand, the fundamental question is the selection of the appropriate mix of goals. The data suggest no solution in themselves. Even after the quantitative data is collected and organized, the pressing question still remains: what does one want to do? Sophisticated costing may provide considerable assistance in revealing the tradeoffs among various alternatives—but the ultimate decision, the choice of the strategic approach to the problem, remains a matter of faith that will always elude rigorous quantification. The attempt has been made to apply the methods of operations research to this class of problems and—when this is done—as frequently as not the term "systems analysis" is tagged on. But the two classes are not amenable to the same type of treatment. The chief danger in the first sort of problem is that of mathematical miscalculation or mishandling of the data, whereas in the second the danger is something akin to consumer ignorance. (p. 298)

Later in the same article Schlesinger pointed out that:

> System analysis provides assistance to logical thinking, but it can never "solve" problems in the same sense that operations research can. (p. 299)

Unfortunately, this idea for a clear distinction has not been widely accepted. At present, the only distinction usually made is that systems analysis deals with "higher level" and operations research with "lower level" problems. This points in the right direction in that the higher ranks in government or other organizations are more likely to set objectives; however, confusion remains. Nevertheless, given any problem, the more objectives there are and the more they conflict, the larger the number of parameters and factors to be considered, the more need for reliance on judgment and intuition, and the less the dependence on quantitative analysis and computers, the more likely the work will be labeled systems analysis rather than operations research.

In the absence of an agreed on definition, Quade and Boucher (1968) attempted to characterize systems analysis as

> . . . a systematic approach to helping a decision-maker choose a course of action by investigating his full problem, searching out objectives and alternatives, and comparing them in the light of their consequences, using an appropriate framework—insofar as possible analytic—to bring expert judgment and intuition to bear on the problem. (p. 2)

Systems analysis, since it generates and presents information in such a way as to improve the basis for decision-makers to exercise their judgment, has the same purpose as policy analysis. As a consequence the terms are often used interchangeably with the choice depending on the context. The term policy analysis is more likely to be used, however, when political and social factors predominate, as they are likely to do in public policy decisions.

> as chance and risk, with which to predict and compare the outcomes of alternati decisions, strategies or controls. The purpose is to help management determine i policy and actions scientifically.

Operations research thus seeks to use scientific methods to assist deci makers in getting the most out of available resources. Unlike science, its pur is not merely to understand or to predict, but to maniuplate the real world r effectively. Both depend on model building as Helmer (1963) makes clear:

> . . . Both the exact scientist and the operations analyst tend to make use of what sometimes called a mathematical model of the subject matter; in the case of tl scientist such a model is apt to be part of the well-confirmed body of our scientif knowledge, whereas an operations research model is of a more tentative, *ad ho* character. In other words, even if the current status of science provides no well-esta lished theory for the phenomena to be dealt with by the operations analyst, the latt must nevertheless construct a model as best he can, where both the structure of tl model and its numerical inputs may be based merely on intuitive insight and limite practical experience by the analyst himself or by whatever expert advisers on tl subject matter may be available to him. As further insights accrue and mo experimental data become available, the operations analyst has to be ready to discar his first model and replace it with an improved one. This tentative procedur dictated by pragmatic considerations, is thus essentially one of successive approxim tion. In this regard, operations research has a status similar to that of the so-calle inexact sciences, of which medicine, engineering, and most of the social sciences a examples.
>
> Therefore, in comparing operations research with an exact science, it is wit regard to exactness that operations research falls short, but not necessarily wit regard to the scientific character of its methods. . . .

The term operations research is commonly used very loosely–narrowly fined, it refers to an attempt to apply mathematics or logical analysis to h client improve his efficiency in a situation in which it is clear what "r efficient" means; broadly defined, particularly by operations researchers tl selves, it refers to all quantitative policy analysis. This practice is spreac hence the distinctions made here may soon disappear.

SYSTEMS ANALYSIS

When operations research came to be applied outside the military fo immediately after World War II, the term was interpreted in its narrow se being confined to studies of "low-level" problems where the decision-maker a clear objective in mind. The term systems analysis then began to be applie broad "higher level" studies that looked into aspects that OR workers usually considered "given" (the objective, for instance) and accepted mo that seemed to some hardly scientific. Before very long, however, the terms v often employed interchangeably.

To avoid confusion, it was suggested (as we do here) that the term operati research be confined to efficiency problems and "system analysis" to probl of "optimal choice." James R. Schlesinger (1963) explains:

The distinction made earlier, even though fairly arbitrary, has an historical basis. Policy analysis developed largely as an extension of systems analysis. Hence we consider systems analysis to be an imperfect or specialized form of policy analysis, one that does not emphasize the distributional consequences of costs and benefits and pays insufficient attention to implementation and to political and organizational effects.

Systems analysis or the systems approach makes no pretense of providing a complete theory of systems. That sort of thing is reserved for General Systems Theory (Bertalanffy, 1968). The term is also used differently in other contexts, e.g., computer science or office management.[7]

COST-EFFECTIVENESS

Cost-effectiveness is a form of systems analysis in which the alternative actions or systems under consideration are compared in terms of two of the consequences, dollar or resource costs and the effectiveness, associated with each alternative. The effectiveness of an alternative is measured by the extent to which that alternative, if implemented, will attain the desired objective.

The preferred alternative is usually taken to be either the one that produces the maximum effectiveness for a given level of cost or the minimum cost for a fixed level of effectiveness. On occasion, but this sometimes leads to error, for instance, if the scale of the operation is not specified, the preferred alternative is taken to be the one for which the ratio, C/E, of cost to effectiveness is least.

Effectiveness measurement almost always presents a problem. It must be measured on a scale that depends on the nature of the goal and that may require the use of a proxy. For example, in a cost-effectiveness evaluation of educational programs to improve reading performance the effectiveness of the programs could be directly related to the goal by measuring effectiveness on a standardized reading test. If, however, the cost-effectiveness analysis were concerned with evaluating various programs to train paramedics and the goal were improved health for the population they were being trained to serve, one would be forced to use a proxy. An example would be to compare the various alternatives, something like the number of mortalities averted, because improved health itself is difficult to measure directly and involves factors besides death. The costs—properly the opportunities foregone—can ordinarily be represented somewhat more satisfactorily by the nominal or resource costs.

Whereas we are often able to use cost-effectiveness to rank competing alternatives for the same goal,[8] we cannot use it to compare alternatives that seek different goals—to decide, say, the best over-all use of our money when we have several long-range objectives in mind. That requires something more, e.g., that there be a way to compare the worth or benefit for a particular cost of

[7] For further discussion of the confusion in the use of the term systems analysis, see Hoos (1972).

[8] Provided the competing alternatives can be made comparable in either cost or effectiveness.

achieving a certain effectiveness in accomplishing one goal with that in accomplishing another.

COST-BENEFIT ANALYSIS

One possibility for guiding choice between programs designed to accomplish widely differing tasks would be to measure the benefits and costs in the same units in all programs, so that the difference between the benefits and the costs could be calculated for each program and compared with the corresponding difference for other possible actions. In practice this means expressing both the benefits and the costs in monetary units, dollars for example, a process that often must be done very arbitrarily and that leads to the neglect of certain benefits and certain costs. When this is done, we have a specialized form of systems analysis that is called cost-benefit analysis.

Cost-benefit analysis is, in theory at least, a much more powerful tool for decision-making than cost-effectiveness. It can be used, for example, to choose between such diverse alternatives as allocating funds to constructing a dam with irrigation, flood control, and recreation as goals or to a screening and pretreatment program to reduce mortality from heart attacks. If the projects are roughly on the same scale with respect to cost, it is merely a question of choosing the project for which the benefits exceed the costs by the greater amount.[9] The great disadvantage of cost-benefit analysis is that it is very hard to perform satisfactorily. Prest and Turvey (1965) put it this way:

> . . . One can view cost-benefit analysis as anything from an infallible means of reaching the new Utopia to a waste of resources in attempting to measure the unmeasurable.

Ideally, all costs and benefits should be identified, converted to monetary units, and taken into account in the evaluation. This means costs and benefits for the life of the project, not just for the immediate future. An attempt also should be made to consider the indirect consequences resulting from the project—the so-called externalities, side effects, and spillovers—such as the roadside business that is ruined because the new lake diverts traffic or the beach that increased shipping renders unusuable for bathing. One procedure is to estimate the installation costs of the project and the times at which they occur and then to add the estimated annual operating costs. The average value of the output or primary benefits and their probable future market value are also estimated. The more intangible "secondary" benefits and costs are then estimated year by year. Finally, a discount rate is assumed and the time streams of costs and benefits are discounted[10] so as to obtain their present values. Projects with benefits greater

---

[9] Sometimes the ratio of benefits to costs is used but this can lead to error.

[10] Exactly as money is discounted. If the rate is r, a cost, C that need be paid until n years from now is equivalent to only $C/(1 + r)^n$ today.

than their costs are then considered for approval, or the ratio of benefits to costs may, with some consideration of relative scale, be used to rank projects in order of desirability.

The aim of cost-benefit analysis is to maximize "the present value of all benefits less that of all costs, subject to specified restraints" (Prest and Turvey, 1965). It can thus be used to promote economic welfare. But whose welfare? One of the defects of cost-benefit analysis is that the format does not consider the distributional effects of costs and benefits. In the field of water resources where cost-benefit analysis developed, federal participation could take place, as specified in the Flood Control Act of 1936, only "if the benefits to whomsoever they may accrue are in excess of the estimated costs." The question of who in the society should bear the costs of a project and who should reap the benefits is thus left for some other means to determine.

Cost-benefit analysis has other defects, associated with such difficulties as the choice of a discount rate and the quantification of many elements, for instance, attempting to estimate the cost in dollars of killing off fish or animals that have no economic value. The same such considerations must naturally be taken into account no matter what analytic approach is used, but in cost-benefit analysis they are more apt to be buried in the computations and treated rather arbitrarily by the analyst. For further discussion, one should consult Prest & Turvey (1965), Mishan (1971), McKean (1958), or Wildavsky (1966).

Alternative Sources of Advice

There are, of course, other means than analysis for providing help to a decision-maker who has to arrive at a choice between alternatives.

One is pure intuition, with or without divine guidance. The intuitive process is to learn everything possible about the problem, to live with it, and let the subconscious provide the solution. This approach can in no sense be considered analytic because no effort is made to structure the problem or to establish cause-and-effect relationships and operate on them to arrive at a solution. But one may use help; read tea leaves or inspect the entrails of a sheep as the Romans did, or even support intuition with analysis. Later we shall see that analysis *must* be supported by intuition.

Between pure intuition on the one hand and policy analysis on the other, there are intermediate sources of advice that are at least partially analytic, although the activities involved are ordinarily less systematic, explicit, and quantitative. One alternative is simply to ask an expert for his opinion. What he says can, in fact, be very helpful, the more so if it results from a reasonable and impartial examination of the facts with due allowance for uncertainty and if his assumptions and chain of logic are made explicit so that others can use his information and reasoning to form their own considered opinion. In other words, if he bases his advice on whatever analysis he can do, that advice is likely to be superior to what he might give on intuition alone. But an expert, particularly an unbiased expert, may be hard to identify. And an expert's knowledge and opinions are likely to be more valuable if they can be used in direct association with those of other experts. This suggests the use of a group of experts forming a committee or panel or other consensus device. Unfortunately, such groups seldom do very much analysis or make their reasoning explicit. Their findings are usually obtained by bargaining, and here personality and prestige often outrank logic. There are, of course, methods of using committees, particularly panels of experts, that are systematic and can be made part of the analytic process. The Delphi process in which questions are answered anonymously, discussed in Chapter 12, is one.

Another alternative is the approach sometimes known as "muddling through," a sort of trial-and-error process in which naturally occurring feedback from what actually happens, supplemented by limited analysis, serves to provide the help. This approach encompasses the procedures by which administrators and policymakers have long gone about making decisions–using analysis on parts of their problem, taking remedial steps rather than innovative ones, moving away from ills rather than toward definite objectives, seeking vague goals sequentially. Advocated as a substitute for policy analysis, muddling through is discussed in the next section.

Incrementalism

As an alternative to systems analysis and the systems approach, Charles E. Lindblom (1959) suggests "disjointed incrementalism," a refinement of a process he had earlier referred to as "muddling through."

Although Lindblom would agree that some or possibly considerable analysis is necessary to policy-making, he has grave doubts about the practicality of placing analysis in any sort of a dominant role. Based on an extensive examination of the political process in federal, state, and local institutions in the United States and to a limited extent elsewhere, he finds that process so fragmented and complex, incorporating the interaction of executive, legislative, and judicial branches, political parties, pressure groups, and citizens, that analysis can have only marginal effect on change. The difficulties encountered in formulating operational goals, lack of agreement on values and criteria for choice, and even resistance to analysis by interest groups lowers the quality of public policy analysis and makes recommendations for anything except minor changes politically infeasible. According to Lindblom (1968), what is feasible politically is

> . . . policy only incrementally or marginally different from existing policies. Drastically different policies fall beyond the pale.

Lindblom assumes that policymaking is serial, that it proceeds through long chains of political and analytical steps, with no sharp beginning or end and no clear-cut boundaries. He may believe it should be this way. In these circumstances, Lindblom suggests that analysis not be used to help seek major changes but merely marginal or remedial ones. Thus only those alternatives that differ incrementally from current policy and from each other need be examined; to do more would be a waste of energy. An analyst need not even examine all the consequences of even these limited alternatives for there will be other analyses conducted simultaneously elsewhere. Analysis becomes more a device to help a decision-maker by contributing to his bargaining power than a means to help him by providing him with sufficient information for him to decide what decision should be made.

Kathleen Archibald (1970) compares systems analysis with incrementalism in the following manner:

> While both approaches recognize that certain alternatives and consequences have to be omitted from an analysis, systems analysts attempt to make those omissions rationally or strategically, while Lindblom is content to have them made quite arbitrarily. The incrementalist feels he can afford to make only minor changes and to make mistakes because policy-making is serial and fragmented. Problems are never solved; instead some analysis is done, a decision is made, unanticipated adverse consequences show up, more analysis is done and more decisions are made to remedy the adverse consequences, etc., *ad infinitum.* The incrementalist feels he can ignore consequences at will because if those ignored should prove damaging to certain groups such groups will press for new analyses and new decisions. He arbitrarily excludes some consequences from his analysis because he depends on "politics" in a pluralist democracy to provide, eventually, whatever degree of closure is feasible in an obstreperous world.

Policy analysis and systems analysis also differ from disjointed incrementalism in maintaining a different view of the analyst's role and of an ideal decisionmaking structure and in putting greater stress on economic rationality. To again quote Archibald (1970):

> Three somewhat random comparisons may be helpful in clarifying the difference between systems analysis and incrementalism in operational implications.
>
> (1) Incrementalism assumes that the costs of analysis and delay (the costs of delaying action, that is, not necessarily the costs of delaying a "solution") are greater than the costs of error; systems analysis tends to assume the opposite. Nuclear strategy is a field where the costs of error are perceived by most people as high; it is not surprising that systems analysis made a name for itself in this field. People are not as aware of the possible costs of error in domestic fields, nor are the costs usually as high, so systems analysis is likely to have a tougher time competing with political pressures when working on domestic problems.
>
> (2) Systems analysis is more likely to act on the notion of "integrity of design," that is, the systems analyst can be expected to sometimes say, "The consequences of such and such a compromise are likely to defeat the major intent of the program, so if that compromise is a condition of its acceptability, it's better to have no program at all." If the systems analyst is wrong, he may sometimes throw out the baby with the bath water; on the other hand, the incrementalist will sometimes save the bath water while forfeiting the baby.
>
> (3) The soul of systems analysis shudders at what J. R. Schlesinger . . . has called foot-in-the-door techniques, that is, commitment to beginnings without much consideration to total costs and eventual benefits should the foot stay in the door nor to the costs of reneging on what looked like a commitment if the door should be slammed shut. Disjointed incrementalism rather applauds such techniques, pre-

sumably on the assumption that survival of the fittest somehow holds for policy beginnings.

The ways in which analysis and bargaining interact are being increasingly discussed and there is a growing recognition of how they can complement each other as well as of how one may subvert the best features of the other. Lindblom's view of the policy-making process has been more influential than this view of the analyst's role. Thus a synthesis between systems analysis and incrementalism, which some would call policy analysis, seems to be moving in the direction of the analyst taking more account of internal and external politics and being more aware of the partial nature of any analysis, rather than in the direction of the analyst welcoming a partisan role and becoming more sanguine about arbitrary omissions. . . .

Two other aspects of an incremental approach should be mentioned. What is not "politically feasible" today may be tomorrow or next year or the following year. In any event, if no one looks at "politically *in*feasible" alternatives now, we may never know the real costs of the purely "political decisions" we take. Also, just looking at small changes without examining the base from which these changes originate, may not enable us to properly assess the implications of these changes.

The view of the policy process expressed by Lindblom is more realistic than that which seems to be assumed in many studies. The two approaches are quite compatible and policy analysis, in growing from systems analysis and operations research is beginning to improve its impact by paying attention to the incrementalists' view of the policy-making process and modifying its practices somewhat.

In summary, even if one does not believe that better and more policy analysis can contribute significantly to the improvement of public decisions, there is a definite need for it. In fact, without analysis, or at least partial analysis, that is, looking into the various courses of action and working out what is likely to happen, we have no real scheme for guidance. Without analysis, one may, for instance, be forced into the dilemma of the intolerable result. One may have to argue that I must take this new course of action, A, because the present course is not working–without asking how likely it is that A will work, or how much it will cost if it does, or what damage will be done if it fails. For if I do what I am now doing, it will surely fail. The results will be intolerable. There is another strategy that offers some chance of success. Therefore we should change to the new–no need to ask the cost, for we have no choice.

References

Ackoff, R. L. & Sasieni, M. W., *Fundamentals of operations research.* Wiley, New York, 1968.

Archibald, K. A., Three views of the expert's role in policy-making: Systems analysis, incrementalism and the clinical approach. *Policy sciences,* 1970, **1**, (1) 73–86.

Bertalanffy, Ludwig von, *General systems theory.* George Braziller, New York, 1968.

British Operational Research Society, *Operational research quarterly,* 1962, **13**, (3), 282.

Helmer, Olaf, *The systematic use of expert judgment in operations research,* The Rand Corporation, P-2795, September 1963.

Hoos, Ida R., *Systems analysis in public policy: A critique.* University of California Press, Berkeley, California, 1972.

Lindblom, Charles E., *The policy-making process.* Prentice-Hall, Englewood Cliffs, New Jersey, 1968.

Lindblom, Charles E., The science of "muddling through." *Public administration review,* 1959, **19**, 79–88.

McKean, R. N., *Efficiency in government through systems analysis.* Wiley, New York, 1958.

Mishan, E. J., *Cost-benefit analysis: An introduction.* Praeger, New York, 1971.

Morse, Philip M. (ed.), *Operations research for public systems.* M.I.T. Press, Cambridge, Massachusetts, 1967.

"New York City–Rand Institute Research in 1970–1971," *Operations Research,* **20**, (3), May–June 1972, 474–515.

Prest, A. R. & Turvey, R., Cost benefit analysis: A survey. *The economic journal,* 1965, **15**, (300), pp. 683–735.

Quade, E. S. & W. I. Boucher (eds.), *Systems analysis and policy planning: Applications in defense,* American Elsevier, New York, 1968.

Schlesinger, J. R., Quantitative analysis and national security. *World politics,* 1963, **XV**, (2), 298.

Snow, C. P., *Science and government: The Godkin lecture at Harvard, 1960.* Harvard University Press, Cambridge: 1969, p. 1.

Szanton, Peter, Analysis and urban government: Experience of the New York City–Rand Institute. *Policy sciences,* 1972, **3**, (2), 153–161.

U.S. Department of Health, Education, and Welfare, Office of Assistant Secretary for Problem Coordination, *Motor Vehicle Injury Prevention Program,* August 1966.

Wagner, Harvey M., *Principles of operations research with application to managerial decisions.* Prentice-Hall, Englewood Cliffs, New Jersey, 1969.

Wildavsky, Aaron, The political economy of efficiency: Cost-benefit analysis, systems analysis, and program budgeting. *Public administration review,* 1966, **XXVI**, (4), 292–310.

Chapter 3

THE BASIC FRAMEWORK AND A HOMELY ILLUSTRATION

Background

The idea of using analysis to help someone (who may be the analyst himself) make a choice or a decision is not new[1] and the sort of inquiries that need to be made (although possibly not expressed formally) have been known and practiced ever since man first realized the limitations of his resources. This chapter presents a basic framework for what is sometimes called the rational or analytical approach to decision-making and illustrates its procedures by an example of systems analysis familiar to almost everyone, a problem in economic choice—the purchase of a family car.

When the term *systems analysis* was first used, it referred to the study of an existing or prospective but clearly recognizable system, in particular, to a complex grouping of men and equipment put together with a clear mission in mind. The continental air defense system consisting of radars, antiaircraft, interceptors, control centers, and bases with men to man them was such a system. The purpose of the study was to help decide whether the system should be procured or developed by investigating the behavior of the system, by suggesting improvements in its configuration, and by comparing the improved system with other systems that might carry out the same mission, in terms of performance and the resources required to procure and operate over the life of the system. Technical, operational, and economic feasibility and desirability were emphasized; the investigation of political and organizational acceptance was left to others. Even inquiry into the desirability of the mission was not at first made part of the study. In brief, classical systems analysis dealt with problems of economic choice in a fairly narrow sense, usually one in which a decision to go ahead had already been made and all that remained was to choose the "best" from the set of alternatives.

Success soon brought opportunities to apply the approach to broader and less well-defined problems, forcing a number of changes in the way things had been done. Difficulties with objectives and criteria appeared almost immediately, causing the emphasis to shift from "how to do it" to "what should be done." Societal issues such as those that occur with problems found in the urban

[1] One congressional committee located the origin of cost benefit analysis in the Garden of Eden (U.S. Senate, 1970).

environment also brought difficulties with implementation, with the target groups, with spillovers, and with the distribution of costs and benefits. The basic framework for helping a decision-maker in a problem of choice, however, did not change. The core remains that of classical systems analysis. It is this basic framework which is the subject of this chapter.

Analysis designed to help or influence public or private policy decisions, if it is to realize its full potential, can do more than merely aid in discovering a course of action. The real goal of a decision-maker is not just to discover and make the right decision; it is to get that decision accepted by others and then implemented without being vitiated. Analysis can help also with problems of acceptance and implementation. This chapter, however, is directed toward discovery, to illustrate how one finds an alternative which is in some sense "best" or preferable. Helping with acceptance and implementation are tasks of broader policy analysis that are considered in later chapters.

Concepts and Principles

In an abstract or general sense, what needs to be done when faced with a problem of choice should become obvious to anyone who pauses to think about it. One must first determine the *goals or objectives*—what it is he wants to accomplish—so that he can search out the various *alternatives* or possible means of attaining what he wants. He should then investigate each alternative to determine as well as he can the *impacts* or consequences that will follow from its choice. In narrow problems, the resources required (or costs) and the extent to which the alternatives will accomplish the objectives are the dominant impacts. In a problem of any complexity, finding the impacts requires that the analyst explicitly work out one or more schemes called *models* for helping him make these determinations. In addition, he needs a *criterion* or standard by means of which he can rank the alternatives in order of preference, using the information he has uncovered about their impacts.

To illustrate these concepts and how they are related to decision-making, suppose you are asked by your boss to carry out an analysis to help with the following problem: Determine the best route for him to use in driving from home, A, to place of work, B.

This is a problem of choice; the five elements underlined above might be interpreted as in Figure 3-1.

To solve the problem as interpreted in Fig. 3-1, one would lay out various routes on the map or model and, by inspection or measurement select the shortest. Although this solves the problem as interpreted in Fig. 3-1, this solution might, or might not, be accepted as "best" by the decision-maker. For instance, he might not like the criterion, preferring a route within minimum travel time to one of minimum distance. To determine such a route would require a more elaborate model, one that took into account traffic congestion, signal lights, road surface, and so on. Or, when he translates the number of miles required into dollars or gallons of gasoline, he may find it excessive and decide to reformulate his problem, say, by substituting "traveling" for "driving" and by changing the criterion to "lowest cost." In this case other forms of transporta-

OBJECTIVE:	To drive from home to work.
ALTERNATIVES:	The various routes from A to B.
MODEL:	A map of the area showing streets and freeways.
IMPACT:	The number of miles in the route.
CRITERION:	The shortest route possible.

FIG. 3-1. A simplified choice analysis.

tion must be considered. The point to be made even with this simple example is that an analysis is at most an aid to decision-making; it can provide the decision-maker with information and perspective but can seldom, possibly never, tie everything down so tightly that the rational decision-maker has no choice but to take the solution from the analysis.

The five key factors or elements–objectives, alternatives, impacts, criteria, and models–are always present in a problem of choice although they may not be explicitly labeled as such in the report or presentation. Of these five, models–schemes designed, appropriate to the problem, to estimate or predict the major consequences of alternative actions–are central to every analysis. A model, broadly defined, is a simplified representation of the phenomenon in which we are interested. It may range in form from an elegant computer simulation, or a set of mathematical equations, or an elaborate game played by the decision-makers and their deputies to a few assumptions and rules of thumb for operating with them. What little theory we have about policy and systems analysis is mostly centered on models and model-building.

To further clarify and make more concrete these concepts and to give some idea of procedures, I will attempt to illustrate how one might go about carrying out a cost-effectiveness analysis by tracing through a realistic but hypothetical example.

An Illustrative Example

Assume you have been asked to help a family with a decision to purchase a new family car. How would one apply the procedures we have been talking about to help them decide what car to purchase?

In form, a procedure to help with decision-making here is basically no different from one that the Federal Aviation Agency might go through before procuring a new air traffic control system or an airline before ordering a new fleet of aircraft. In the course of demonstrating with this example how one might go about carrying out an analysis of cost and effectiveness, I hope to show that

1. There is nothing mysterious about the concepts and procedures and no reason why they should not be used when appropriate by all decision-makers.

2. A thorough cost-effectiveness analysis of a problem even as simple as this one can result in a very complex process.
3. The cost of analysis can sometimes outrun the cost of error.
4. It is important to make objectives and assumptions explicit.
5. It is desirable to take a "systems view" of what might at first glance appear to be a relatively straightforward problem.
6. Considerations that cannot be handled quantitatively are present and must be taken into account.
7. In the final decision, the judgment of the decision-makers will still play the major role.

This example also illustrates a truism—it is hard to do a good analysis, but it is easy to do a better one.

Let me outline the steps one might take in tackling this problem. One could begin work at various points; one good place to start would be with clarification of the problem.

1. CLARIFICATION OF THE PROBLEM

For this, among other things, one needs to find out if the decision to purchase has already been made irrevocably or whether there are circumstances under which the alternative of continuing to use an old car or some form of public transportation would be acceptable. Also, any constraints must be discovered—for example, a price above which the purchaser can or will not go no matter how cost-effective the purchase might turn out to be. Or are certain cars barred because they do not have automatic transmission, or power steering, or some other accessory?

In general, a most vital step in the application of cost-effectiveness or systems analysis to someone else's problem is a detailed investigation of their assumptions. If the analysis is to be useful to others, the assumptions must be clearly stated. Unfortunately, the concise format of the typical study report is not conducive to clear presentation of the assumptions on which the outcome is based. In the case of some public reports, the situation is even worse because no attempt is made to be precise with respect to assumptions.

A next step might be to determine what they hope to attain with the decision being considered.

2. IDENTIFICATION OF THE OBJECTIVES

In general, the decision to purchase a new car may have many motivations. Among others, it might be to (A) enable the wife to drive in a carpool, taking her children and those of neighbors to school; (B) provide a reliable and safe vehicle to drive to work over freeways during rush hours; (C) serve as a status symbol; (D) make the family vacations more enjoyable—by having a four-wheel drive or by being amphibious or by serving as a camper. In every analysis, it is necessary to be quite precise in defining the objectives. If the four purposes listed really include all the important uses one intends to make of the car, it is quite likely that a satisfactory compromise can be found. But, if the real reason for wanting a new car is to be able to vent a little steam on weekends by roaring through the countryside at high speed over winding roads or for the pleasure of driving along mountain roads with the top down, these reasons had better be included or the results of the analysis will almost certainly be disappointing. Unless such factors are considered, the study of cost and effectiveness may show that a Volkswagen bus is the best buy when the head of the household really had his mind on a Mercedes 450 or a Jaguar XKE. Outcomes that are far off the mark are seldom the fault of the analytic techniques, but are frequently the result of not having the right objectives.

To be definite, assume that in addition to transportation at an acceptable cost the objectives are to obtain a car that will provide: (A) comfort with room for one adult and five children; (B) safe and reliable transportation; and (C) a clear indication of status. We now need to determine ways to measure the attainment of these objectives.

3. MEASUREMENT OF EFFECTIVENESS

Having decided on the above objectives, since different cars will certainly attain the various objectives to different degrees, we need some way to assess these differences. That is, we need a way to measure effectiveness of the various alternatives with respect to attainment of the objectives. This may take extensive investigation.

Of the three objectives, (A), (B), and (C), comfort and status are almost completely subjective in nature. Safety and reliability are less so but in practice may have to be handled in the same fashion.

When buying cars, most of us do this in a relatively subjective fashion, relying primarily on our own judgment, intuition, and experience. We might also get a little more information by talking to friends, and more precise information and some analysis by checking *Consumer Reports* and other publications. Given enough time and money we could go still further with safety and reliability. By analyzing data in Highway Department and insurance company records, we should be able to establish the comparative frequency with which a car of particular type breaks down or is involved in accident with injury on the freeway while carrying someone to or from work. But we would soon reach a stage where the cost of analysis would far exceed the cost of error. (An airline interested in a $12 million jet might carry a search for data much further.)

One way to handle subjective assessment of the attainment of an objective such as status would be to ask the decision-maker to assign a number between 0 and 100 to each alternative to represent his best estimate as to its percent effectiveness in attaining the objective in question. If there are several decision-makers, the median of their estimates could be used. Objectives A, B, and C might be handled in this way. The cost can be measured more objectively, say, in terms of the annual cost associated with owning the car or the present value of the stream of future costs suitably discounted.

We also need a standard for choice.

4. DETERMINATION OF A CRITERION

While there is no fully satisfactory way to measure the attainment of the three objectives on a single scale, in practice, there are numerous ways to proceed. One possibility would be to establish a minimum level of effectiveness for A, B, and C and then select the car with the least total cost from among those that exceed the minimum effectiveness level on all three standards.

Another approach that is not completely unreasonable would be to combine the measures for the three objectives into a single "payoff" function by weighting them in a way to reflect how well each alternative or car accomplishes its desired purposes. For example, we might use the notion of an effective cost, i.e., if a car were 100% satisfactory in all its uses, its effective cost would equal its actual cost; if it were only 50% satisfactory, its effective cost would be twice its actual cost. The supposition would be that it would take twice as much money as we would actually be spending to really get the job done. In other words, effective cost might be defined as actual cost divided by the overall satisfaction rating, expressed as a fraction. Thus, if a car cost $10,000 and were only 25% satisfactory over-all, its effective cost would be $10,000 divided by 0.25 or $40,000.

Assuming we decide to use effective cost, a decision must first be made as to the relative importance and interrelationships of the three purposes in seeking the new car. For this, suppose it is decided that the three goals A, B, and C are equally important and that they are independent (that the degree of satisfaction with respect to one of the purposes can be evaluated without knowing the degree of satisfaction for the other purposes). Then the over-all satisfaction rating can be taken as the product of the three ratings (expressed as decimals) and the effective cost as the actual cost divided by that product.

Assume we decide on effective cost as the criterion for ranking the alternatives, the car with the lowest effective cost being ranked first.

We should not be fully comfortable with this criterion. We would prefer a way to translate the combined worth of A, B, and C into dollars so that we could rank the alternatives in terms of their excess of worth over their cost. A ratio does not take absolute magnitude into account–it may lead us to a car that would barely run provided its cost were almost zero. For that reason we should set some bounds, e.g., we do not consider any car unless the product of the satisfaction ratings (expressed as decimals) exceeds 0.5 and unless the initial cost is under $20,000. Moreover, we should use the rankings obtained in this manner with discretion, looking at them to see whether the higher satisfaction is worth the higher cost or conversely and basing the final decision directly on judgment.

5. DETERMINE THE ENVIRONMENT

Having settled on the objectives of the system and a scheme to measure their attainment, one can consider next the operational environment. Let us say that the system must operate primarily in fair weather, with only occasional rain, snow, ice, or very hot weather. Weather, however, is only a small part of the environment. Who will be the operators? Does one of them have a tendency to strip gears when shifting a manual gearshift? What other systems will be operated in parallel? (For instance, if the addition of the new car means there are now to be two cars, the new one may achieve some of the desired purposes without directly accomplishing them itself.) What is the logistics situation? That is, are there repair facilities and spare parts available? Is there adequate garage space? All these factors are part of the environment and are probably a much more important part than the weather. Only when the environment is completely defined will there be a rational basis for the evaluation of the utility of the system parameters such as automatic transmission, air conditioning, body style, color, radio, etc.

6. INVESTIGATE THE ALTERNATIVES

Next we might define the alternative systems. They must be examined for feasibility and cut down to a reasonable number. Objective B, for instance, eliminates all cars that will not seat at least six people; initial price and the availability of service may eliminate others. We decided earlier to eliminate all alternatives for which the product of the three satisfaction estimates for A, B, and C came out below a standard which we took to be 0.5.

7. FORMULATE MODELS

To compare the various alternatives we need to devise a scheme or model for computing the "effective" cost. To do this, we first need a model for the cost before it is modified for effectiveness.

To work out the cost model, assume we have a car W with purchase price \$$P$, which we plan to use for t years and then sell or trade in for \$$R$. Assume also that, for some fixed number of miles to be driven, the annual operating cost (fuel, tires, servicing, insurance, repairs, taxes, and so on) is \$$A$ per year. Because costs for different alternatives occur at different intervals, to compare the total costs we must discount all costs to the same date, say to the date of purchase.

Since the use periods are different, what we can compare are the cost streams for the various alternatives, defined as follows. Assume that car W is purchased for \$$P$, used for t years, then traded in on a new car of the same type for a total cash outlay of \$$(P-R)$; this car is used for t years, then traded for another of the same type again for \$$(P-R)$, and so on. This yields the cost stream which we discount to the time of purchase.

The present value of the cost stream for alternative W will then involve \$$P$ paid at the start, \$$A$ paid each year for t years, and the purchase of a new car for \$$(P-R)$ at the end of every t years.

The discounted value of this cost stream will then be

$$P+\frac{A}{1+i}+\frac{A}{(1+i)^2}+\frac{A}{(1+i)^3}+\dots \qquad +\frac{P-R}{(1+i)^t}+\frac{P-R}{(1+i)^{2t}}+\frac{P-R}{(1+i)^{3t}}+\dots$$

Because the two series are geometric series, they can be summed, giving the formula for the discounted cost stream:

$$P + \frac{A}{i} + \frac{P-R}{(1+i)^t - 1}$$

To obtain the model for the "effective" cost, this sum can be divided by the product of the satisfaction ratings, yielding

$$\text{Effective cost} = \frac{P + \frac{A}{i} + \frac{P-R}{(1+i)^t - 1}}{(S_A \cdot S_B \cdot S_C)}$$

where S_A, S_B, and S_C are the combined satisfaction ratings for objectives A, B, and C, respectively, expressed as decimals.

8. COLLECT THE DATA

For simplicity, assume that only three alternatives survive the initial screening, and that the relevant cost information is as in Table 3-1. (These figures are not intended to be very realistic, but are presented merely to make an example.)

TABLE 3-1.
Cost Data

	Car *X*	Car *Y*	Car *Z*
Purchase price (*P*, dollars)	4,000	7,000	13,000
Estimated use period: (*t*, years)	3	4	6
Value at end of period: (*R*, dollars)	1,000	1,600	5,000
Annual costs: *A* (dollars) (includes fuel, repair, taxes, insurance, etc.)	1,450	1,600	1,550

Assume further that for objectives A, B, and C, the cars have satisfaction ratings as in Table 3-2.

TABLE 3-2.
Satisfaction Ratings

Purpose	Car *X*	Car *Y*	Car *Z*
A. Comfort	.94	.89	.85
B. Safety and reliability	.90	.90	.92
C. Status indication	.61	.80	1.00
Combined satisfaction rating	.52	.64	.78

9. CARRY OUT THE COMPARISON

Substituting the cost figures from Table 3-1 in the cost model, assuming a discount rate of 6%, we have:

	X	*Y*	*Z*
Present value of cost Streams at purchase	43,900	54,200	58,000

The effective costs of the three cars are then

$$X = 84{,}400 \quad Y = 84{,}700 \quad Z = 74{,}300$$

Before coming to any sort of a decision or even suggesting one on the basis of these or similar calculations, a "sensitivity" analysis should be carried out. This is a procedure for determining the effect of errors in the inputs or in the model relationships.

10. EXAMINE THE ANALYSIS FOR SENSITIVITIES

The results of the effective cost calculations derive from some fairly arbitrary procedures and, as a result, are quite dependent on the assumptions involved. This is, of course, a basic characteristic of all decision-making processes.

Consider, for example, the satisfaction rating of the various alternatives for status. This is highly subjective; the possibility that the rating of car *X* might be raised by adding some fancy accessories and going to the "top-of-the-line" should be examined. Even at the addition of something like $500 to the purchase price, an increase in the status satisfaction rating from 61% to 75% would reverse the order of preference.

The ratings in Table 3-2 are just best estimates and the range of uncertainty was not specified. Suppose the complete results of the rating procedures gave three estimates for each objective—a low estimate, a "best" estimate, and a high estimate.

TABLE 3-3.
Satisfaction Ratings in Percent
(Lowest Possible—Best Estimate—Highest Possible)

Purpose	Car *X*	Car *Y*	Car *Z*
A. Comfort	90- 94- 95	80- 89- 95	82- 85-100
B. Safety	85- 90- 93	80- 90- 95	90- 92-100
C. Status indication	50- 61- 63	75- 80- 85	100-100-100
Combined ratings	.38-.52-.56	.48-.64-.78	.74- .78-100

A simple and common form of sensitivity analysis is the so-called "worst case" analysis, an *a fortiori* sort of approach which is a familiar tool in engineering. To verify preference for car Z over car X, the worst case analysis simply assigns all the lowest possible ratings to car Z and the highest possible ratings to car X. The effective cost would then turn out to be

$$\frac{43{,}900}{(.95)(.93)(.63)} = 78{,}900 \quad \text{for car } X \text{ and}$$

$$\frac{58{,}000}{(.82)(.90)(1.0)} = 78{,}600 \quad \text{for car } Z$$

The preference for car Z is then strengthened. Now the worst case analysis is not really satisfactory for a problem such as this where there are many variables with questionable estimates. Some of the cost data are probably very questionable too in this kind of a problem; it is most improbable that all of the uncertainties with respect to car X would turn out to be favorable to X and that all of those for car Z would turn out to be unfavorable. So even if the worst case analysis failed to verify the choice of car Z, your enthusiasm for car Z might be only slightly dampened. However, your position is somewhat strengthened if the worst case analysis does not reverse your preference.

Although this is somewhat of an aside, suppose that in this attempt to choose among cars the decision-maker refused to assign any numbers as status ratings, claiming that this was so subjective he could not do so. If he later decided to purchase Z in preference to X on the basis of the calculations involving the other data, it would mean that he was implicitly deciding that the status rating of Z was at least 43% greater than the percent satisfaction with the status indication of car X, whether he admits it or not; that is, the rating for Z divided by the rating for X is greater than 1.43.[2] If the decision-maker is willing to agree with the rest of the procedure, and if he believes that his decision is correct, he is stuck with the assumption of greater than 43% difference because that is the only number missing and we can easily back-calculate to see what the number had to be in order to throw the decision one way or the other.

TABLE 3-4.
Present Value of Cost Streams for Varying Discount Rates

Rate	Car X	Car Y	Car Z
2%	126,000	153,000	154,000
6%	43,900	54,200	58,000
18%	16,700	21,600	26,300

[2] If the previous effective cost of X, $43{,}900/[(.94)(.90)X]$, equals the effective cost of Z, $58{,}000/[(.85)(.92)Z]$, where $100X$ = percent satisfaction with X as status indicator and $100Z$ = percent satisfaction with Z as a status symbol, then, setting the two effective costs equal we have $Z/X = 58{,}000/[(.85)(.92)] = 1.43$. That is, Z is 43% greater than X.

Another factor to which this analysis and, in fact, most policy studies are sensitive, is the assumption regarding the rate of discount. We took this to be 6%; if it were high, a decision-maker would tend not to tie his money up in an expensive car but to invest it. For example, if the rates were as in Table 3-4, the changes in the present value of the cost streams would be considerable.

The effective costs would then change proportionally and car *X* would be preferred to car *Z* at 18%.

TABLE 3-5.
Effective Costs for Varying Discount Rates

Rate	Car *X*	Car *Y*	Car *Z*
2%	241,000	238,000	197,000
6%	84,400	84,700	74,300
18%	32,200	33,800	33,700

11. CONSIDER DEFICIENCIES IN THE ANALYSIS

In this analysis, as in almost every analysis, many factors and alternatives have been left out. We have, for example, not considered renting a car or using public transportation for some of the purposes. We have neglected spillovers–the addition of a second car brings gains such as greater flexibility in transportation and possibly diseconomics such as the need for a second place to park.

Other considerations we have not taken in account are possible constraints. Suppose the downpayment required on each car was 10% of the purchase price. If the total available for a downpayment was $700, it would not matter that *Z* had been shown to be the best choice because one could not buy it anyway. Constraints such as this frequently simplify the procedure since they allow the easy exclusion of certain cases that would otherwise be suitable candidates for analysis. Constraints are often extremely important in public policy problems for they often appear as legal restrictions that may be purely arbitrary but extremely difficult to get changed or to bypass.

There are other deficiencies in the analysis. For instance, we assumed a set use period for each car. In reality, repairs are not distributed uniformly in time over the use period of a car and their frequency strongly influences how long one wants to keep a car. Also, they do not occur uniformly for every car of the same make and model. Furthermore, a satisfaction rating such as status depends on the age of the car and current fads; we have not taken such factors into account. These considerations could be taken into account by means of more complicated models that make the decision to replace depend on the condition of the car. To do so, however, would require an immensely more complicated analysis and cost of the analysis would again become a factor to consider.

We might also go in the opposite direction and use a simpler model for the cost. For example, we might have computed an annual cost as

$$\frac{P-R}{t} + A + Pi$$

where the last term represents the annual income lost by having money tied up in the car that might otherwise be invested at rate *i*. For different rates, this cost model would give Table 3-6, corresponding to Table 3-4,

TABLE 3-6.
Annual Costs for Varying Rates

Rate	Car *X*	Car *Y*	Car *Z*
2%	2,530	3,090	3,140
6%	2,690	3,370	3,660
18%	3,170	4,210	5,220

Effective annual costs, given in Table 3-7 would correspond to those in Table 3-5.

TABLE 3-7.
Effective Annual Costs for Varying Rates

Rate	Car *X*	Car *Y*	Car *Z*
2%	4870	4830	4030
6%	5170	5270	4700
18%	6100	6580	6700

We note that for each rate the cars maintain the same order of preference in Table 3-7 as in Table 3-5. Also, previously, when the status rating estimate for *X* was raised to 75% and the cost to $4500, the order of effective costs between *X* and *Z* was reversed; the same thing happens with this cost model.

12. SUMMARIZE AND, WHERE APPROPRIATE, RECOMMEND

Up to this point, the analysis is, in outline, conceptually identical with the choice process used by most successful decision-makers, or the analysts who work for them. The issues necessitating the decision are clarified and defined, certain assumptions are made, conflicting desires are weighed against each other, the various options are identified, the pros and cons of each option are investigated, and a scheme is devised to compare and rank the options. The typical decision-maker may be less inclined to record his assumptions and procedures explicitly or in quantitative form, but the process is the same.

In the light of the information developed by the analysis (and possibly on the basis of a recommendation by the analyst that attempts to take into account things that we left out of the analysis) a choice can be now be made which, in some sense of the word, is the best of all those considered. One final step remains—a record of what was done should be made.

Did this analysis really tell the decision-maker what car to select? No—but this is the case with all policy analyses. The decision-maker has better information than he had before. Some of it is fairly firm—say the costs in this example—other parts are far less so—the satisfaction ratings, for instance. But these latter are important for they give him an idea of the preferences of his constituency. He should thus be able to make a better decision than, without the analysis, he otherwise would have been able to make.

References

U.S. Senate, *Planning Programming Budgeting,* Subcommittee on National Security and International Operations, Henry M. Jackson, Chairman, 91st Congress, 2nd Session, Washington, D.C., 1970, p. 10.

Chapter 4

A BROADER FRAMEWORK

The successful application of analysis to policy problems is to a large extent an art. There are principles and procedures that offer considerable guidance but they are not so firmly grounded in theory or experience that they can be followed by rote. One must frequently do things that, if challenged, he cannot really justify and that depend to a large extent on intuition and judgment. What this implies, I think, is that the methodology per se is less crucial to the worth of the product produced than the skill and judgment of the analyst who applies it. An approach that may produce valuable insights when taken by one analyst may produce faulty or misleading conclusions when taken by another.

Policy analysis and systems analysis, like engineering, seek to see that things are done well and cheaply. In the confidence their results engender, however, they are far behind. Except for the most narrow cost-effectivenss comparisons, policy analysis lacks the strong analytical foundation and proven "rules-of-thumb" found in most fields of engineering. The late Leonard J. Savage[1] pointed this out:

> If a city hires an engineer to design a bridge, it may perhaps have his work checked by another engineer, but the city fathers will not presume to study his report with a view to seeing for themselves whether the proposed bridge is likely to collapse. They believe, with more or less reason, that the field of civil engineering is sufficiently well developed and a licensed engineer so likely to be firm in his science that his judgment in this matter is overwhelmingly better than their own. Similarly, they will trust the authority of their engineer that the clearance, carrying capacity, safety, and durability of the proposed bridge cannot all be increased without an increase in cost. The trade-offs among these values might in principle concern the city fathers, and in special cases they will. But, by and large, there will be none among them capable of or feeling the responsibility for going deeply into these matters. Where, however, a defense system for the nation is concerned, tradition cannot be relied upon; for there is too little of it. The trade-offs between the various values involved are properly felt to be high concerns of the nation and the immediate responsibility of high government officers. Of course, even in the case of the bridge, there are some trade-off considerations that the city fathers cannot dodge. They must make some decisions that depend on their estimates of the political temper of the city and of its probable future growth—decisions for which the civil engineer has no particular competence.

[1] Unpublished communication (1963).

With judgment and intuition by both analysts and decision-makers playing such a large role, far more than in engineering, one cannot expect policy analysis to serve as a tight mechanism for choice. Nevertheless, we approach decision problems as if such a mechanism were operating, using the framework outlined in the previous chapter for a simple problem of choice. Although few if any public decision problems fit this framework very closely—objectives cannot be reconciled, or values cannot be integrated to produce a preference ordering, or a complete comparison of alternatives is excessively burdensome, or some other difficulty arises—this is the most practical approach to take.

In providing analytic assistance to help a decision-maker (an individual or an agency) with a decision, our assumption is that he wants to do the best he can. Furthermore, we assume that for all states of the world that might come about he can decide, possibly with help from the analysis, the state he wants to reach and, based on the information available from the analysis and elsewhere, the "optimal" way to try to get there. If there are several decision-makers and they do not agree on the goal they want to reach, analysis may still be able to help but sometimes only by carrying out separate analysis for each goal. The characteristics that determine what is optimal, that is, what is most desirable or satisfactory, are determined by the decision-maker, not the analyst. The analyst may be able to point out the lowest cost way or the least risky way or what is preferable according to any given criterion but the choice of the criterion itself is the decision-maker's.

The Elements of Analysis

In the previous chapter we listed and illustrated five important elements or factors that must be considered[2] in facing, or in helping someone facing, a choice or decision of consequence. By a decision of consequence we mean one in which it is important that the best decision possible or at least a very good decision be made. In this section we discuss these elements and their role in detail.

The five most important elements are discussed below in an order in which they might logically be considered in tackling a problem.

1. The Objectives. The objectives are what a decision-maker seeks to accomplish or to attain by means of his decision. Often the most difficult task for the analyst is to discover whether or not the objectives, often stated or implied by the decision-maker in such general and abstract terms as to be ambiguous, are really the objectives that are wanted; if not, the analyst should investigate and get agreement on what the objectives should be. Unfortunately, in practice, despite its importance, a thorough inquiry into objectives may not be possible; the client may simply not countenance the inquiry. Alternatively, the decision-

[2] But not necessarily all in the same study. A policy analysis may help a decision-maker by considering just part of his problem, for example, the feasibility of certain alternatives he thinks he has.

maker may be multiple, say a legislative body, and information on objectives have to be inferred from actions taken.

2. The Alternatives. The alternatives are the options or means available to the decision-maker by which, it is hoped, the objectives can be attained. Depending on the particular question, they may be policies or strategies or actions of any sort. Alternatives need not be obvious substitutes for each other nor perform the same specific functions. Thus, education, recreation, family subsidy, police surveillance, and low-income housing (either alone or combined in various ways) may all have to be considered as alternatives for combating juvenile delinquency. In addition, the alternatives are not merely options known to the decision-maker at the start; they include whatever additional options can be discovered or invented later.

3. The Impacts. The designation of a particular alternative as the means of accomplishing the objective implies a certain set of consequences; we speak of these as the impacts associated with the decision. Some of these are benefits and contribute positively to the attainment of the objectives; others are costs, negative values in the decision, things the decision-makers want to avoid or to minimize. In addition, there may be other impacts associated with an alternative (often factors over which the decision-maker has no control) that, while they have little effect, positive or negative, on the attainment of the desired objective, must nevertheless be considered in the analysis. These are the spillovers or externalities, which may affect the attainment of certain of the decision-maker's (or the public's) other objectives. By broadening the objectives, the externalities can be internalized or made part of the study.

In the narrow sense, costs are the resources required to implement an alternative that is no longer available for other uses once it is implemented. In the broadest sense, costs are the "opportunities foregone"–all the things we cannot have or do once we have a particular alternative.

Many but by no means all costs can be expressed adequately in dollars or other quantitative terms. Others cannot. For example, if the goal of a decision is to lower automobile traffic fatalities, the delay caused to motorists by schemes that force a lower speed in a relatively uncrowded and safe section of road will be considered a cost by most drivers. Such delay not only has a negative value in itself, which may be expressed partially in dollars, but it may cause irritation and speeding elsewhere and thus lead to an increased accident rate or even to a contempt for law, a chain of consequences that can be very difficult to quantify.

4. The Criteria. A criterion is a rule or standard by which to rank the alternatives in order of desirability. It provides a way to relate objectives, alternatives, and impacts. A frequently used example would be: Given a fixed task, rank first the alternative that can accomplish it at the least cost.

5. The Model (or Models). The heart of any decision analysis is the existence or creation of a process that can predict or at least indicate the consequences that follow the choice of an alternative. That is, if an alternative were to be selected for implementation, a scheme or process is needed to tell us what impacts would be generated and to what extent the objective would be attained. This role is fulfilled by a model (or by a series of models, for it may be inappropriate or absurd to attempt to incorporate all the aspects of the required process into a single formulation).

The term model comes from the physical sciences. And, in fact, in policy analysis, in systems analysis, or in operations research, the most used models, on the whole the most useful, and often the only type even considered, tend to resemble "scientific" models. That is, they consist of a system of logical relationships that attempt to express the processes that determine the outcome of alternative actions by means of a set of mathematical equations or by a computer program. Our current capability to design valid and reliable models of these types is limited, particularly for public policy questions where social and political considerations tend to dominate. Here, in a sense less satisfactory models, that depend more, and more directly, on judgment and intuition and are thus not as precise and manageable and provide less definite feedback, may have to be used to predict the consequences of choice.

An explicit model, scientific or otherwise, however, introduces structure and terminology to a problem and provides a means for breaking a complicated decision into smaller tasks that can be handled one at a time. It also serves as an effective means of communication, enabling the participants in the study to make their judgments in a concrete context and in proper relation to the judgments of others. Moreover, through feedback–the results of computation in an analytic model or the countermoves in a game–the model can help the analysts and the experts on whom they depend to revise their earlier judgments and thus arrive at a clearer understanding of their subject matter and of the problem.

These secondary characteristics of a model–the separation of tasks and the provision of a systematic, efficient, and explicit way to focus judgment and intuition–are of crucial importance for they provide a route for tracing out the major consequences of choice when adequate quantitative methods are not available.

Three sorts of inquiry involving these five elements must be carried out during the analytic process. They involve (1) values and the determination of goals and criteria, (2) research and the identification of alternatives, and (3) evaluation and the comparison of alternatives.

Determination of Objectives and Criteria. A thorough investigation of what the objectives of the man or institution you are trying to help really are (as opposed to what he may at first say they are or, without some analysis, what they may appear to be) is needed. Therefore, insofar as his charter allows, the analyst should inquire into values and ultimate goals. To find out how effective the various means or alternatives are in achieving those objectives it is necessary to determine a way to measure their effectiveness, which again involves values. And finally, for helping the decision-makers decide which alternative to choose, criteria for ranking them in order of preference need to be determined.

An analyst can help here, and help a great deal, but he must remember that objectives and criteria for public goals must be set by the policy-makers responsible to those who are ultimately affected by the decisions that follow his study.

Identification of Alternatives. Analysis to help with a public policy decision implies research to find ways to accomplish the objectives. This involves an attempt both to search out existing alternatives and to invent or design entirely new ones. Research, invention, and intuition are also needed to discover or predict the conditions and constraints under which the various alternatives must perform. They are also needed to build models to identify the impacts of the possible decisions.

Comparison of Alternatives. After possible alternatives are identified, they must be investigated thoroughly and examined for feasibility (legal, economic, technical, political, and so forth). Their availability and the risks associated with them must be estimated. A way must be found to measure how well they accomplish the objectives. Finally, they must be compared so that a choice can be made.

In these inquiries one should try to look at the problem as a whole, not just at its separate parts. Thus, if our investigation seeks to reduce crime through an increase in policy activity, we should consider the related activities of the other public agencies that affect the situation–courts, corrections, welfare, probation, and so on. The analysis should also consider changing training, communications, technology, work hours, the possibility of using auxiliaries of various kinds–everything that might influence the outcome. In addition to the direct effects, one needs also to investigate the spillovers–the costs and benefits to those not directly concerned–what increased policy activity might do to the quality of life in the area, for instance.

Unfortunately, broad inquiries of this sort may have to be done very sketchily; time may be too short or the client may not allow it.

Looking at the entire problem may sound like common sense, but if taken literally, is impossible in practice, because everything in the world is connected in some way with everything else. To perform analysis, some considerations, usually many considerations, must always be left out. The determination of the boundaries–what to leave out and what to hold constant–is largely a matter of judgment supported by very rough analysis. The point is that we should at least think about the entire problem and deliberately decide what aspects we are going to tackle or include and what to leave out. It is also important for the analyst not to pretend that he has treated the whole problem.

For an analysis to take place, someone must have or anticipate a problem, that is, must be dissatisfied with some aspect of the current or projected state of affairs and want to consider a decision in terms of altering it. Although he may not realize it, he may not be clear at the start as to what the problem really is or why he is dissatisfied. And finally, when he has a proposed solution or just further information, other aspects to consider may occur to him and he may see the problem differently. For this reason alone, the process of analysis would be an iterative one.

In practice, things are seldom tidy: too often the objectives are multiple, conflicting, and obscure; no feasible alternative is adequate to attain the objectives; the predictions from the models are full of crucial uncertainties; and other criteria that look as plausible as the one chosen may lead to a different order of preference. When this happens one must try again. A single attempt or pass at a

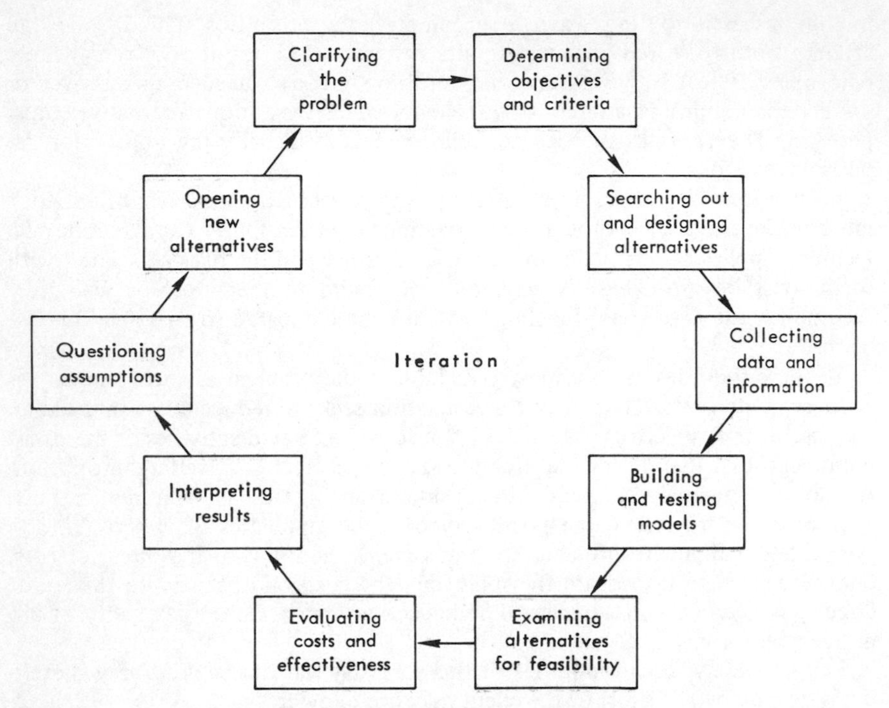

FIG. 4-1. The iterative nature of analysis

problem is seldom enough. Successful analysis depends upon a continuous cycle of formulating the problem, selecting objectives, designing alternatives, building better models, and so on, until the client is satisfied or lack of time or money forces a cutoff. The process is pictured schematically in Fig. 4-1 where some of the major activities are shown.

The Activities in Analysis

The process of analysis, what we do first and what we do next, depends on the problem and its content but no matter what the problem, five activities are involved: Formulation, Search, Comparison, Interpretation, and Verification. They are undertaken originally in the order listed but they may soon have to be carried out simultaneously as the iterative nature of the analytic process forces us to reconsider what we have done before.

An analysis starts with the process of Formulation, an attempt to clarify the objectives and define the issues of concern and to limit the problem. We then, or to some extent simultaneously, go into a research phase that I have called Search. Here we look for data, relationships, and new alternatives. We investigate possible alternatives and build models to discover their impacts in order to have a basis for comparison.

In the Comparison we use the probable impacts, that is to say, the costs, benefits, and other consequences likely to follow from each choice of alternative to compare and rank the alternatives by means of various criteria. Where Formulation ends and Search or Comparison begins is not very clearly marked. At any stage we may turn up information which causes us to go back and reformulate, search for additional data, change our models, and reevaluate.

The fourth activity, Interpretation, is exclusively a matter of judgment and intuition. It can take place at any stage but most obviously following the comparison. Using the information and insight obtained from the work up to and including the evaluation, the analyst derives conclusions and, possibly, suggests courses of action that he feels should be preferred under various conditions. The decision-maker should be very much involved here because he almost always has information and insight not available to the analyst. Decision-makers and political leaders are likely to be keenly aware of constraints that the context imposes, that must be taken into account in formulating policy.

These latter considerations are not always evident to the professional analyst for he may enter a situation at a late stage in its development, without the political background information of those who have struggled with its problems. He may proceed, even though consulting contenuously with his client, to develop a set of objectives and a rational program for achieving those objectives by ignoring political and other considerations that seem to him irrelevant or trivial—as well as, of course, those of which he is unaware, often because the client did not regard them as of sufficient importance to mention. Some such considerations may include local interests and prejudices, personalities, habits, prejudgments, rivalries, and other very real but unquantifiable human factors. It is in this activity that all of these and other elements that are involved in creating a political consensus, and which so often constrain political decisions, must be considered by the client and the analyst (even though he has done so earlier) before formulating policy.

Finally, one would like a Verification of the conclusions. The preferred scientific way would be by controlled experiment. Experiments, however, cannot be carried out before most public policy actions are taken, although they are much more frequent now than in the past and becoming more so. There are many other reasons but frequently time alone precludes experimentation. Hence experimental verification must usually be omitted or done vicariously, say, by simulation. Often we have no better way to verify our conclusions than to expose our analysis to the criticism of others.

As for the decision that follows the analysis, in many cases, even long after that decision has been made and the course of action that was selected has become history, we may still have no way to tell whether the best action was chosen. There is even a class of problems—a comparison of land-based with sea-based strategic missile systems would be an example—in which verification may be possible in principle, but the consequences of an actual test would certainly involve too high a cost. Thus, if we wanted to compare the damage that two missile systems, say Minuteman and Polaris, could do to the Soviet Union, we could, at best, use simulation to devise a vicarious experiment. Hence Verification for the study as a whole may have to be done crudely or not at all.

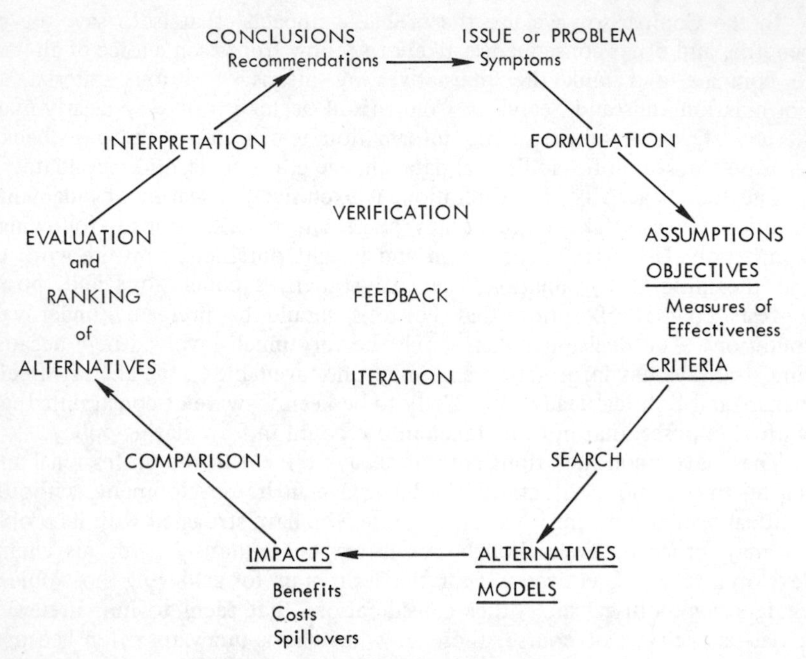

FIG. 4-2. A paradigm for policy analysis

A second diagram, Fig. 4-2 which appears above, and a simple example[3] may help to clarify the relationships between these various activities and the elements of a decision problem.

Let us say a problem is perceived as one of excessive turnover in a particular program, say a manpower training program. Whereas this may not be what is wrong with the program but only the symptom, to those operating the program it is *the* problem–excessive turnover with the constant influx of new personnel may mean inefficient operation and incomplete training for those who complete the program. To the operators the goal of any analysis may be obvious–reduce the turnover. The difficulty is to find out how and how much. To the level above that at which the turnover is taking place, however, the problem may look entirely different. They may feel they should first address the issue: Does a high turnover really have an adverse effect on the program operation? It may be that the field for which the trainees are being prepared is an exacting one, requiring considerable dedication. Convincing people early that they do not belong in the field, offering more opportunity to others who may be more dedicated, and avoiding the expense of carrying people along who are likely to become discouraged as soon as they start to work, may be a virtue of the program.

Thus the nature of the problem and what constitutes a solution may depend

[3] This example and a simpler version of Fig. 4-2 are from an unpublished paper by Kathleen Archibald.

on the decision level at which it is considered. At the upper level, for the first iteration, the questions are: What is it we wish to remedy? Are there other ways of remedying the problem besides reducing turnover? Suppose this go-around confirms the initial definition of the problem, i.e., that it is important to reduce turnover. An analysis can then be carried out with reduction of turnover taken as the objective, alternative ways of reducing the turnover rate can be identified and evaluated, and finally one designated as the "best" possibility for reducing turnover.

Public policy problems are usually complex and, as Fig. 4-2 indicates, feedback is everywhere. Objectives cannot be really set until everyone has a fair idea of what can be done; new alternatives cannot be designed until the hidden deficiencies in the original possibilities are uncovered; the boundaries of the investigation are influenced by how well the proposed solutions behave; and the true nature of the problem may not become clear until we have what appears to be a solution. And then feedback from the decision-maker may imply further iteration.

We have now said about all that we can say in the abstract with regard to "theory" or general principles for tackling the broad scale, ill-defined, complicated problems associated with public policy decisions. Most of the ideas involved are no more than common sense, but actually carrying them out can be extraordinarily complex and difficult. Moreover, they are not newly discovered; the concepts and procedures for carrying out an analysis to make a decision or to provide advice to someone else for that purpose have been known and practiced for a long time.

To keep the discussion from getting too abstract and to show the procedures clearly, we will use a realistic, but hypothetical, example to illustrate how one might go about carrying out an analysis to assist a resource allocation decision.

Assume the following situation. You have been assigned the task of supervising a study to provide an evaluation of possible future systems to provide passenger transportation along the heavily traveled routes between fairly large cities, something like 200–500 miles apart. The results are to enable government officials to make informed decisions on the provision of funds for research and development of a preferred mix of transportation systems for such routes to be operational in 10–15 years. The decision-makers are interested in such questions as: What are the advantages and disadvantages of R & D investment in various mixes of air and ground systems? What are the trade-offs between improved service and environmental impacts? Suppose also that your charter is broad, i.e., that you are not barred from spending funds to investigate particular systems or certain aspects of the problem as may sometimes happen.

The plan is to say a few words about each of the five activities in this context and then outline briefly how one might go about the required analysis.

FORMULATION

Formulation implies an attempt to isolate the questions or issues involved, to fix the context within which these issues are to be resolved, to clarify the objectives, to discover the major factors that are operative, and to get some feel for the relationships among them. At the start, these relationships may be extremely hypothetical, for empirical information and knowledge is likely to be in short supply; but the attempt at statement will help to make the logical structure of the analysis clearer. In a sense, formulation is the most important stage, for the effort spent restating the problem in the different ways, or redefining it, clarifies whether or not it is spurious or trivial and may point the way toward a solution.

Good policy analysis should seek to establish the boundaries of the issue under investigation where thought and analysis show them to be and not where off-the-cuff decision or convention, whether established by government jurisdiction, academic tradition, or industrial practice, would have them be. This may not be easy to do. Convention, jurisdiction, tradition, and practice may hamper the effort. The attempt to find a solution of one problem may require the solution of others. For example, urban air pollution cannot be adequately controlled without affecting many other aspects of city life including local transportation, and controls there may affect housing and jobs. Conversely, a comparison of transportation systems must taken into account their effect on pollution. Systematic investigation of these linkages may alter perception of the issues, expose hitherto unobserved relationships, and show the way to new opportunities.

In these inquiries, as we said earlier, one must try to look at the problem as a whole, not just at its separate parts. In transportation, for example, we must be concerned that the movement of people is not merely a matter that affects such things as miles of freeways and the location of airports, but one that affects the values of society—how many acres of recreational land it is willing to relinquish to rights-of-way or what noise level it will tolerate, for instance. Land usage, property values, air pollution, safety, must all be taken into account. The fact that many of these concerns elude quantification encourages the analyst to ignore them, but they are part of the problem and must come into the analysis even though it is purely through judgment. Sometimes good proxies or substitutes can be found for the real factors we want to quantify and failing that, qualitative arguments can be used to good advantage.

The process of formulation is highly subjective. We must, for example, consider what evidence will be meaningful and significant to the decision-maker we are trying to help. Thus, in our transportation example, the context should tell us that it will be sufficient to consider such things as highway additions, dedicated bus lines, air shuttle service, but that we do not have to go to far out schemes that might involve such things as 300-mile subways that penetrate mountain ranges or interurban rocket ships.

Many practical questions arise in formulating a transportation study such as this. Is there much freedom of choice in connection with the routes between the two cities? To what extent must each alternative system be designed? What use is this study going to be put to? Will it be the basis for a decision, or is it merely an input to further consideration of the problem? Are the sponsors ready to put up money to actually implement a proposed scheme that offers immediate improvement or are they more interested in supporting a lagging sector of the economy?

The proper formulation is important and a systematic approach through some fairly formal device such as an "issue" paper may be desirable. One reason is that, until the problem has been defined and the issues clarified, it may not be clear that the study effort will be worthwhile. A more extended discussion of the process of formulation is the subject of Chapter 5. Constructive suggestions are also found in (Checkland, 1972).

SEARCH

The search phase is concerned with finding the alternatives and the data and relationships on which the analysis is to be based. It is usually more productive to look for additional alternatives (including blends or modifications of those we already have) than it is to look for more precise schemes for comparison. Obviously, if we have no alternatives and no ideas about finding any, there is nothing to analyze or to choose between. If in the end we are to prefer a particular course of action, we must have discovered earlier that such a course exists.

One aspect of analysis that should be strongly emphasized (and it is better to be done in this section than at any other because search is the activity that depends most on talents that the analyst may not have himself–scientific, economic, legal, engineering, or whatever is required) is the role of the component or supporting study–scientific, engineering, political, etc.–in policy analysis. A sensible answer to a complex question requires a great many facts, insights, and judgments about the question, and the supporting studies are designed to provide them.

Thus, in our transportation comparison, the feasibility of a ground system may depend critically on the travel time. This, in turn, will depend on the route; for high speed, it may require avoiding excessive curvature and intermediate stops. Land use, land value, tax yields, all these must be taken into account in considering routes. Somebody connected with the study must have more than superficial knowledge of these things, for unless we have the facts and relationships reasonably in accord with reality, the results will be limited, possibly useless. Also, certain kinds of technical error, although trivial, can discredit the entire study. For instance, calculating in cubic yards in a field where the measurement is customarily in barrels. Even though simple conversion factors exist, the "old hands" may conclude that the analyst really doesn't have the necessary background. These supporting studies very often absorb most of the man-hours invested in the study.

As a consequence, an analyst's success may often depend on the technical competence found in his associates in engineering, sociology, cost analysis, and elsewhere. This book should not give the impression that broad analysis is done solely by generalists, academic types who work with their feet on the desk. This may sometimes, but rarely, be the case. Typically, it takes more than abstract thought to result in a good study. It takes research on the part of lawyers, mathematicians, anthropologists, and other specialists who have to travel, investigate, search for information and persuade the holders to part with it, and practice their professional skills.

Some needed information may be hard to come by, or may not even exist. The actual attractiveness of a new transportation system cannot be predicted with any great degree of accuracy or certainty by purely theoretical study. Systems that look attractive on the drawing board may not be so attractive to the riders. For major, critically important, policy issues–for example, the appropriate configuration for a National Health Insurance scheme–we might supplement the purely "paper" analysis by program experimentation. (When many billions of dollars are involved, it may pay to spend a few million on experimentation.) Even though properly conducted experiments are no longer rare, policy analysis must depend a great deal more on informed judgment than, say, advertising or construction engineering.

Every system belongs to a hierarchy. There are subsystems for every system and there are wider systems of which the one under study forms a subsystem. Invariably it happens that the relationship of the system to the wider system is hazy and unclear at the beginning. Hence a great deal of clear thinking will be necessary to fill in sufficient detail so that proper account can be taken of the interaction between the system and the wider system when formulating objectives and identifying costs and benefits. Thus transportation systems are imbedded in the system of economic and social life in the area, and it is important to determine how this broader system reacts, in order to evaluate the way the various transportation systems impact on it.

MODELING

To compare the costs and other impacts associated with different ways of operating or configuring a future system, it is necessary to estimate the performance of the system over a wide range of different conditions. To do this, the standard approach is to build a model of the system and its environment. For this model we would prefer a quantitative description of the behavior of the system, which can be used to predict its performance over the relevant range of operating conditions and real-life environments. If we are lucky, the same model can be used to project the outcomes that follow from the choice of a competing system. If not, we will need another model for that purpose. In their crudest form, such models could consist of a set of tables or graphs. At a more sophisticated level, they might be written in mathematical language, say as a set of algebraic or differential equations or as a computer program.

Model building is not a cut-and-dried process but rather a highly creative activity, possibly an art, certainly more than a craft. It is, of necessity, an iterative and adaptive process in which one moves from a state of little knowledge to one of greater knowledge. To design or evaluate a system, many different types of models may need to be developed. Experience and good judgment are required to decide which type of model should be used in any particular situation. Jenkins (1969) describes the process as follows:

Model building is the cornerstone of any scientific activity and, as such, plays a very important role in systems engineering. . .

However, there are important differences between the approach required for model building in systems engineering compared with that required for other scientific subjects. For example, in physics and chemistry, model building is almost an end in itself since the objective is to subsume as many facts as possible under the umbrella of a single model. In systems engineering, on the other hand, the end objective is to *optimize* the performance of a system and hence model building must be subservient to this objective. Thus a systems engineering team must:

(1) Ensure that model building is carried out *with a sense of purpose,* that is to enable a system to be designed as cheaply and efficiently as possible.

(2) *Tie together* the various specialisations that may be needed for building models of the subsystems. In particular, the systems team will always be emphasising to the specialist model builders that the important thing is to model those areas to which the overall economic criterion is sensitive.

(3) *Ensure that work is concentrated where it is most needed.* Because it takes an overall view, the systems team will be in a much better position to pin-point those areas which are cost sensitive and, if necessary, to ask for more work to be done in those areas. Conversely, it will be in a better position to stop activity in those areas which are not cost sensitive. For example, in a plant design project an overenthusiastic chemist may believe that it is necessary to predict a certain physical property to say within 1%, whereas from a system view point it may only be necessary to predict it to within 5% or even 10%. *As a general rule, models should be kept as simple as possible.*

(4) *Decide when the model is adequate* for the purposes for which it is needed, namely to predict overall system performance to a sufficient degree of accuracy. Model building is a fascinating activity and people, if left to their own initiative, tend to be carried away, tend to become emotionally involved and tend to waste time and money by overelaborating the model. By contrast, the systems approach requires discipline to ensure that when the model is adequate, model building is terminated and the optimisation stage begun as soon as possible.

(5) If the model is to be used for planning, see that an *effective dialogue* is conducted between the systems team and the managers who will use the model. Experience suggests that in such areas as production, investment and corporate planning, that this *dialogue must start whilst the model is being built* and not when the model is handed over. This is because models must be thoroughly understood by management if they are to be used effectively and moreover, because it is management that makes the decisions, their involvement at an early stage can often lead to suggestions resulting in substantial improvements and simplifications in the models themselves.

Of course the search for data can be endless because, in principle, the uncertainties of most planning problems can never be completely eliminated. When should theoretical analysis begin? What portion of the effort should be devoted to empirical research? Much effort can be and often is wasted in gathering the wrong data for failure to do the necessary theoretical homework first. On the other hand, much effort is also wasted in applying sophisticated analytic techniques to inadequate data—trying to make silk purses out of sow's ears.

Physical experiments and data gathering in general are expensive. Making plans and decisions in the face of uncertainty, even if aided by the best possible systems analysis, can also be very expensive. In the end, it comes down to a matter of judgment and successive iterations. We supplement the data we have by sensitivity testing; that is, we systematically but arbitrarily vary various parameters working out how these changed values affect the results. If the results seem to depend too much on uncertain data, we go back and gather some more.

COMPARISON

The ideal is to do more than to prepare a comparison of the alternatives so that the differences and similarities stand out; it is to rank them according to one or more criteria so that the decision-maker's choice is made easy. The situation is at its simplest when the decision problem is to choose one from a closed or at least a restricted set of alternatives, say because a previous decision has already been made to do just that.

Systems analysis has used two principal conceptual approaches to rank the alternatives. These work well when we are comparing alternatives where there is a clear-cut way to measure the level of accomplishment or a good proxy, something that occurs with systems that are very similar. The first is to fix the task or the level of effectiveness, and then seek to determine the alternative which is likely to achieve this level of effectiveness or accomplish the task at the lowest cost. The second approach is that of a fixed budget. For a specified cost level or budget to be used in the attainment of the given objective, the analysis attempts to determine which alternative will produce the highest effectiveness. A cost-effectiveness approach of either sort is useful when the relative merit of a number of competing proposals is under investigation; as we remarked earlier, it does not help much when the question is one of absolute merit—say, in deciding whether to allocate resources to adult education or to highway construction.

The fixed budget approach can still be useful for questions of absolute merit even if we cannot measure accomplishment in any clear-cut way. For example, we can "normalize" the alternatives with respect to budget dollars and then develop a table of comparative effectiveness measures. By inspecting this table and exercising their judgment on how these measures, which may be qualitative as well as quantitative, compare, decision-makers may be able to decide on the basis of their own subjective criteria which type of program they prefer.

In our transportation example we might use the fixed-budget approach to compare the alternative systems for intercity short haul transportation. If we try to compare these systems at a fixed level of effectiveness, however, we run into difficulty. Transportation effectiveness involves a meld of so many different aspects that to configure the systems so that they attain the same level of effectiveness may be impossible. In this circumstance we may be able to use some aspect of effectiveness as a base for a comparison—equal capacity, for instance. There are cases, however, where it would not be sensible to try to consider everything at the same level of capacity. In our example, for instance, we might want to consider supplementing highway transportation between our two cities with conventional aircraft. In comparing this with a high speed rail system, we would not necessarily want the rail system to be designed to carry the same number of passengers as the combined highway-air system. The fixed-budget approach thus is more generally useful. But often-times alternatives come in "lumps" of different sizes, such that working out a fixed budget comparison is not practical.

In many public decisions, say those involving a resource allocation, the decision-makers must judge whether or not a given undertaking is worth the cost. When this has to be done, the most common approach is to express the benefits and costs associated with each alternative in dollars as a function of time, discount the future benefits and costs at some appropriate rate, and then compare the alternatives on the basis of the present value of the net benefits. Alternatives can also be compared on the basis of the internal rate of return, i.e., that rate of discount which reduces the present value of the net benefits to zero. This is the "classical" Cost-Benefit Analysis approach–something that is hard to execute well in analyzing today's complex policy issues.

A fundamental difficulty is that in many public projects it is hard to classify every impact as either a cost or a benefit let alone find an acceptable way to express in dollar the benefits from such amenities as increased comforts in travel or the costs such as the need to move or destroy an historical monument. There are also other dimensions of interest to the decision-makers–for example, the costs may be paid and the benefits received by different sets of people. There is no foolproof way to bring these distributional impacts into the cost-benefit format.

For these and other reasons it may not be possible for the analysts to prepare an unambiguous ranking of the alternatives. When this occurs the best scheme may be to list the characteristics and impacts of the various alternatives and leave the task of ranking to the judgment and intuition of the decision-makers.

The Goeller "scorecard" is an attractive way to do this. For the transportation example we have been discussing, such a scorecard might look something like the one shown in Table 4-1.

The ideal scorecard presents the entire spectrum of impacts both good and bad, with an indication of who gets the benefits and who pays the costs. It is no more than a matrix in which the alternatives being considered are listed in one direction and the impacts in the other. Color coding may be added for emphasis.

No method of comparison is likely to answer all questions the decision-makers will want to consider for there are almost always aspects of the problem that have not been measured or even thought of by the analyst. The scorecard facilitates the asking of such questions.

INTERPRETATION

With this activity the real world gets back into the analytic cycle again (in the event it may have been forgotten by the analyst). It should go on throughout the entire analytic process but it is intensified at the stage at which the analysts consider what to tell the client in the ways of conclusions and how to respond if they are asked what action to recommend. Reaction with the decision-makers provides the impetus to try to correct the always imperfect mappings of reality that the models represent. The outcomes obtained from various models used

TABLE 4-1.
Sample Scorecard[a, b]

Impacts	Base (CTOL) Case	VTOL Case	TACV Case
Transportation Service Impacts			
Passengers (millions yearly)	7	4	9
Door-to-door trip time (avg.)	2 hr	1.5 hr	2.5 hr
Door-to-door trip cost (avg.)	$17	$28	$20
Airport congestion (% reduction)	0%	5%	10%
Financial Impacts			
Investment costs ($ millions)	150	200	2000
Net annual subsidy ($ millions)	0	0	90
Economic Impacts (peak year)			
Added jobs (thousands)	20	25	100
Added sales ($ millions)	50	88	500
Community Impacts			
Noise (thousand households)	10	1	20
Air pollution (% all emissions)	3%	9%	1%
Petroleum savings	0%	–20%	+30%
Households displaced	0	20	500
Land taken (acres)	0	25	8000
Taxes lost ($ millions)	0	0.2	2.0
Landmarks destroyed	None	None	Fort X
Distributional Impacts			
% low income trips taken on this mode	7%	1%	20%
% of noise-impacted households who are low income	2%	16%	40%

[a]From Goeller (1974).
[b]*Abbreviations:* CTOL = conventional takeoff and landing aircraft; VTOL = vertical takeoff and landing aircraft; and TACV = tracked air-cushion vehicle.

must now, if not earlier, be interpreted in the light of practical, real-world considerations.

A solution to a problem that has been simplified and possibly made amenable to calculation by idealization and aggregation is not necessarily a good solution to the original problem. Even if the model and its inputs are excellent, the conclusions proposed may still be unacceptable. The reason is obvious. Major decisions in the field of government policy are part of a political as well as an intellectual process. To achieve an acceptable solution, considerations other than those of direct cost and effectiveness are important: morale, tradition, political acceptability, organizational behavior. Government decisions have a major impact on our domestic economy and on public morale. The men who must somehow integrate these factors with the results of the study must necessarily deal with much that is uncertain—and, of course, the actions they decide to take based on the study results may differ widely, depending on how everything omitted is weighted. They may also draw entirely different inferences from his conclusions than the analyst.

Analysis can be helpful in reaching a policy conclusion even when the objectives are not agreed upon by the policy-makers but it is most helpful when they are. In public policy in particular, and in many other cases as well, objectives are seldom, in fact, agreed upon and this disagreement may not be uncovered by the analyst, no matter how hard he tries, until the decision-makers are presented with the consequences that are likely to occur—a look at the scorecard, for instance—and they realize the full implications of what the analysis assumed to be wanted. The choice, although ostensibly between alternatives, now may turn out to be really between objectives. Nonanalytical methods must be used for a final reconciliation of views. Although the consequences computed from the model may provide guidance in deciding which objectives to compromise, such decisions are not easily made, and in the end judgment must be applied. Another iteration may or may not be implied.

It is important for the user of the analysis to distinguish what the study actually shows from any recommendations or suggestions for action that the analyst might make on the basis of what he, the analyst, thinks the study implies, or in the light of his values, feels should be done. Some experienced and successful users of analysis hold that an analyst should not even make his conclusions known. For instance,

> Simply said, the purpose of an analysis is to provide illumination and visibility—to expose some problem in terms that are as simple as possible. This expose is used as one of a number of inputs by some "decision-maker." Contrary to popular practice, the primary output of an analysis is not conclusions and recommendations. Most studies by analysts do have conclusions and recommendations even though they should not, since invariably whether or not some particular course of action should be followed depends on factors quite beyond those that have been quantified by the analyst. A "summary" is fine and allowable, but "conclusions" and "recommendations" by the analyst are, for the most part, neither appropriate nor useful. Drawing conclusions and making recommendations (regarding these types of decisions) are the responsibility of the decision-maker and should not be pre-empted by the analyst. (Kent, 1967)

Analysts must certainly be careful about making recommendations that are biased by their personal objectives and values, including their methodological biases. There are, however, cases in which it is likely to be very helpful to have the analyst do such things as point out "dominance" in the rare cases when it exists or draws conclusions in terms of major alternatives, e.g.;

- If your value system (utility function) is such and such, then you should prefer D or H because. . . .
- On the other hand, if your utility is *this,* then A or F should seem preferable.

When new minds—the decision-maker's, for example—review a problem, they bring new information and insight. Even though the results obtained from the model are not affected, recommendations for action made on the basis of the model outcomes may be. A model is only an indicator, not a judge. Whereas the analysis may compare the alternatives under a great many different assumptions, using various models, no one should expect the decision in a complex situation to be made solely on the basis of these comparisons—and the same would hold even if an immensely more complicated version of the study were to be carried out.

When should an inquiry stop? It is important to remember that much more so in problems of public policy than in the private sector, inquiry can rarely be anywhere near exhaustive. It is not only almost always out of the question to collect—much less process and interpret—all the information that one would need for analysis so exhaustive that one has complete confidence in the results, but in addition public pressures for answers is great. Thus inquiries are partial, and the decision-maker must get along without the full advantage of all the potentiality of policy analysis and scientific methods. Inquiries cost time and money. This is not to say that such costs cannot sometimes be ignored; the point is rather that paradoxes arise if we allow ourselves to forget that almost all inquiries must stop far, far short of completion either for lack of funds, time, or a justification for spending further funds or time on them.

For these reasons an analysis is usually far from finished when it is presented to the client, or even later when it is published.[4] There are almost always unanswered questions that could be investigated further, particularly when the need for reporting or the lack of further support requires a cutoff. And the decision-maker's questions and reactions will usually make an extension of the study desirable.

Because analysts must so often present their results before they are fully ready, they may be wrong on occasion for this reason alone. They may also be wrong for many other reasons. But one cannot do useful work in the field of policy analysis unless he is willing to accept a certain amount of uncertainty and run the risk of being wrong.

VERIFICATION

Whether or not the decision-maker made a right decision based on an analysis can never be verified. One cannot judge by what actually happens for there are always circumstances beyond his control. What we do have some hope of

[4] It is likely to be even more incomplete, at the time the press and those who may feel threatened begin to speculate about the results.

verifying, of course, are the results obtained from the model or models, particularly when the situation is stable.

For instance, suppose the analysis seeks to determine better traffic patterns and controls in a central business district. Various alternatives might be tried out by means of a computer simulation and the one that seems best selected and implemented. Then unless something happens not taken into account in the model (e.g., property used for a parking lot is sold and immediate construction of a large forty story office building started) the predictions from the model can be verified from experience.

Sometimes the results from a paper study can be tested by experiment, provided the policymakers are willing to finance the experiment before a full-scale program is started. But even this may not provide verification—the experiment may not be properly designed or analyzed. But frequently experimentation is precluded because the decision-makers may feel obligated to act or politically motivated to act long before an experiment can be designed, performed, and evaluated.

The Process of Analysis

The process of analysis, how we go about tackling a problem depends, as we said earlier, on the nature of the problem. To make more than general statements about the activities involved, we need to outline an actual example. For this purpose we take the transportation example introduced earlier about which we have been making general remarks.

There are, of course, almost as many ways to approach the problem as there are analysts. One possibility might be somewhat as outlined below.[5]

No matter what approach is used, three things need to be done at an early stage.

The objective of the study is to help with research and development; therefore, funding decisions regarding future short-haul transportation systems, we need to consider the criterion problem and decide on a basis for ranking the possible alternatives. We also need to investigate what these alternatives might be and to design the various mixes of systems in which they are to be compared. Also, because the time period in which the new systems can be available lies ten to fifteen years in the future, we need to investigate the environment or environments in which they are to operate and in which they must be compared.

A procedure for carrying out the analysis may be presented schematically as in Fig. 4-3.

Transportation systems have so many impacts on the environment and affect the economy in so many ways, including certain groups of people (far more and in different ways than others) that the criterion problem is a difficult one. Goeller (1974) groups the impacts into the following categories:

Transportation service impacts occur to the users of the system. They include

[5] This illustration is based on a hypothetical example discussed in Goeller (1974).

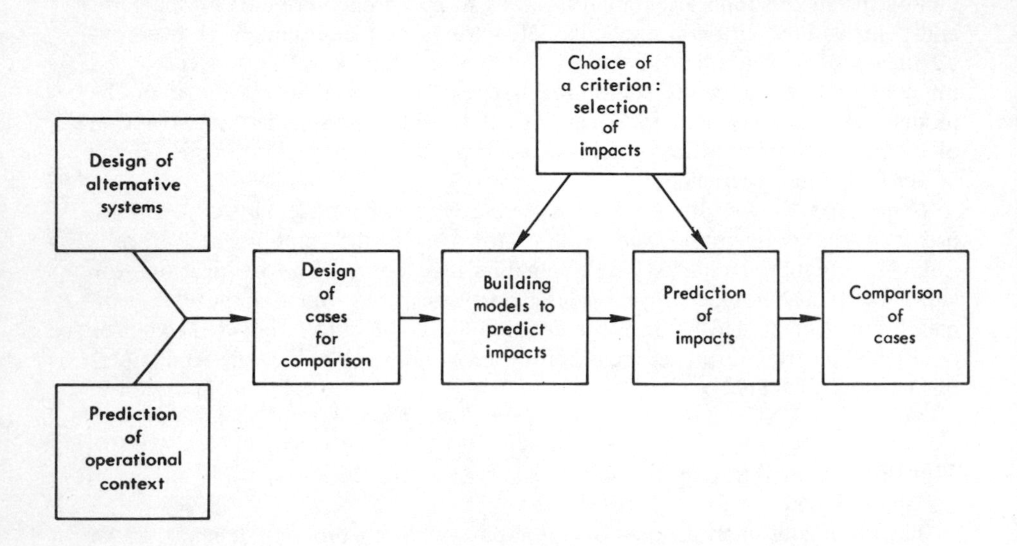

FIG. 4-3. Stages of the analysis

travel volumes, the door-to-door trip times and costs, and the changes in air and surface congestion at existing airports as their traffic is diverted to other airports or modes.

Financial impacts occur to the operators and society. These include investment costs for vehicles and facility construction, and any annual subsidies needed to operate the system (in addition to the revenues collected from fares).

Economic impacts involve changes in the income and employment in the region being studied. They include added jobs as a result of the construction of the system (primarily in the peak year) and added sales as a result of the direct expenditures on the system and the indirect effects of the respending by the people and firms directly employed on the project.

Community impacts involve changes in the activity patterns, tax base, and environment, which would occur to the various communities in the region as a result of the construction and operation of the transportation systems. Specific examples include the number of households annoyed by excessive noise, the amount of air pollution, the savings in petroleum consumption, the households displaced by system construction, the amount of land taken and the resulting tax losses to the community.

Distributional impacts consider how the various aggregate impacts are distributed among different social groups (e.g., income) and locations. These include the percentage of total trips taken by low income households that are made on a particular mode (showing whether the mode is disproportionately unattractive to the low income group) and also the percentage of the households annoyed by excessive noise who are in the low income group (showing whether a particular mode gives disproportionately high noise exposure to the low income group).

Because there are so many impacts, which are so diverse and hard to value in dollars, it seems desirable to use the scorecard approach mentioned earlier, that is, to present these impacts to the decision-makers in their natural units to be ranked, rather than attempting to devise some other means of solving the criterion problem.

Now consider the alternatives. One mode that will surely constitute a part of any 1985–1990 short-haul system is some form of automobile and bus transportation. Other modes, some novel, will complement and compete with the use of the automobile and bus.

A reasonable approach is to define a base case consisting of the most likely mix of transportation systems for the time period, under the assumption that existing trends and plans continue. In addition to the automobile–bus mode, the base case would include an advanced CTOL (conventional takeoff and landing) aircraft operating from existing regional airports. We might then introduce further alternatives by taking variations on the base case. These might be obtained, for example, replacing the CTOL aircraft with VTOL (vertical takeoff and landing) aircraft for all short-haul flights. Another variation might be to introduce a TACV (tracked air-cushion vehicle), which rides on a thin cushion of air on a concrete guideway or track.

Once a reasonable set of alternatives has been defined, the idea is to compare them, find out what patterns of impacts the decision-makers prefer, and then introduce further variations, literally "fine-tuning" the systems to produce something the decision-makers regard as optimal.

The comparison results depend on the environment in which the systems operate. For this purpose, one, two, or three regions can be specified. This essentially fixes the physical and geographical data. For the period 10 or 15 years ahead, population shifts, economic condition, travel patterns, and so forth must be predicted for each region—and, possibly, because the uncertainties are considerable, more than one forecast should be made.

We are now ready to work out the impacts. This must be done on a case-by-case basis—a particular alternative (mix of transportation modes) viewed in the context of a particular regional forecast (e.g., 1987 VTOL with auto access to airports evaluated in the Northeast Corridor with competing auto and bus modes assuming a minimal population growth).

We now need to develop models for working out the impacts in each case. To give an idea of what may be involved, consider Goeller's (1974) description of what is involved in predicting the noise impacts:

Assume that we have predicted the pattern of CTOL operations at a specific California airport in 1985. (This would be but one component of a full assessment of the noise impacts of one alternative case.) This pattern, the individual vehicle noise levels, the geometry of the airport, and other data would be inputs–along with a noise-annoyance criterion[6]—to a noise-exposure computer program. Rand's noise-exposure program generates a noise contour; anyone residing within the contour is exposed to a noise level that exceeds the specified annoyance criterion. This noise contour in turn becomes an input to another computer program which superimposes it on a pattern of fine-grained population data.[7] The output of the latter operation is a count of households within the contour, i.e., households impacted by excessive noise. Obviously, this process can be readily repeated by the computer for each CTOL airport with short-haul operations. The resulting household counts, when aggregated over all airports in the alternative, quantify our selected measure of noise impact.

Having developed the required models, measures of the various impacts are obtained and the set presented to the decision-makers for their inspection (Table 4-1). This alone cannot be expected to complete the analytic work, for the decision-makers will have questions and will want additional comparisons made and further systems and mixes of alternatives investigated. After their questions are answered, however, the decision-makers should be in a far better position to make the required decisions than they would have been without the analysis.

References

Checkland, P. B., Towards a systems-based methodology for real-world problem solving, *Systems Engineering,* 1972, **3**, (2), 1–30.

Goeller, B. F., *System impact assessment: A more comprehensive approach to public policy decisions.* R-1446-RC, The Rand Corporation, Santa Monica, California, to appear.

Jenkins, G. M., The systems approach. *Journal of systems engineering,* 1969, **1**, (1), 3–49.

Kent, Glenn A., Maj. Gen., On analysis. *Air University review,* 1967, **XVIII**, (4), 50.

Quade, E. S., (ed.), *Analysis for military decisions.* North-Holland Publishing Company, Amsterdam, 1964.

[6] The noise-annoyance criterion specifies the noise level that people will tolerate before they become "annoyed." It could be derived from the results of attitude surveys taken near busy airports or from measurements of the noise level where ordinary speech becomes difficult.

[7] For this purpose, Rand has used the 1970 First-Count Census, whose areal unit is the "block group." A block group, on the average, contains about 300 households and is 1000 feet on a side.

Chapter 5

INITIATING THE ANALYTIC PROCESS

A policy analysis is generally initiated when someone perceives a problem exists. A decision-maker, for instance, may realize he needs help—he may have noticed that something is going wrong with one of his programs and he may need help to decide whether to continue the program or how to improve it. Alternatively, an analyst or some other person may suggest to the decision-maker that something needs looking into. The latter may then decide to have an analysis done. To this end he may consult with an individual analyst or an analytic unit within his agency or he may contract with someone on the outside, a university faculty member, for instance, or a management consulting firm. Sometimes it is the outsider who suggests the analysis. But no matter how the process is started, it is not always clear at the start what the problem is or at what level it should be tackled or even that there is a problem.

Defining the Problem

When beginning to work on a problem, the analyst should interrogate the decision-maker or whomever it was that commissioned his analysis and all other persons within the organization or associated with the problem who seem likely to be able to help. In particular, he needs to ask such questions as

1. How did the problem arise? Why is it a problem?
2. Who are the people who believe it to be a problem?
3. If it involves implementing a decision made higher up, what is the chain of argument leading to that decision?
4. Why is a solution important? If an analysis is made, what will be done with it? Will anybody be able to act on the recommendations?
5. What should a solution look like? What sort of solution is acceptable?
6. Is it the right problem anyway? Might it not be just a manifestation or a symptom of a much larger or deeper problem? Would it be better to tackle this larger problem, if there is one?
7. Analytic resources are always limited. At this stage, does it seem that there would be a return from the study effort that would be justified or would this analytic effort be better applied elsewhere?

By questions of this and other types and their answers, a clearer picture should begin to emerge regarding the nature of the problem, its scope, and the benefits likely to result from an extensive analytic effort.

A major job may be to determine what the decision-maker really wants to accomplish. Ultimate goals may be easy to state but more immediate objectives that lead toward them are harder to determine. Someone has to look out for the public good and although this is a responsibility of the man who makes the decision, it may be very difficult for him to determine what that public good is; even if he knows it abstractly and ultimately, it may be difficult for him to pick out shorter-range targets that really carry him where he should want to go. Time spent on this aspect of the problem is well spent; there is no point in looking for the best way to achieve the wrong objective.

It is easy for an analyst to accept a client's view of what is wanted without further discussion and then to set about searching for feasible alternatives and gathering data without giving a thought as to whether the stated objective, if attained, will contribute to solving the problem under attack. Because policy studies have resulted in rather important changes, not only in how the policy-maker carries out his activities, but in the objective of his policy itself, it would be self-defeating to accept without inquiry[1] the client's or sponsor's view of what the problem is. A case in point would be the New York City–Rand Institute's work on rental housing. This changed the objective of the city housing policy from one of attempting to maintain the supply of available housing solely by new construction to one of helping to maintain the supply, in addition, by keeping old housing from leaving the market.

But how is the analyst to determine a better formulation of the problem or to know that another formulation is superior? His only possible advantage lies in analysis. He examines alternative formulations and the answers and decisions they imply. Using the few facts and relationships that are known at this early stage and assuming others, the analyst simply makes an attempt to solve the problem. It is this attempt that will give him a basis for better formulation. He frequently has some idea as to possible solutions to the problem, for otherwise he probably would not have been asked to work on it. A systematic approach helps. One such is the "issue paper," a form of inquiry that was developed in association with one of the many attempts to implement Planning–Programming–Budgeting.[2]

Preparing an Issue Paper

The preparation of an issue paper is essentially nothing more than a formalized approach to problem definition. It attempts to identify what the problems at issue really are, to isolate the fundamental objectives involved, to suggest

[1] Unfortunately not all clients are willing to support the necessary inquiries here. It is my belief, nevertheless, that something can usually be done.

[2] State-Local Finances Project [1968].

alternatives and appropriate measures of effectiveness, and to identify the population subgroups that are affected. It lists the government agencies and other organizations concerned with various aspects of the issue. It finds and lists the resources currently or that can readily be applied to the problem. It stops short, however, of an investigation and comparison of the impacts of the various alternatives—for otherwise it would be the analysis itself.

An issue paper is supposed to be as complete an assessment of all that is currently known about the problem or issue as the readily available data will allow. The original idea of an issue paper was to explore the problem at a depth sufficient to give the reader a good idea of its dimensions and the possible scope of the solution, so that it might be possible for a decision-maker to conclude either to do nothing further or to commission a definitive study looking toward some sort of an action recommendation.

An issue paper could, of course, stand by itself as a description of the problem area in order to provide an improved perspective by management on the problem. But in the sense we wish to use it, it is to set the foundation for an in-depth policy analysis by acting as the first phase.

We would like an issue paper to answer such questions as:

1. What is the magnitude of the problem? How widespread is it now? How large is it likely to be in future years?
2. Toward what public objective should programs for meeting the problem be directed? How can estimates of progress toward the objectives be made? What proxies might be used if the estimates do not seem to be directly measurable?
3. What specific activities are currently being undertaken by the government that are relevant to the problem? What alternative programs or activities should be considered for meeting the problem.

In order to be reasonably certain no aspect is overlooked, it is a good idea to systematically work through a standard format even though all the required pieces of information are not available and those that are available may not be of the desired accuracy or dependability. The format presented below was modeled after one used in association with the implementation of a PPBS system in the State of Hawaii. In addition to helping with problem formulation, such a format can help to provide management perspective.

The main sections of a standard issue paper format might be as follows.

A. Source and Background of the Problem
B. Reasons for Attention
C. Groups or Institutions toward which corrective activity is directed
D. Beneficiaries
E. Related Programs
F. Goals and Objectiveness
G. Measures of Effectiveness
H. Framework for the Analysis
   1. Kinds of Alternatives

2. Possible Methodology
3. Critical Assumptions

I. Alternatives
1. Description
2. Effectiveness
3. Costs
a. Year by Year
b. Public, Private
4. Spillovers
5. Comments on Ranking
6. Other Considerations

J. Recommendations that might follow

K. Appendices (as needed).

The original from which this format was adapted was designed primarily for use with a PPBS system. For problem formulation alone one may wish to omit certain sections. In some cases it may be desirable to add categories; for instance, one dealing with constraints—for example, political constraints which might preclude a particular alternative. An explanation of the various sections listed above and the sort of material that ought to be included within them is in order.

A. SOURCE AND BACKGROUND OF THE PROBLEM

This section asks for a short, clear description of the problem, issue, or situation that is proposed for analysis, where it came from, what it symptoms are, and why it is a problem. It should discuss such questions as: What seems to be the real problem? What appears to be the cause of this problem? To what extent are these causes known?

B. REASONS FOR ATTENTION

This section is inserted to give others the reasons why the situation needs attention and requires analytic resources at this particular time. It helps to justify the expenditure of public funds on analysis.

C. GROUPS OR INSTITUTIONS TOWARD WHICH CORRECTIVE ACTIVITY IS DIRECTED

Exactly who or what should be considered the target of a program or decision to eliminate a problem is not always clear or unambiguous. Essentially it is that at which the activities are aimed; the target might be agricultural pests, or narcotics peddlers, or the welfare rolls, or factories discharging pollutants. The idea is to spell out the specific population if it is other than the general public. Their general characteristic, such as age group, race, income class, special needs, and geographical location should be identified. If a health program were being considered, the target group might be those with a particular illness or those who were in a high-risk group. We would also want to distinguish between the numbers of those who are actually in the target group and the numbers who might potentially be included if certain alternatives were undertaken. For example, if the program contemplated a change in, say, welfare eligibility rules, the actual target group would be composed of all those who would normally be expected to claim the benefits under the new rules.

D. BENEFICIARIES

These would be those who stand to benefit by the effort to achieve the objectives. For a program to eliminate coyotes, sheep ranchers would be the beneficiaries; for one to eliminate air pollution, the general public. The people here may be the same as those in the target group, but again they may not. For example, whereas criminals would be the target group for a law enforcement program, the victim and potential victims would be the beneficiaries of such a program. For a drug-control program, the users might be both the target group and the beneficiaries. Alternatively, the target group might be the suppliers. In all cases we would like quantitative estimates of the numbers in the various beneficiary and target groups affected. The question of who gets the benefits and who pays the cost is central to the resolution of almost all public policy problems and issues. Hence it is particularly important to have these groups clearly in mind and to alert the decision-makers to whom they are.

E. RELATED PROGRAMS

This section should list the specific activities currently being undertaken by others that affect the problem. Such programs should be identified, and, to the extent possible without great effort, costs should be found and the impacts estimated on the target and beneficiary groups. An indication of the number currently being served by these programs in each population group identified earlier is helpful. Projections based upon whatever plans are known are also helpful. It is important to make the list complete: federal, state, city, and county, and private sector programs should all be identified because obviously a new effort to achieve the same or essentially similar objectives should be integrated and coordinated with those of all the other agencies having effect on the problem or issues.

F. GOALS AND OBJECTIVES

Unless the objectives at issue in the decision are carefully, clearly, and precisely spelled out, it is very likely that the rest of the analytic effort will not be directed properly. Measures of effectiveness may be misconceived; incorrect and ineffectual alternatives are likely to be designed; and resources will be invested without favorably affecting the problem that generated the analysis in the first place.

An issue paper should suggest the objectives toward whioh programs for meeting the problem should be directed. The fundamental purpose is to look beyond the immediate physical output to call attention to the ultimate goal toward which the solution is directed. Any possible negative effects on other government programs should be identified.

As we will see in more detail in the chapter on objectives, there is no simple procedure for insuring that objectives will be directly and perceptively specified. There are some techniques, however, which may be helpful. At this stage the only technique really available is to involve several other people in the discussion and resolution of objectives. They need not be and, in fact, probably should not be, involved with the substance of the problem under analysis. They should be critical, skeptical-minded outsiders. A thoroughgoing exploration of the issues with such people will substantially increase the likelihood of getting objectives that point toward the real goals.

G. MEASURES OF EFFECTIVENESS

Measures of effectiveness attempt both operationally and quantitatively to measure the degree to which the objectives are being achieved. The basic difficulty is to ensure that the measures of effectiveness reflect in an accurate way the achievement of the ultimate good effects intended. In general, measures of effectiveness relate to the immediate and physical measurable capabilities or effects of the particular system, piece of equipment, or program being considered. The hope here is to suggest measures that seem to be good proxies.

H. FRAMEWORK FOR THE ANALYSIS

This section suggests how the complete analysis could be approached methodologically if the issue paper were to lead to a full-scale study. The applicability of various techniques such as linear programming or computer simulation might be discussed or the merits of considering equal cost versus equal capability systems considered. Of course, at this early stage the methodology can only be spelled out very tentatively because until we actually get into the problem and have lived with it for a while, we will not know which schemes have much chance of being successful.

The discussion of the kinds of alternatives to be considered should include consideration of their level. The highest level of choices concern what to do, including the related questions of why, for whom, and when? If questions of this sort are involved, they must be considered first before questions of how.

It is important that the major assumptions should be laid out for all to see. This is very important because these assumptions are what determine the nature of the analysis. An otherwise good analysis may turn out in the end to be unsatisfactory simply because somewhere along the way a hidden, and hence unquestioned, assumption was made that turned out to be untrue or otherwise inappropriate.

I. ALTERNATIVES

At this early stage an attempt should be made to identify the set of alternatives, even though some of them may seem, a priori, unorthodox, impractical, too costly, technically infeasible, and so forth. It will be time enough later when the analysis is well started and when most of the available data has been collected to do more than make "top-of-the-head" judgments about which alternatives are attractive ones and which are not. At the start we want to emphasize a free wheeling, wide open, unconstrained look at all of the possibilities. At least up to a point, but one must consider that too wild a proposal may turn off the client.

At the beginning it is probably adequate to deal mainly with the pure alternatives and simply suggest, where it seems obvious, the possibility of mixed solutions or combinations of parts of the various pure alternatives.

For each alternative certain items of information are helpful. Among these items are the following.

1. *Description*

A brief description of how the alternative works, the personnel and equipment required, the facilities needed, the policies and technologies used, the agencies involved, and the activities carried out will help in making judgments about its advantages and disadvantages.

2. *Effectiveness*

A quantitative but largely judgmental estimate of how effective the alternative is likely to be in meeting the objectives as determined by any of the suggested measures of effectiveness may be useful in considering whether to continue the analysis.

3. *Costs*

Some estimate of the total cost usually can and should be given in the issue paper. This will, of course, be refined greatly later. One of the important causes of faulty analysis is a failure to identify all of the cost elements that should be attributed to a given alternative. In particular, those costs that are not going to turn up until some time in the future are sometimes missed. It may, in fact, be a good idea to disaggregate the cost, by cost categories. That is, assign nominal costs to various categories such as investment and operating, and, as far as possible for each category separate the costs roughly with respect to the various government agencies and elements of the private sector where these costs are going to be paid.

4. *Spillovers*

Spillovers (side effects or externalities) are those unintended effects that spill over onto persons or things outside the problem or issue which is being directly analyzed. At the start an effort should be made, even if the estimate must be a purely qualitative one, to identify the significant spillovers associated with each alternative and note who bears their costs or enjoys their benefits.

5. *Comments on Ranking*

An issue paper can do no more than make a very tentative comparison of alternatives. The strongest comment could be that certain alternatives appear to be potentially more attractive than others and should receive priority in their investigation and exploration. The general nature of the results may become apparent, however. A forewarning that a new program is likely to be required, or taxes raised, or that a certain group will be much disturbed may be very helpful.

6. *Other Considerations*

Any point not previously discussed relative to alternatives would be discussed at this point in the issue paper. The things that might warrant explicit treatment are questions of risk and uncertainty, social and moral constraints, if any, and possibly, matters of political feasibility. It is not expected that these would be handled in an exhaustive fashion at this stage, although we would try to do so later. Nevertheless, it is important that the problem areas of this type related to each of the alternatives be clearly identified and at least the direction of their effects on the attractiveness of the alternatives.

J. RECOMMENDATIONS THAT MIGHT FOLLOW

About the only recommendations appropriate to an issue paper or that can be made at the formulation stage are: (A) drop the whole subject as not worth further analysis; (B) continue the analysis, but on a low priority basis; or (C) undertake or continue with a full-scale study. Occasionally, however, so much information may become available during the preparation and preliminary design of the study that one of the alternatives may be found to be so unmistakably superior that purely on this basis alone one will be warranted in making an action recommendation. In this case, the preliminary design to initiate the study in fact becomes the study.

K. APPENDICES

The formulation stage of the study will not have an appendix. However, an issue paper might have one as the purpose of an issue paper is to produce a document that is short enough and clear enough to be read in its entirety by executives and legislators who have to make up their minds whether to continue or not and on what scale. Because we might want certain items to be checked by another analyst, the presentation of the necessary material can be handled by using the appendix section for references, footnotes, back-up tables, and charts, raw data, extensive calculations, special exhibits and so on.

Organizing the Work

SELECTING THE STAFF

Once the problem has been defined, the way it is to be tackled must be laid out. Policy analysis, in most cases, requires an interdisciplinary activity. For this, it is desirable to set up a study team or project team. This may be difficult. Most organizations do not have on their staffs all of the skills that may be required for a complex problem. Even when the staff is large, it may be necessary to supplement their knowledge with outside consultants–economists, sociologists, and specialists of various kinds–transportation engineers, pollution experts, etc. In fact, few organizations have a large enough staff to make a team of more than two or three. Hence most policy analysis is probably done by a single analyst. In that event, the analyst or analysts must make the best of the situation.

For a large project the staff might consist of something like the following:

1. *The project leader* is ideally an experienced analyst who has some knowledge of the area being tackled, in most circumstances the more the better.[3]
2. *The project team* is a group of analysts or specialists of various sorts, preferably from several different backgrounds, who can devote a large fraction of their time to the project. A multidisciplinary team is valuable not only because of the variety of knowledge and skills they bring to the problem, but because of their differing perspective. A problem looks different to an engineer than it does to an economist, a sociologist, or a mathematician, and their different approaches can contribute greatly to finding a solution. The number on the team depends on the issue being examined; if it is large, a steering group of not more than four or five is a good idea.
3. *Consultants* are specialists in the important aspects not represented on the team. Most of these might not work on the project full time and some of them might have no role other than that of an expert who answers a question when asked. A few might be assigned subprojects–to build a model to predict the impact of traffic noise on property values, for example.

Our transportation example in Chapter 4 would probably require an economist to carry out the analysis of supply and demand as affected by each particular transportation system, a cost analyst to consider the resource impact of such systems, and engineers or scientists to calculate the environmental impacts–noise, pollution, etc. We also want someone to consider the political organizations and institutions that will be involved in implementing any system. And in addition, of course, we would like to have access to transportation engineers and designers in order that we may configure the systems that are to be considered, plus operations researchers, computer programmers, and mathematicians to build the various models and to program the models for the computer.

The project leader should try to ensure that the team is given broad terms of reference and that they have access to any information or person required. Moreover–something which is often neglected–*the project team should apply a systems approach to the conduct of their own activities in order to ensure that the work is carried out systematically and in the time allowed.*

As far as reasonable, problems should be tackled in the order in which the results are likely to be needed. One must always remember that even partial

[3] But not someone so caught up in the current conventional wisdom that he cannot have an original thought.

analysis can be very valuable and, hence analysts, when called on, as happens frequently, to report on the project before it is completed, should try to be in a position to respond with something useful.

LAYING OUT THE TASK

To guide the project, an outline or work statement with estimated times and personnel assignment is helpful. For the transportation example of the previous chapter something like the following might serve:

Work Statement

1. A number of alternative transportation systems for intercity passenger traffic will be defined and investigated for the 1975–1985 time period including systems that:

a. Emphasize highway additions
b. Emphasize dedicated bus lanes
c. Emphasize air shuttle service
d. Emphasize new passenger service on existing (as opposed to new) rail facilities
e. Emphasize high-speed ground transport such as a tracked air cushion vehicle employing new facilities
f. Emphasize an extension of the subway and elevated system from the largest of the two cities.
g. Blend the above systems designed to be mutually complementary and reinforcing

For each system, a time-phased design will be developed.

2. Several alternative regional scenarios of future demand will be defined.

3. The impacts (benefits and disbenefits) that are significant to the decision will be identified and considered in the study. The list of impacts investigated may vary during the course of the study, as a result of interactions with government officials and the public.

4. For each case (defined as a particular mix of transportation systems operating in a given regional scenario) an estimate will be made of the impacts produced by the system. The exact list of impacts will be developed during the study (item 3 above), but the following list is representative.

a. *Service Impacts*
Door-to-door travel time and cost
Congestion reduction
Passenger volumes carried
b. *Cost Impacts*
System investments

Cost per passenger mile
Annual subsidy required

c. *Economic Impacts*
Changes in income and employment (as a result of the construction and operation of the systems)

d. *Community Impacts*
Amount of land taken
Number of households displaced
Number of households exposed to excessive noise
Amount of energy consumed
Amount of air pollution

e. *Social Impacts* (the distribution of benefits and disbenefits by social group)
What fraction of the travelers are:
high-, medium-, low-income?
What fraction of those exposed to excessive noise are:
black, white, or other?
low-, middle-, or high-income?

5. The alternative systems will be evaluated and compared in various ways.
6. Oral presentation material will be prepared and a report written.

The process of formulation is highly subjective, depending greatly on the judgment and intuition of the analyst. Any work statement such as that above leaves many aspects of the problem still to be decided. For example to what extent should we deal with the problem of the interface, that is, investigate how the systems would tie into the nationwide transportation network? To what extent should we assume private ownership? If new roadways or tracked vehicles are to be considered, what criterion should govern route selection? How should the systems that we are going to compare be defined? Among other things, what capacity should they have, how fast should they go, how many intermediate stops? Whereas it might be reasonable to consider that highways or railways could handle all the traffic between the two cities, what capacity should we assume for the air shuttle system?

Choosing the Approach

In any study, the problem as viewed at the start seldom remains static. Interplay between a growing understanding of what is involved and what might be involved in the future with what is known at the start forces a constant redefinition. For example, initially, high-speed ground transportation might seem a very attractive alternative to freeways; but once preliminary routes are laid out and the number of households of various types displaced, or affected by a high noise level, becomes clear, a scheme which might have seemed politically feasible at the start may no longer appear so.

Primarily as a result of discussion and intuition, the original effort to state a problem should suggest one or more possible solutions or working hypotheses. As the study progresses, these original ideas are enriched and elaborated upon—or discarded—and new ideas found. The process of analysis is an iterative one. Each hypothesis serves as a guide to later work—it tells us what we are looking for while we are looking. In a sense good analysis consists of a series of "hypothesis testings." It is not a mistake to start with an idea as to the solution; the error is to refuse to abandon such an idea in the face of mounting evidence.

It is important to recognize that anything going on in one part of an activity or system is likely to affect what goes on in other parts. Thus in our transportation investigation, if we increase the capacity of the air shuttle system, the problem of access to the airports may become a significant factor by a change in the demand for some other system, such as airport buses. The natural inclination in problem solving is to select a part of the problem and to analyze it separately, to reduce the problem to one that looks manageable. We must do this in policy analysis but we recognize that when we do divide a problem we are sacrificing something. Ultimately, we have to solve the problem that really exists. This calls for us to extend the boundaries of the problem studied as far as required (largely a matter of judgment) to determine what interdependencies are significant, and to evaluate their combined impact.

It is, however, necessary to suboptimize. Inevitably, not all decisions can be made simultaneously, or at the highest level, or by one individual or group. Some decisions must be delegated to others, others must be delayed. Decisions as to objectives, however, must be made first but these may have to be revised later for, as is obvious, one cannot always decide what one wants to do until one has some idea of what one can do.

Analysts and decision-makers must always consider actions that pertain to only part of a complex problem. We would like, for instance, to minimize simultaneously door-to-door travel time, systems investment, cost per passenger mile, annual government subsidy required, amount of land taken from agriculture, number of households displaced, pollution generated, and noise level, etc., but this is beyond our capabilities. Other choices are set aside temporarily, possible decisions about some things being neglected, and specific decisions about others being taken for granted. Breaking up the problem in this way brings with it, on the one hand, a number of advantages. Models can be less aggregative, more detail can be taken into account, and they may for that reason yield more accurate predictions. In contrast, suboptimization of this sort can introduce errors because objectives and criteria for subproblems can so easily be chosen in ways not consistent with those for the full problem.

Even for small-scale problems, the number of factors under consideration at any one time must be reduced until what is left is manageable. To consider in detail the complete range of possible alternative ways available ten years hence to provide transportation between two cities is impossible. The mass use of bicycles, horse-drawn carts or rocket vehicles probably does not belong in our comparisons; these and other possibilities have to be discarded somewhat arbitrarily.

Fortunately, the vast majority of alternatives will be obviously inferior and can be left out without harm. The danger is that some alternative, better than those covered by the analysis, might also be left out. Thus, although constraints must usually be imposed to reduce the number of alternatives to be examined, this should be done by a preliminary analysis, not by arbitrary decree. Moreover, such constraints should be flexible, so that they may be weakened or removed if it appears in later cycles or iterations that their presence is a controlling factor.

At some time we must decide on a specific approach to our problem. Uncertainty is the essence of our difficulties, particularly uncertainty about future environments and contingencies. Here we often may have almost no idea about the value of parameters and little more about the distribution of possibilities that may have to be handled. For our transportation example it is necessary to forecast what the situation will be a number of years from now when the new transportation systems might be installed. Will growth continue as it has in the past or will some circumstance that we do not know about cause population or other factors to change? For such reasons we must look for an approach that offers a hope of producing something constructive in spite of the great uncertainties that might exist. That approach is to make excursions and to examine the alternatives under many differing assumptions, to consider what can happen under a wide range of contingencies and circumstances. We call this sensitivity testing. Its importance can not be overemphasized.

A point of view that has evolved from experience with military problems runs something like the following.

Any attempt to determine a sharp optimum or the unique best solution to a problem having a large number of largely indeterminate parameters, some subject to the influence of others, is probably doomed to failure. The goal instead should be to search out and emphasize those policies or systems that are close to the best for the most likely contingencies and, at the same time, are largely insensitive to many uncertainties–specifically, policies that might work well under many widely divergent contingencies that could even give reasonably satisfactory performance under more catastrophic circumstances. This characteristic of a system is what the military man intends when he speaks of "flexibility."

It is helpful here to recall that the primary purpose of policy analysis is to advise a decision-maker, help answer his questions, sharpen his intuition, and broaden his basis for judgment. In practically no case should we expect to "prove" to the decision-maker that a particular course of action is uniquely best. The really significant problems are just too difficult, and there are too many considerations that cannot be handled quantitatively. If we insist on a strictly quantitative treatment, we are likely to end with such a simplified model that the results will be almost meaningless, or arrive at an answer too late to be useful to the decisionmaker.

These observations suggest two rules of thumb:

1. Throughout the inquiry, it is well to look for gross differences in relative costs and effectiveness among the alternatives and, specifically, for differences of

the sort that have a chance of surviving many likely resolutions of the various uncertainties and intangibles. Thus, in comparing future systems, the questions to address is what systems have a clear advantage, rather than precisely how much better one is than another.[4] Something of this attitude no doubt underlies the motto of the Systems Analysis Office in the Office of the Secretary of Defense during the time of Alain Enthoven: "It is better to be roughly right than exactly wrong."

2. All comparisons should be made with the uncertainties in mind. Before a choice between policies is recommended a careful investigation of the relevant range of contingencies should be made, particularly of those that involve major changes in the environment such as political acts or actions that depend on human caprice.

Gathering the Data and Information

The discovery and design of a successful public policy may depend upon uncovering a great deal of information and data. An issue paper outlines the problem in a way that can be very helpful in telling what is needed. But the data and information are not necessarily easy to acquire.

Suppose, for example, we are investigating what to do with mental patients discharged from state hospitals or institutions but still needing some follow-up care. Various possibilities might be considered ranging from placing these people in private homes in residential areas to such institutions as half-way houses or intermediate care facilities under close state supervision. Where does one go for information on such a problem?

A first source is probably an expert, someone who has worked with mental patients or written articles about their care. Such people are often biased and hold fixed opinions on the subject but in talking to them it may be possible to discount their bias, to get a good deal of information, and, in addition, to locate documents. This last is important. It is often assumed that documents can be searched for and located without the cooperation of people but this may not be the case when one is looking for up-to-date or current material. Getting access to the material once it is located may also be a problem.

The gathering of data and information may be, and often is, the most expensive and frustrating stage in any policy or systems analysis. Data will be required not only to provide information about the operation of the policy or

---

[4] One analyst notes that even the discovery that the differences among the alternatives are insignificant can have a considerable value to the decisionmaker. "This is especially true if sensitivity analyses have been made . . . [and] the final results are still within relatively narrow ranges. Given results of this kind, the decisionmaker can be less concerned about making a mistake regarding the quantitative aspects of the problems, and he may then feel somewhat more comfortable about focusing more of his attention on the qualitative . . . considerations" (G. H. Fisher, *The Analytical Basis of Systems Analysis,* The Rand Corporation, P-3363, May 1966).

system that is to be designed or built, but also to make forecasts of the environment in which it will have to operate.

The requirements for efficient information and data collection are stated by Jenkins (1971) as:

(1) Clear thinking about the problem so that relevant sources of information can be tapped.

(2) Ability to communicate in speech and in writing so that people are stimulated into parting with information and into volunteering a new information.

(3) A grasp of statistical techniques so that the significance of the data can be appreciated and used in subsequent model building and decision making.

The data needed to analyze a problem area or system are defined quite directly by the model that is designed to handle the prediction problem. Therefore it is useful to specify the model fairly early; it will clearly and unambiguously indicate the sort of data needed and thus can lead to more efficient organization of what can be an extremely laborious job. In fact, if it turns out that the model as originally specified demands data that cannot be furnished, some reworking of the model to accommodate what is available may be sensible. Sometimes lack of information about a particular variable can be handled by assuming a set of values for it—often a "high," "medium," and a "low" value—and then running the models with these values held constant. It is important to let the model determine the data needed, otherwise there is a tendency to try to gather all remotely usable facts, hoping to come up in the end with those that prove to be needed. Fortunately, complete enumerations are not always required. Very often judicious sampling will suffice. But even this can be expensive in time and dollars.

Research for a policy analysis is likely to develop into a contest involving locating relevant sources, finding ways to gain and maintain access to those sources, accumulating information that is useful in obtaining further information, and then preparing for the contingencies that result from the political nature of the problem.

Policy analysis data are seldom primary data. What the policy analyst uses most often is data that has been collected by others. He may have to interpret it, criticize it, and put it together in various ways, using people to lead to other people and documents, and documents to lead to other documents and then to people. Such items as agency reports, legislative hearings, even statutes and regulations, usually require somebody to locate them. The analyst has to ask leading questions but be tactful. He also has to be careful not to antagonize people and possibly to stay away from certain individuals. Also, with public documents, the nominal author may not be the real author and the best information may come from the later source. Some of these practices are foreign to academic research. For example, finding out what one official is doing or thinking by asking another rather than by asking the man himself. Of course, the one may be biased about the other, but then one is always biased about himself.

Policy analysis becomes a political resource. It is a weapon of persuasion even if the tone is neutral. If it can influence decisions, it has power. Informants are sensitive to the political implications of what they say. The best leverage for getting information is knowledge. Because knowledge increases over time, it may be a good idea to interview the most difficult source last. To protect against criticism that a study is politically motivated it is desirable to touch base with anyone or any institution that might try to undermine what has been done by claiming to have been ignored. It is a good idea to quote such parties, if at all possible. In fact, very frequently in a report an analyst will raise an objection that has been made by others to his work or conclusions in order to say something about it and thus bring out points that might not occur to the decision-makers until later.

References and Notes

State and Local Finances Project, *A First Step to Analysis: The Issue Paper,* PPB Note 11, The George Washington University, July 1968.

Jenkins, G. M., The systems approach. *Journal of systems engineering,* 1969, 1, (1), 3–49.

Chapter 6

OBJECTIVES AND CRITERIA

Importance

> Has it ever occurred to you that to design a bungalow is a problem different in kind from designing the first atomic bomb and basically much more complex? The difference is that for the Manhattan project the criteria of success were simple and compatible, whilst for even the smallest bungalow they are multiple and conflicting. Scientific questions are 'why?' questions. Technological questions are questions of 'How?' or 'How best?', where the criteria of 'best' are given. But the simplest policy question is a question of what *is* best, where the criteria of 'best' are multiple and conflicting and their relative weight is just what has to be decided. (Vickers, 1971)

When a decision-maker or public body calls on analysis to help make a decision or choose a policy, it is with a purpose in mind, some objective or goal that the decision or policy to be chosen is supposed to accomplish. Unfortunately, such goals or objectives, may not be clearly stated or even fully perceived by the decision-maker, let alone be unambiguously communicated to the analyst. In addition, even for the individual decision-maker, and certainly for a composite one, the goals are likely to be multiple and they may often be conflicting.

Consider, for instance, the goals of primary and secondary education. They include increasing knowledge and skills, possibly even mental ability; instilling values in keeping with those of society; improving character; freeing parents to work by keeping children in school; changing the social structure by providing for mobility; and other items.

These different goals are to be achieved, if at all, at different times, some immediately, others–such as providing educated mothers who will in turn motivate their children to study–occurring in the next generation. A number of these goals are intangible and cannot be quantitatively measured; some even defy qualitative formulation. Others conflict; education should promote both social stability and social change, both satisfaction and the fulfillment of social needs, both desire to achieve (for the gifted) and contentment with self (for those with low ability).

In addition, what might be regarded as a desirable goal or benefit by some people, say, changing the social structure, is certain to be regarded as a disbenefit by others. No policy output can be evaluated unless it is determined whether "more" or "less" of the given output is good—percentage of criminals in jail, number of highway miles paved, number of individuals on the welfare roles—all can have this indeterminancy.

If analysis is to help a decisionmaker, it is necessary to determine with as much precision as possible what he wants to achieve. With multiple objectives, this may require determining in detail how much of each he is willing to give up in order to obtain more of another. With an individual it may be possible to determine these trade-offs with satisfactory accuracy, say by asking the right questions, but when the decisionmaker is the state or a public body, we are usually unable to do this. Hence, for this reason alone, even if we are able to determine the best solution possible to the problem we have tackled, it may not be "good enough."

Selecting an operational objective or task, one that a policy can be designed to achieve, is seldom an easy task. Even if the goal appears obvious, it may be so vague that the effort to formulate it operationally may lead to misinterpretation. Consider an elementary school principal who would like to know whether the program is operating properly and accomplishing its purpose or whether it should be altered. Before he can really find out, he will have to state the goals of providing an education in some operational form. For example, the goal of developing reading skill to the extent of each pupil's ability might be formulated as "to achieve a mean score for his students in the top X% on the standard reading screening test." But this may not give the true picture and other attributes may have to be measured.

It is crucially important to choose the "right" objective, more important than it is to make the very best choice from among the alternatives. A wrong objective means the wrong problem is being tackled. The designation of the wrong alternative may merely mean that something less than the "best" policy is being recommended. Frequently, when the uncertainties are great, we can do little better than demonstrate that a suggested action is in the given direction anyway. If the recommendations from the analysis are not going to be acceptable to the decision-maker because they are designed to help him with something he is not trying to achieve, it is absurd to spend a long time figuring out how to obtain them most efficiently.

The key to the analysis of public policy is the identification of policy and organization goals. To be "rational" in our choice of actions, we have to know what we want to achieve. Until it is decided what a policy or program is supposed to accomplish, information about alternatives, costs, and effects has no particular value except insofar as it helps to firm up the goals. But that, of course, is extremely important, for knowing what can be done may be extremely helpful in deciding what one should to try to do.

Economic analysis, operations research, and management science usually assume that the decision-maker knows what he wants. One way policy analysis and systems analysis are distinguished from these disciplines is that they consider the problem of what ought to be done as well as how to do it.

A thesis of this chapter is that, in many important problems, the objectives cannot be taken as given. As one of his major tasks the analyst must help the decisionmakers to clarify and ultimately define the objectives they have in mind before he can do much to help them determine which alternative to select. And in order to decide among the latter he needs to find criteria that will reflect those objectives.

In classical systems analysis we considered the choice of objectives but in a very narrow sense. We usually had relatively similar alternatives to compare and the problem was more to select an objective that would help with that comparison than it was to guide the policy-maker in what he wanted to do. Thus, for example, NASA might have a number of space missions to perform without sufficient funds for all. The problem was to choose a specific objective that would enable us to rank the feasible missions in order of preference. We did not consider the broader question of whether the resources devoted to such missions should better be used for other far different goals, say, improving urban transportation or medical care.

Some Difficulties with Objectives

Legally, there are few requirements for policy-makers to make clear statements of their goals. On the other hand, in many circumstances, there are clear political advantages in being ambiguous about, and even silent, concerning, goals and objectives. In fact, the tendency to state objectives in broad statements of direction rather than as a precise target, as Hovey (1968) remarks,

> . . . is probably attributable to a hesitancy by any participant in the political process to accept anything less than perfection as his publicly stated objective. It probably also reflects a reluctance by those participants to subject their actions to the precise tests available when performance is promised in specific terms within specific time periods. The political repercussions of a small increase in aircraft accidents are more easily borne by a President committed to a safe and efficient civil aviation system than by a President who has committed his prestige to "the reduction of the civil aviation accident rate by 10 per cent annually over the next 5 years."

As a typical example of broad statement, President Johnson indicated his thinking on objectives of his major programs as follows (Weekly Compilation, Oct. 1966):

> I should like for this period of the 20th century to be remembered as the period when we produced more food to feed more people—because food is the necessary sustaining ingredient for all the other things—the period when we spent more money and more effort on educating more people; the period when we spent more time and more dollars on providing health for our bodies; the time when we did more planning and added more acres for conservation, recreation and beautification."

Public policy objectives are usually the product of a political process, not of an individual decision. Here it may be easier to agree on an action than on a statement of purpose; the motives of the individual participants may be far different. Thus the United States Congress may establish a Civil Rights Commission but the decision to do so may reflect far different desires by the various factions involved. For example, quoting Hovey (1968) again:

> . . . the decision may reflect (a) a desire to get on with civil rights by group A, (b) a desire to forestall stronger legislation by group B, (c) a desire to get more facts on the subject before deciding on the merits by group C, (d) a desire to find jobs for one's relatives by group D, and (e) a desire to logroll for support for a group of hometown projects by group E.

In stating objectives, officials do not always reveal what they really want, sometimes because they feel they must maintain a position or front, sometimes because it might mean loss of support, sometimes (maybe most often) because they do not know what they want. Hitch (1960) put it this way:

> Even in the best of circumstances ignorance and uncertainty about high level objectives make reliance on official definitions a precarious procedure. We know little enough about our own personal objectives. There are doubts about the therapeutic value of psychoanalysis, but no doubt at all that it has revealed to surprised patient after patient that his *real* motives for action bear little relation to what he believed were his motives. National objectives can only be some combination of distillation of the objectives of people who comprise (or rule) the nation; and we should learn to be as skeptical and critical of the verbalizations and rationalizations that pass for national objectives as we have learned to be of apparent or claimed personal objectives. No lower order of caution and sophistication gives promise of success.

For public policy issues, the question of *whose* objectives are the relevant ones is also important. Certainly it is some subset of the citizens–those of today or tomorrow.[1] The decision-maker or his agency may not be the one whose objectives we should be considering. The decisionmaker is merely the person or organization with the authority and responsibility for changing the system. Those served by a government agency usually are not in that position but their goals may be the relevant ones. The philosophy of the public policy analyst cannot be that of the engineer who depends on the customer to specify what he wants. The role of the analyst in helping a client who has goals and values that the analyst feels contradict the basic values of democracy and human rights is not at all clear. There may be circumstances when the analyst, recognizing that he has clients beyond the immediate one, should not continue to provide advice (see also Chapter 18).

How Analysis Can Help

How does an analyst help identify objectives? There is certainly no reason to assume that he is better qualified to select goals than those who hire him, yet

---

[1] Here, the question of "time preference" is of fundamental importance. Some people (and *most* organizations) have very high discount rates; that is, they put very much more "weight" on objectives for *now* and the *very near* future, than on objectives for the distant future. Others, however, may stress the future and heed objectives regarding "future generations."

somehow he must obtain one or a most a limited set to use in his analysis. To help he has only one tool–analysis, supported by his judgment. Hitch (1960) put it this way:

> Nothing but rigorous, quantitative analysis can tell us whether some objective makes sense or not–whether it is feasible, how much it will cost. Nothing but rigorous analysis can reveal the conflicts between objectives. Nothing but inventive and ingenious analysis can uncover means or systems that contribute to several objectives, or that function in a wide variety of relevant circumstances, or that satisfy influential people or organizations with quite different views about higher objectives.

The fundamental reason why analysis is needed to clarify objectives is the following, well stated by Hitch (1960):

> First it is impossible to define *appropriate* objectives without knowing a great deal about the *feasibility* and *cost* of achieving them. And this knowledge must be derived *from* the analysis.

People do not know and may not able to decide what they really want, simply because the alternatives and their implications are not fully revealed to them. Beckerman (1968) offers an example:

> A simple example will illustrate this point. Suppose an economic advisor were asked, by a community of monks, to draw up a plan that would increase their incomes and hence enable them to provide, every evening, better services for the dead. One might then submit a plan which, say, involved providing better facilities for tourists to visit the monastery and one would draw attention to this possible objection to the plan. But the monks might then say that, far from this being an objection, they would on reflection, be delighted to have more tourists since a greater flow of tourists was desirable for its own sake in that it permitted greater scope for religious conversion. In the latter case one should then, logically, revise the whole plan, for if, after all, it appears that one of their revealed objectives was the provision of inducements to tourists, the original plan was clearly an inefficient plan since it did not allow for this additional objective. The new plan might then, say, incorporate a steel band as an attraction for tourists, but the monks might object to this on the grounds that they attached even more importance to silence than to services for the dead. This newly revealed independent objective would require a new plan that made provision for sound-proof facilities in their establishment at the expense, perhaps, of both services for the dead and tourists. This process can obviously continue a long way, if not indefinitely; at each stage new and hitherto unrevealed objectives coming to light, some of which would force them to appraise more deeply than hitherto their real order of preferences.

Although there is widespread belief that goals should and can be set independently of the plans to attain them, there is overwhelming evidence that the more immediate objectives are–possibly more often than not–the result of opportunities that newly discovered or perceived alternatives offer, rather than a source of such alternatives. For instance, the objective of landing a man on the moon did not arise until technology made it a feasible attainment.

Analysis can provide more than information on least-cost means of attaining a given objective. It can give data on costs or prices of achieving different objectives, or sets of objectives, through several alternative systems, so that the decision-maker is better able to choose among the possibilities on the basis of his system of preferences.

One technique to uncover objectives is to confront the man who must act with a list of alternatives and their consequences as determined on the basis of analysis and then ask him if he would be willing to follow the course of action implied by each of the alternatives, were further analysis to indicate that it should be preferred. If he says, "I can't do this for the following reasons," then we may get a better idea of what he would like to accomplish.

What should be done when no unique objective can be selected, the degree of attainment of which is the sole measure of performance or success, but when several objectives are present, some of which can be shown to be in direct conflict with others? This is the common situation in public life; an agency is told to cut costs and stay within the budget and at the same time to hire more minority staff members and provide better service, or to supply more housing for the public and simultaneously eliminate slums.

One possibility is to relate the real objectives to an economic base, to do somehow what is best in terms of net dollar returns. This may not be too hard within an industrial firm. But for a government agency, translating the benefits from the services of, say, the patent office or the FAA into dollars alone is harder and not very satisfactory. Nevertheless, this cost-benefit approach has been the dominant one in the public sector.

If objectives compete—that is, if people disagree on objectives or if an individual cannot determine what his objectives are—one can try to find a higher or more general objective on which there is agreement. One way to find such a higher objective is to look at the next higher organizational level. (Unfortunately, sometimes this may come down to a mere "do better"—and it is hard to know what this implies until we are able to trace out the effects.) It then may be possible either to carry out the analysis with the higher level objective or, by examination of the original objectives for consistency with the one agreed on, to make a choice.

For another thing, we can examine each objective to see if it is important only as a means to another objective; if it is, then we can eliminate it. For still another, we can examine each alternative to see if the attainment of any of the objectives would be unaffected by a choice among the alternatives. Again, if it is, we can eliminate that objective as a factor.

We may be able also to convene the people who disagree as a committee or panel and possibly get them to compromise and agree on a group objective through discussion or some other approach[2] for arriving at a group position.

One thing we cannot do is construct some unique group objective from all the individual objectives by a formula which automatically weighs all separate ones; both theory and experience demonstrate that this cannot be done satisfactorily.

[2] For example, by majority vote or by use of some process such as Delphi (of Chapter 12) where the median or other average is used as the group opinion.

To illustrate, let me quote Herman Kahn (1960), whose explanation I find to be particularly good:

> Consider a committee formed of three people, Tom, Dick, and Harry, as shown in the table below. Tom has the set

GROUP CHOICES[3]

Chooser	*Preference*
Tom	A to B, B to C, A to C
Dick	B to C, C to A, B to A
Harry	C to A, A to B, C to B
All	A to B, B to C, C to A

> of preference relations A to B and B to C. Because Tom is a rational person, I do not have to ask him which he prefers between A and C. If he prefers A to B, and B to C, then he prefers A to C. Similarly, Dick prefers B to C, and C to A, and therefore B to A. Lastly, Harry prefers C to A, A to B, and therefore C to B. Now if we ask this committee to vote for their preferences between A and B, we note that they give two votes, Tom's and Harry's for A over B. Similarly, if we ask them to choose between B and C, B will win out over C by two to one. We might then conjecture that since the committee is composed of reasonable people it should not be necessary to ask which they prefer between A and C. We ought to be able to deduce that they prefer A to C, since they prefer A to B and B to C. However, being experimentally minded, we might just go ahead and ask the committee which they prefer between A and C and find, to our amusement or horror, that two to one, they prefer C to A. This is most disturbing. If one is a rigid type, it may be extraordinarily distressing. It turns out that it is perfectly proper to be disturbed because, even after analysis, there seems to be no way in principle (and very often in practice) to make this committee act reasonably—unless we accept a rule of autocracy and delegate the decision making to one of them, to a dictator. Even this may not solve our difficulty. Some dictators, though they have the external appearance of being integrated, are really committees. In any case, there is no subtle voting method for a committee that is even mildly democratic that seems to avoid the difficulty.

Multiple and Conflicting Objectives

If efforts to agree on a single objective fail, we are faced with essentially two ways to proceed. One is to attempt to determine a preferred policy by working with such schemes[4] as

- ○ Establishing a system of relative values and tradeoffs among the objectives;
- ○ Setting up a preference ordering of the objectives and then optimizing in sequence;

---

[3] This "paradox of voting" seems to have been first pointed out by E. J. Nanson. The reference is given by Kenneth J. Arrow in *Social Choice and Individual Values,* Wiley, New York, 1951, p. 3.

[4] Elion (1972) contains a good discussion; this section owes much his exposition.

- Converting all except the most important goal into constraints, by agreement as to the minimum level of attainment acceptable.

The other possibility is to give up the idea of determining any sort of "optimal" policy and turn to "satisficing,"[5] that is to determining not the best solution to the problem, but one that is "good enough."

First, consider schemes that allow us to seek an optimal solution. As we have noted earlier, a policy analysis can seldom, if ever, represent a true optimization. The best we can do is find the most desirable or satisfactory alternative, under given circumstances. Full optimization would require simultaneous consideration of all possible allocations of one's resources among all the alternatives; taking into account the effect of exogenous events and their expectations. The world is too complicated; one simply cannot consider all contingencies and all alternatives.

In order to trade off one objective for another, we need specific agreements as to the quantitative amount that it would be acceptable to give up in, say, objective A in order to obtain some specific increase in attaining objective B. By this approach, multiple goals can be reduced, through scaling factors, to a weighted sum that represents the aggregate objective.

Cost-benefit analysis is an example of how this may be done indirectly. Here the benefits and costs of any proposal are quantified (usually by making heroic assumptions!) and expressed in monetary terms. The dollar values then provide a system of weights. Because the objective now becomes that of determining the system that yields the greatest return in dollars, a goal such as reduction in noise from a transportation system is automatically traded off, in dollars, for an increase in speed. The difficulty, of course, is to successfully express these factors in monetary terms.

When we are unable to obtain agreement on a set of relative weights, it may be possible to rank the objectives in some order of preference. For example, in a recent transportation study in laying out the routes for the ground systems, it was decided the goals in order of preference were, (1) maintenance of maximum design speed, considering curvature, steepness and safety, (2) low cost for land acquisition and elimination of grade crossings, and (3) minimum socioeconomic impact in noise and households displaced.

After the objectives are ranked, an "optimal" solution is determined with respect to the first-ranked objective, then with respect to the second, and so on.

In this approach it is assumed that, after optimizing for the first goal, there is still room for maneuver to search for the second. If this freedom does not exist, then the analyst can explore the flexibility that can be introduced by deviating from optimality with respect to the first. The process then becomes essentially another way of expressing trade offs, except that instead of posing the question in the form "How much speed are you prepared to give up for lowering the cost by $10 million?," it is stated as "What deviation from the optimum speed are you prepared to tolerate to allow more choice with respect to cost?"

---

[5] A term suggested by Simon (1961) to describe the behavior of a manager who does not seek the best solution to a problem but is satisfied with a solution that is "good enough."

By determining the minimum level of performance that would be acceptable for certain goals, and considering these levels as constraints, we may make it possible to find an optimal solution for the remainder. This approach at least insures that these goals are not simply ignored.

Satisficing (Simon, 1961) is an alternative approach. If one cannot reconcile conflicts between goals, then a way to proceed is to give up the idea of obtaining the "best" solution to the problem, set lower bounds for the various goals that, if attained, will be "good enough" and then seek a solution that will at least exceed those bounds. The question of what is "good enough" is largely answered by individual judgment or by a consensus of opinion.

The satisficer is not looking for a unique solution and he does not need to consider how goals are to be ranked or how conflicts between them can be resolved. He does have to worry to some extent that the performance standards specified are not too "tight," for then no solution may be possible.

The philosophy behind optimization is the hope to determine the best solution possible to the problem in the given circumstances. The satisficer, on the other hand, holds that in real world problems there are too many uncertainties and conflicts in values for there to be any hope of obtaining any sort of optimization and that it is far more sensible to merely set out to do "well enough" (but preferably better than has been done previously). To ensure this, he sets his constraints according to his best judgment and seeks a feasible solution.

The Relation of Objectives and Criteria to Effectiveness[6]

Let me now amplify the list of elements of a decision problem (Chapter 4) by inserting something we call an effectiveness scale, which, in turn, we use to define effectiveness. Assume we have reduced the problem to one with a single goal. Thus

Objective:	What we desire to achieve
Alternatives:	Competitive[7] means for achieving or moving toward the objective
Costs:	What must be given up to acquire each alternative
Effectiveness Scale:	Scale indicating degree of achievement of goal
Effectiveness:	Position on effectiveness scale reached by each alternative (determined from the model)
Criterion:	Statement about cost, effectiveness, risk, timing, etc. by which the alternatives are ranked

To illustrate, consider Los Angeles County, where there are large numbers of elementary school pupils of Mexican descent whose English is at best frag-

[6] This discussion follows L. D. Attaway, Chapter 4 in (Quade and Boucher, 1968).

[7] In the sense that if we do A, then we cannot do B. An alternative may contain complementary and supplementary elements; i.e., another alternative might be one-half A and one-half B.

mentary. An objective of one school program is to teach them better English. Alternatives might be special classes, supplementary programs, separation from other Spanish-speaking students, etc., or various combinations of these means. An effectiveness scale on which to indicate achievement might be the number of words in the English vocabulary at the end of grade two. The *effectiveness* of the program might be a number, the average number of words making up the English vocabulary of the pupils in that particular program at the end. A criterion for choosing a program might then be the maximum effectiveness obtained for a given number of teacher hours per pupil; or, better,[8] for a given cost for the program.

The rationale for an effectiveness scale is straightforward. Clearly, without such a scale on which the position of an alternative will indicate its ability to achieve the objective,[9] quantitative evaluation of alternatives would be next to impossible. The scale is a yardstick, along which we place our alternatives by means of some analytic or subjective technique of measurement; this position indicates the alternative's effectiveness. Now, people sometimes want to substitute the term "criterion" for "effectiveness scale," or replace "scale of effectiveness" with "measure of effectiveness." But to keep the following remarks unambiguous, and to preclude some of the semantic difficulties often met in similar discussions, we will use the terminology and definitions just given.

Considerations in Selecting a Criterion

Given that the cost and effectiveness of each alternative can be determined separately, one is still faced with the problem of how to choose among them. We need a test–a rule of choice or criterion. In principle, the criterion we want is clear enough: the optimal system is the one that yields the greatest excess of positive impacts (attainment of objectives) over negative impacts (resources used up, or costs and externalities or "spillovers" that reduce effectiveness). In practical problems, this ideal criterion can seldom be applied; there is no generally acceptable way to subtract dollars spent from lives saved by a medical program or from enemy targets destroyed in a military operation.

To illustrate how to go about selecting a criterion, let us consider a series of examples.

Consider an example based on our previous discussion of teaching better English to elementary school students of Mexican descent. Suppose two programs have been proposed and that on the basis of cost and effectiveness they are predicted to compare as in Fig. 6-1.

Assume that both programs are designed for the same grade levels and are

---

[8] Using teacher hours as a cost measure implies no way to evaluate other inputs, for instance, Spanish speaking teacher aides or electronic devices of various kinds.

[9] Many policy issue questions may require more than one effectiveness attribute to indicate achievement of the objective. In the example of this section, a concern with how the vocabulary increase is distributed among the students might also be important.

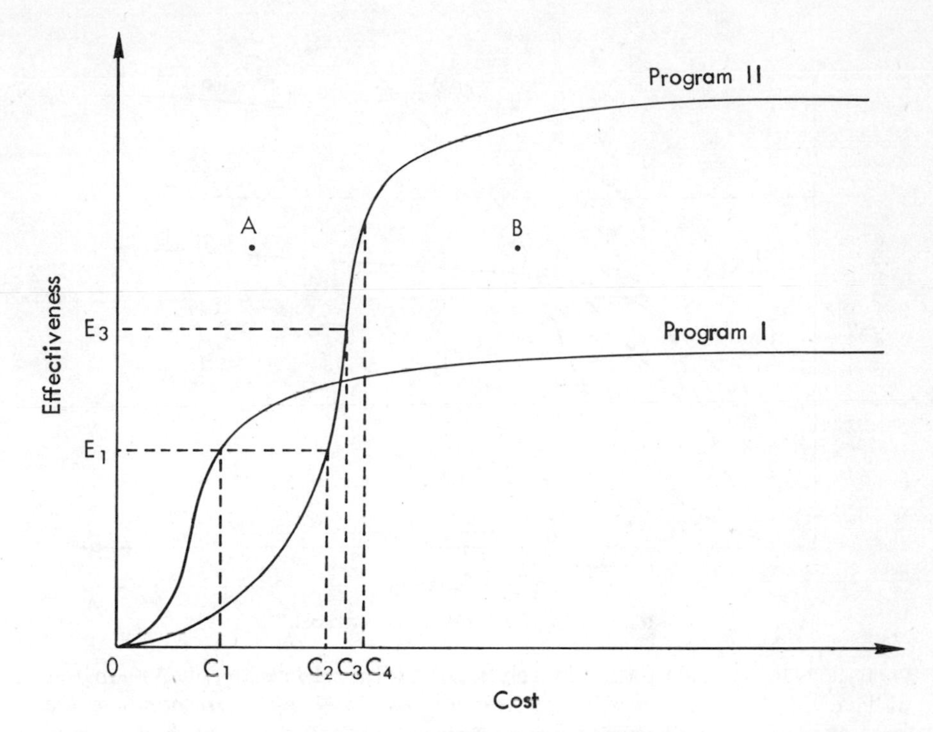

FIG. 6-1. A cost-effectiveness comparison.

subject to essentially the same amount of uncertainty. Can we decide between them? Not without some additional information, for example, how much cost is permissible or what level of effectiveness is adequate. Without straining, Program I achieves only a modest level of effectiveness (E_1), but does so at one-half the cost of Program II. If the level E_1 is adequate, why not select Program I and thereby minimize cost? Indeed, quite often the budget for a program will be limited by decree to some level such as C_2, in which case the first program is the obvious choice. In contrast, the goal of the proposed program may be to achieve some new level of effectiveness, such as E_3, no matter what the cost within "reason." Then Program II is obviously the choice.

The point to be made is that, *in general, it is not possible to choose between two alternatives just on the basis of cost and effectiveness* data such as those shown in Fig. 6-1.[10] Usually either a required effectiveness must be specified and then the cost minimized for that effectiveness, or a required cost must be specified and the effectiveness maximized.

Clearly, preliminary analysis of cost and effectiveness should influence the selection of the final criterion or basis for choice. For example, if C_3 is truly a reasonable cost to pay, then the case for C_4 is much stronger, in view of the

[10] Unless we have dominance—see Program III in Fig. 6-2.

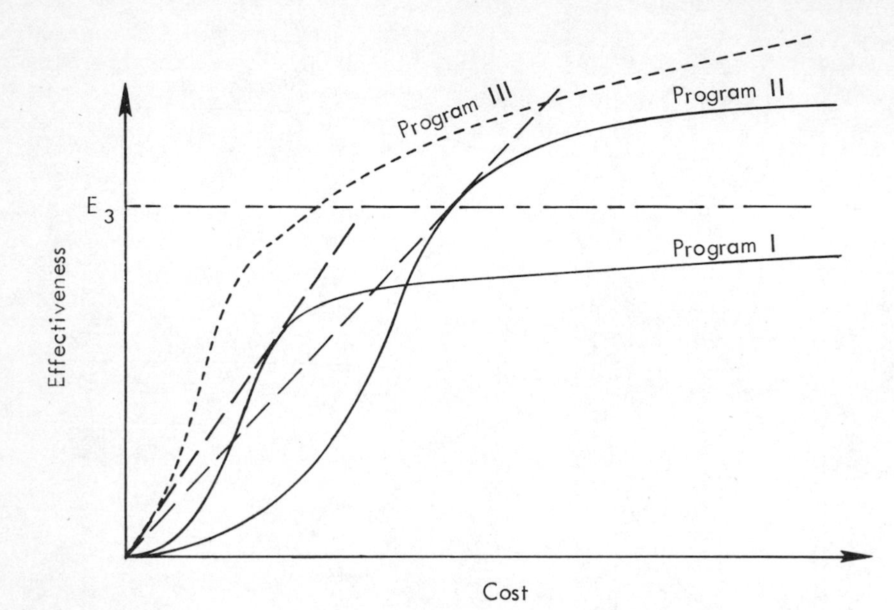

FIG. 6-2. Ratios of effectiveness to cost.

great gains to be made for a relatively small additional investment. As a matter of fact, this approach of *setting maximum cost so that it corresponds to the knee of the cost-effectiveness curve* is a very useful and prevalent one, since very little additional effectiveness is gained by further investment.

On the other hand, one should not specify *both* the proposed expenditure and the effectiveness; this overspecifies the criterion, and can result in asking for alternatives that are either unobtainable (point A in Fig. 6-1) or underdesigned (point B in the same figure). An extreme case of criterion overspecification is to require maximum effectiveness for minimum cost. These two requirements cannot be met simultaneously, as is clear from Fig. 6-1, in which minimum cost corresponds to zero effectiveness, and maximum effectiveness corresponds to a very large cost.

Somewhere in the middle are criteria that apparently specify neither the program cost nor the effectiveness. One widely used criterion calls for maximizing the ratio of effectiveness to cost. This seems to be a workable criterion since, in general, we want to increase effectiveness and decrease cost. Nevertheless, we can see by examining Fig. 6-2 that it has a serious defect. Since the effectiveness-cost ratio for either alternative is simply the slope of a line drawn from the origin to a given point on the curve for that alternative, and since, in this example, the ratio obviously takes on a maximum at the knee of the curve, our choice between the two alternatives seems to be settled at once. Thus, Program I is clearly preferred with this criterion. However, if E_3 is the minimum level of effectiveness acceptable (in our example, say, the vocabulary required for next years program to be worthwhile), then Program I is the obviously unsatisfactory.

The point to be made here is that unless the decision-maker is completely unconcerned about *absolute* levels of effectiveness and cost, something that probably never happens, then a criterion such as this, which neglects them, must be avoided.

In some cases, it is possible to escape this need for specifying either the required cost or effectiveness by expressing both cost and effectiveness in the same units, for instance, in dollars. In our language instruction example, this would mean assigning a dollar value to the attainment of various levels of proficiency in the English language. This could be done; it might have to be accomplished on the basis of some fairly arbitrary assumptions but proficiency in the English language in the United States clearly has an economic value. We could then have a cost-benefit comparison, for it would be possible to subtract cost from effectiveness and take as the criterion the maximization of this difference. But we can seldom put the data in a form adequate to express effectiveness and certain costs[11] satisfactorily in monetary units without a good deal of work and a large element of arbitrariness. Hence, with a single goal and a limited number of alternatives to consider as in our example, the cost-effectiveness approach is usually more direct.

It sometimes happens that selection between alternatives is easy. A case of this is shown in Fig. 6-2, in which Program III is more effective than any other at every cost. In such a case, which is extremely rare, it is clearly advantageous to select III, which is said to dominate I and II at all levels of investment and effectiveness. Note that it is still not permissible to overspecify the criterion and require maximum effectiveness for minimum cost, for minimum cost still corresponds to zero effectiveness for Program III, and so forth. Even though dominance designates the third program as preferred, the required level of effectiveness must be specified before the preferred level of investment can be selected or conversely.

Suboptimization[12]

The selection of good criteria is made difficult in part because problems must be broken down into component pieces or subproblems. It is inevitable that decisionmaking be divided, some decisions being made by "high level" officials or groups, and some being delegated to "lower levels." All decisions cannot be made simultaneously by one official or group, even in a military organization where the authority to do so might be present. In most of the public sector authority is divided and in any given decision there are likely to be many participants who enter at various stages.

A solution, to be really satisfactory to a client, must be within his capability to handle. A bureau chief wants to find out what action he can take to improve

---

[11] For example, the so-called "social costs" (for instance, the cost of human life). To see how these might be handled, see Chapter 9 Appendix and Chapter 7.

[12] The discussion is based on that of R. J. McKean in Chapter 5 of (Quade, 1964).

his situation—not to support an analysis to show the federal government what to do, if congress and the courts approve, to eliminate the difficulty. An attempt at the clarification of issues far beyond the capability of the supporting agency or even its superior at the next higher level in the hierarchy to cope with, is likely to generate little interest and have no impact on policy.

For these reasons, as well as the difficulty of solving the problem, the process of analysis must be broken into parts; alternatives at all levels cannot be analyzed simultaneously. (No connotation of greater or lesser significance should be associated with the terms "higher" and "lower" levels.)

Analysts, like decisionmakers, are frequently forced to devote their attention to courses of action that pertain to only a part of the problem. Other aspects are temporarily shelved, possible decisions about some things being neglected and specific decisions about others being assumed. The resulting analyses, nevertheless, can still *help to find* preferred solutions, or at least policies that are improvements over present or proposed solutions to subproblems. In the language of systems analysis, these are "suboptimizations."

The term suboptimization may have been first applied to policy analysis by Charles Hitch (1952). He comments (Hitch, 1960):

> It might appear then that it would make sense to *begin* with some broad "given" or accepted objectives; to derive from them appropriate local or sub-objectives for the systems problem in hand; and then to design the analysis to maximize, in some sense, the proximate objectives. . . .
> Not only is this a plausible approach; it is in some special cases an acceptable one; it is usually (not always) better than making no systems study at all; and it is frequently, given limitations on available time or manpower, the only feasible approach. I think I was the first to use the term "sub-optimization" to describe this style of operations research (in 1952), and I am no implacable or dogmatic foe of its use. Some of the most rewarding systems studies have in fact been low level sub-optimizations.
> But in general sub-optimizing is not good enough; . . ."

To repeat, whenever we simplify the problem of selecting an alternative by completely fixing certain characteristics that might, in fact, vary, the selection that results is less than optimum. It is a "suboptimization" because we could almost always do better if we allowed some or all of these characteristics to vary simultaneously and had the capability make our selection from the larger set of possible mixes.

Suboptimization brings with it, on the one hand, certain advantages. One is that more detail can be taken into account by analysis in the small than by broader analyses. Models can be less aggregative, and they may for that reason yield more accurate predictions. On the other hand, suboptimization brings great difficulties in the selection of criteria, because lower-level tests can so easily be inconsistent with higher-level criteria.

For example, in the design of a car-parking operation for an amusement park, one size of lot and pricing policy might be derived from a suboptimization in which the criterion is that of maximizing net revenues from parking and another size lot and pricing policy from an analysis that sought to maximize net revenues from the amusement park as a whole. Lack of parking for a few people, who then go elsewhere, might lead to considerable saving in land costs for parking and in salary for attendants, and thus in greater revenue for the lot, but the loss in goodwill, and thus in revenues for the park as a whole, might be serious.

A policy more consistent with the higher-level objective, and thus a better policy for the lot, might simply be to provide parking for all comers–or even to provide it free. We cannot know *a priori,* however, that income from a parking lot should be small relative to the income for the facility it serves. Whenever there is little opportunity for visiting the facility except by private car, a significant general admission charge might be more acceptable if disguised as a parking fee.

Let us take a more extreme illustration. Consider the selection of a camera for a surveillance satellite. If the test we use is based on the minimum cost of achieving some standard of laboratory performance, the camera might be too heavy or too delicate to be suitable for use with the satellite. It is hazardous to "factor out" such a decision from the larger problem, the selection of the surveillance system. It is necessary to at least ponder higher level criteria in order to avoid serious inconsistencies.

Two points for guidance in choosing a criterion, originally stated by C. J. Hitch (1956) are

> 1. The criteria in a low-level problem must be consistent with high-level criteria[13] . . . .
> 2. Beware of criteria that have to be repeatedly hedged by constraints to prevent them from giving absurd results.

As an example of the second point, we can consider any use of a ratio as a criterion. For example, if the criterion used to determine the choice of plans for public buildings is minimum cost per cubic foot, we tend to get huge boxlike structures full of empty space. A constraint on the total cubic footage, the denominator, is necessary to keep them from being even larger. Ratios are particularly treacherous because they ignore the absolute magnitudes of both numerator and denominator. For a further discussion of ratios as criteria, see Hitch (1958).

The determination of a rule to tell you what is "best" is seldom solved even in the most ordinary and prosaic circumstances. Consumers Research doesn't tell you which washing machine to buy–it merely lists dollar costs and some important characteristics and indicators for performance that you should consid-

[13] I am not saying that we must convert low-level OR problems into higher level ones and work only on those. On the contrary, the operations researcher must cut his problems down to workable size. The whole point of my article on suboptimization [Opns. Res. 1, 87 (1953)] is that he can do so and reach valid and useful conclusions by choosing with care criteria that are *consistent* with objectives at higher levels. He need not *solve* global problems in all their complexity in order to recognize their general features, a happy circumstance which makes useful OR possible. (Hitch, 1956).

TABLE 6-1.
Impact Summary for the 1977 Reference Cases[a]

Item	Case: Nominal	A	B	C	D	E
Strategy components						
Fixed source	Nominal	Nominal	Maximal	Medium	Maximal	Medium
Retrofit[b]	D[c]	R	L	M	L	M
Mileage surcharge, ¢/mi	0	0	0	0	0	0
Bus-eligible population, %[d]	70	70	70	70	80	80
Bus headways, peak/off-peak, min	40/40	40/40	20/40	20/40	20/40	20/40
Bus fare, ¢/trip	25	Free	Free	Free	Free	Free
Environmental Impacts						
Worst-day oxidant concentration, ppm	0.13‡	0.08*	0.08*	0.08*	0.08*	0.08*
Number of days oxidant above standard	15‡	1*	1*	1*	1*	1*
Lead standard achieved?	No‡	Yes	Yes	Yes	Yes	Yes
Annual gasoline consumption, million gal	615	618	609	607	604	603
Annual diesel-fuel consumption, million gal	2.4	2.4	3.4	3.4	4.8	4.8
Transportation-Service Impacts						
LDMV mileage reduction, % of uncontrolled	0.0	2.9	4.3	4.3	5.0	5.0
Trips forgone, % of uncontrolled	0.1*	1.0‡	0.9§	0.9§	0.9§	0.9§
Trips by bus, % of all trips	1.1	1.4	2.6	2.6	3.1	3.1
Carpooling, % of home–work trips	5	5	5	5	5	5

Economic Impacts						
Annual strategy expenditures, $ millions						
Fixed-source control	2.8	2.8	5.1	3.2	5.1	3.2
Retrofit	7.7	67.6	51.5	55.2	51.3	55.0
Mileage surcharge	0	0	0	0	0	0
Bus	10.0	10.0	15.0	15.0	23.0	23.0
Net expenditures after transfers, $ millions[e]	20.5*	80.4§	71.6†	73.4†	79.4§	81.2‡
Unallocated income from surcharge, $ millions	0	0	0	0	0	0
Incremental recurring employment	77‡	815§	1041†	1048†	1536*	1543*
Incremental nonrecurring employment	212*	2071‡	1144*	1483§	1144*	1483§
Number of buses	175*	175*	348	348§	490‡	490‡
Distributional Impacts						
General-aviation annual cost, $/aircraft	0*	0*	500‡	0*	500‡	0*
Retrofit procurement cost per household, % of income[f]						
Under $5000						
User-pays financing	0.8*	8.9‡	5.4†	6.5§	5.4†	6.5§
Income-proportional financing	0.2	2.6	1.6	1.9	1.6	1.9
Over $15,000						
User-pays financing	0.1	1.2	0.8	0.9	0.8	0.9
Income-proportional financing	0.2	2.6	1.6	1.9	1.2	1.9

[a]Coding: * = best; † = next best; ‡ = worst; and § = next worst.
[b]See Appendix D for detailed composition.
[c]Strategy currently mandated by CARB.
[d]As percent of total county population.
[e]Surcharge income used to offset bus subsidy to whatever extent possible.
[f]Percent of average annual income for income group, retrofit procurement expenditure made in a single year.

er. There may be quite a problem, however, in deciding which characteristics and indicators are significant to your choice.

In the immensely more complicated problems of public policy we can seldom do much better than take a similar approach. That is, to display the various consequences, costs, and spillovers, associated with each alternative and their sensitivities to changes and uncertainties–the approach we have earlier called the "scorecard" approach–and put the criterion problem squarely up to the decision-makers.

The Criterion Problem

One price we pay for the increase in scope and comprehensiveness of policy analysis over systems analysis and operations research is increased difficulty with the criterion problem. This occurs not only because the analysis is more complex adding more factors to be considered in the decision itself, but also because the number of decision-makers who now become aware that they have an interest in the decision increases. Moreover, when attention is paid to the distributional impacts of the various alternatives on such elements of society as income classes or ethnic groups, representatives of those elements become aware of how the decision to be made affects them.

It becomes extremely difficult to obtain practical and acceptable ways to measure effectiveness and costs when transfer payments are relevant to the decision, and spillovers into the environment and into particular segments of society must be considered because sensitive value judgments have to be made. The cost-benefit approach or any other approach in which the factors that must be considered are aggregated to obtain a simple rule for ranking alternatives necessarily involves a large element of arbitrariness and can generate endless argument. In addition, the use of weights, dollars, or otherwise, ordinarily depends upon many judgments that must be made by analysts,[14] judgments that have to be made on personal values and with highly uncertain data. Moreover, it is difficult to make these judgements explicit to the decision-makers. Also, if there were considerations that were not quantifiable, these would have to be omitted, or at least treated differently.

Another drawback to the single measure is that although it may indicate the over-all worth of one system or policy with respect to the other alternatives, it does not indicate anything about the direction to take if improvements are needed.

The obvious alternative to the single all-inclusive measure is to present each significant impact or consequence on its natural scale, leaving the problem of integration to the judgment and intuition of the decision-makers. This approach is not new but has been developed and used very successfully in a number of

---

[14] In principle, it is possible to get the decision-makers to supply these judgments but the practical difficulties are extreme.

transportation studies, particularly by Bruce Goeller (1974) of the Rand Corporation.

In this approach the impacts, that is to say, the consequences that ensue from the decision to select one of the alternatives–costs, benefits, spillovers, risks, segments of society affected, and so forth–are displayed in terms of the physical units commonly used to characterize them rather than being converted to a scale such as dollars. In some circumstances, if appropriate, some impacts may even be described in natural language.[15] An example of a display of this type is given in Table 6-1 (pp. 98–99).

The study from which this display was taken (Goeller *et al.*, 1973) was undertaken to analyze alternative air pollution control strategies in terms of their impacts, positive and negative, on San Diego, California with the hope of identifying the most promising strategy for implementation. The display was, of course, accompanied by extensive discussion to point out how the risks and uncertainties might affect the ranking of the alternatives.

A scorecard is not only an acceptable method of handling the criterion problem when no clearly dominant single measure can be agreed upon, but it is also a way to protect against or counter the biases of the analyst, which are possibly unconscious. These biases cannot be countered completely, of course, for the analyst still largely decides what impacts to present, and his is the point of view that originally classes them as favorable or unfavorable. But decision-makers can reshuffle the analyst's arrangement and they can call for various sensitivity tests to be run to determine how changes in the assumptions originally made by the analyst affect the results. A scorecard presentation is also something the public can understand. They, like the decision-makers, can ask "what if" questions of the analysts, which, when answered, will show not merely the changes in ranking of alternatives but also the impacts.

As used in the transportation studies, the "scorecard" has been enthusiastically received by a wide variety of decision-makers, ranging from federal transportation and environmental policy-makers to state legislators and local planners.

References

Beckerman, Wilfred, Methodological links between economic and social planning, in *Social planning,* Everett Reimer (ed.). Editorial U.P.R., University of Puerto Rico, San Juan, Puerto Rico, 1968, 101–115.

Elion, Samuel, Goals and constraints in decision-making. *Operational research quarterly,* 1972, **23**, (1), 3–15.

Goeller, B. F., *et al., San Diego clean air project: Summary report* R-1362-SD, The Rand Corporation, Santa Monica, California, 1973.

---

[15] For example, if it were necessary to destroy a historical landmark, it would be identified and the damage described.

Goeller, B. F., *System impact assessment: A more comprehensive approach to public policy decisions.* The Rand Corporation, RM-1446-RC, Santa Monica, California, to appear.

Hitch, Charles J. Comments by Charles Hitch. *Operations research,* 1956, **4**, (4), 427.

Hitch, Charles J., Economics and military operations research. *Review of economics and statistics,* 1958, **XL** (3), 199–209.

Hitch, Charles J., *On the choice of objectives in systems studies,* P-1955. The Rand Corporation, 30 March 1960.

Hitch, Charles J., *Sub-optimization in operations problems,* P-326. The Rand Corporation, Santa Monica, California, 18 November 1952.

Hovey, Harold A., *The planning-programming approach to government decision-making.* Praeger, New York, 1968.

Kahn, Herman, *On thermonuclear war.* Princeton University Press, Princeton, New Jersey, 1960, p 121.

Quade, E. S. (ed.), *Analysis for military decisions.* North-Holland Publishing Company, Amsterdam, 1964.

Quade, E. S. & Boucher, W. I. (eds.), *Systems analysis and policy planning: Applications in defense.* American Elsevier, New York, 1968.

Simon, H. A., *Administrative behavior.* MacMillan, New York, 1961.

Vickers, Sir Geoffrey, Institutional and personal roles. *Human Relations,* 1971, **XXIV**, (5), 433–447.

Weekly compilation of presidential documents, October 17, 1966, p. 1460, quoted in Hovey, 1968.

Chapter 7

EFFECTIVENESS AND BENEFITS

In 1966 the House Armed Services Committee (1536, 16 May) remarked of the Defense Department that their

> . . . almost obsessional dedication to cost-effectiveness raises the specter of a decision-maker who . . . knows the price of everything and the value of nothing.

This remark calls attention to two of the most serious problems in the application of analysis to the public sector. The first lies in the difficulty of determining valid ways to measure the value or the effectiveness of programs and to take their benefits into account. The second, almost as serious, lies in the implication that we know how to calculate the cost.[1]

As we have seen, public statements of objectives are not only likely to be missing but, in the rare cases in which they are announced, they are likely to turn out to be vague, multiple, and conflicting. To compound the problem, even when we have a single, clear-cut, agreed-upon objective, we may be forced to use measures of attainment that are, at best, barely adequate approximations for what we would like to measure.

It is hard to find a measure that even points consistently in the right direction. As an illustration to indicate the attainment of such a standard military objective as deterrence or victory, consider deterrence, for instance. It exists only in the mind—and in the enemy's mind at that. Therefore, we cannot use some scale of deterrence to measure the effectiveness of alternatives we hope will lead to deterrence, for even if there were such a scale we could not apply it. Instead we must use a proxy,[2] such as the potential mortalities that might be suffered if war were to come or the number of square feet of roof area that might be uncovered in a bombing attack. With such proxies, even if a comparison of two proposed strategic bombing systems indicated that one could inflict 50% more casualties on the enemy or destroy 50% more roof area than the other, we could not conclude that the more damaging system would supply 50% more deterrence[3]

---

[1] Since costs are benefits foregone, the same difficulties are there.

[2] A substitute scale, one that can measure at least approximately the extent to which the real objective is attained.

[3] There is, of course, no need that such a criterion be linear.

or, indeed, any more. And it may supply less, for in some circumstances, a strategic force may look so potentially threatening that an opponent may feel he has no choice but to strike before the situation gets worse. Thus the system capable of inflicting the greatest number of casualties may provide the least deterrence.

The measurement of effectiveness and the evaluation of benefits are great stumbling blocks in public policy analysis. What we do here is often woefully inadequate. For example, we measure the success of a city government in traffic management, sewage disposal, and all the other diverse and conflicting activities which it is required to regulate usually by comparing its achievements with what "informed" public opinion thinks it might be able to achieve in these dimensions of its public responsibilities.

Some Difficulties in Measurement

As defined earlier, the effectiveness of an alternative measures the extent to which the objective is attained. If we have multiple objectives that cannot be represented by a single one, we often talk about the benefits of an alternative rather than its effectiveness. By the benefits we mean those impacts or consequences that we regard as favorable or positive, keeping the objectives in mind, as opposed to the costs, which are the unfavorable or negative impacts. To measure the benefits means to estimate their value in terms of a common unit, usually monetary. If the costs are measured in the same units, preference for an alternative is then indicated by the excess of benefits over costs.

We would like to be able to designate that course of action which, with the available resources, would lead to the greatest increase in something like the profits of a group, the satisfaction of an individual, or the military worth of the defense establishment. This prescription, however, usually helps little more than merely saying that we want the best. Nobody knows precisely how the public welfare and well-being are related to the observable outcomes of various courses of action. In practical problem-solving, therefore, we have to look to "proximate" or secondary measures, chosen because they are considered to be positively correlated with and more measurable than some aspects of the primary goals, to reflect what is happening to such concepts as the public welfare or military worth.

Actual measures are thus approximate, practical substitutes for whatever it is we would ultimately like to maximize. For example, the degree to which the school system is decentralized and subject to local control, the amount of money put into the system per pupil, or the percentage of pupils who enter college, have all been assumed to measure the quality of a school system. As a second example, if the aim is to improve the "quality" of arrest procedures, how is improvement to be measured? The following have all been considered:

(i) a survey of a representative sample of citizens
(ii) the percentage of felony arrests that pass preliminary court hear-

ing (or prosecuting attorney hearing or grand jury, if no court hearing)
(iii) the percentage of arrests leading to a conviction (a) for some charge (b) for the highest charge before plea bargaining.

For other examples see *Measuring the Effectiveness of Basic Municipal Services,* The Urban Institute and International City Management Association, Washington, D.C., February 1974.

We would like to be able to designate that course of action which, with the available resources, would lead to the greatest increase in something like the profits of a firm–say the welfare of the public, the well-being of a group, the satisfaction of an individual, or the military worth of the defense establishment.

Behind every public program there is some idea (which, of course, may be very rough and imprecise) as to why it is good and how it contributes to the public welfare. In trying to measure the benefits or effectiveness of such programs, it is important to determine first what we should want to measure without being restricted by our beliefs regarding our current ability to measure things–we should avoid prejudging based on what we may think to be current practice as to what may be measurable and what may not be. Improvements in methods of measurement can be made, particularly as analysts in governments everywhere are increasingly focusing their efforts on such measurement. The whole social indicator movement, for example, is an effort in this direction.

Let me list some of the major obstacles to the measurement of effectiveness and benefits.

First, the problems of benefit measurement are not just technical problems; they are also conceptual problems. It is by no means obvious how the benefits of most social programs should be defined. For example, the federal government supplies money to the states and cities for law enforcement. How can we measure the effectiveness of their programs to fight crime? The first impulse is to use the number of crimes as a measure of effectiveness. This has important difficulties, however. Crime is a very heterogeneous phenomenon–murder, shoplifting, drunkenness, and joy riding may all be crimes, but all are not equally serious.

Another possibility is to devise an "index of seriousness," but thus far this effort has turned out to involve very arbitrary judgments and is not very satisfactory. The index reported annually by the FBI simply adds the reported number of seven index crimes (willful homicide, forcible rape, aggravated assault, robbery, burglary, larceny of $50 and over, and auto theft). This approach has been widely criticized, partly because it attaches equal weight to these types of crime.[4]

One would like to consider a higher-level measure of effectiveness, which might be called "social disruption." Such a measure would bring together the total social costs resulting from crime control efforts as well as those costs

[4] Also because a rise may mean better record-keeping or laxness in preventing crimes as well as greater efficiency in detecting crime. The varying rate of failure to report crimes on the part of the victims also makes it unreliable.

resulting from crime itself. A social disruption function, however, is not so sufficiently well defined as yet that we can operate with it analytically (Blumstein, 1971).

Secondly, there may be no data, or poor data, or we may not know how to aggregate what data we have. Even when the analyst knows what benefits he wants to measure, he may not be able to obtain the measurement. This is particularly the case when we are involved with social programs such as those concerned with health, education, and welfare, which are directed at changing the lives of people in some way.

In order to measure the effectiveness of such programs, we need to find out what actually happens to the people affected by the programs. For this purpose, we usually cannot rely on the routine records of operations and administration as they are kept by most public and private organizations. Such record-keeping is often, if not usually, wastefully and inefficiently performed; it is qualitatively poor and badly adapted to the purpose of effectiveness measurement. Moreover, many of the effects we would like to measure are delayed in time. To get the required data, new statistical activities are probably required.

Third, even if we could conceptualize and measure the benefits of a given program, often we would face a further problem: the fact that the benefits and costs often go to different people. The benefits from a proposed new freeway can be measured by evaluation of time saved by drivers and in reduction of accidents, tire wear, and gasoline used. The obvious costs consist of the expenses of right-of-way, construction, and upkeep, but there are others: noise, pollution, inconvenience to those living nearby and those displaced.

If benefits exceed cost, it may be concluded that the project would be economically efficient. However, the income redistribution consequences may be significant. The owners of motels, gas stations, and stores along the old route stand to lose heavily, but their loss will be "offset" by the gains of others who operate along the new road. The parallel railroad will lose business to truckers whose driving time and costs have been lowered. One town may replace another as a shopping center for many customers whose access to the first town has been eliminated and whose access to the other has improved. Cement manufacturers, sign painters, and highway contractors will enjoy increased income whereas the manufacturers of railroad cars and mass transit may have less. In general, the people who make use of the road or serve its users get the benefits whereas the general taxpayer pays the costs.

A fourth problem involved in assessing benefits in that many of the benefits of government expenditures are not reflected in the marketplace. Public health expenditures may simply lessen pain and anguish or they may extend the life of someone too old to work. An educational expenditure may increase the capacity for cultural enjoyment without increasing earning power.

Proxies, or indirect measures, although necessary, can introduce ambiguity. For one thing, a proxy measure may be used elsewhere to represent some other objective. For example, the infant mortality rate is used to measure the quality of life in a community, or the standard of medical care available, or to rate hospitals. For another, consider the objective of the various federal antipoverty programs, which was to raise the standard of living of those at the lower end of the income scale. The proxy used was the number of families crossing a purely arbitrary level of dollar income called the "poverty line." The problem was that the particular level selected carried different connotations for different people, engendering endless debates over whether the standard had been set at the right point. This took away attention from the number of people moving across the line as a general indicator for evaluating whether the programs were working.

Approaches to Measuring Effectiveness

In spite of the difficulties in finding useful measures of effectiveness, a great deal can be done. Substantial improvements in data gathering are possible. There are new efforts, such as the movement to find social indicators, which may eventually provide us with better means for taking social factors into account. Also, we have as yet made little use of techniques such as rating-point scales and other schemes for pooling judgment to take factors into account that cannot otherwise be qualified.

Our first step, of course, is to decide what we want to measure. To do this, we need to be sure that we know the objective that the analysis is to help achieve. The mark of a good effectiveness measure is that it closely reflects the objective. In public policy analysis, we would also like it to reflect the effect on people of attaining the objective.

A number of inadequate and sometimes absurdly executed approaches to the measurement of effectiveness are in fairly common use. Let me mention these before we go on to discuss better measures.

1. The first inadequate (but not always completely absurd) idea is that program costs can be used to measure program effectiveness. It is surprisingly prevalent. For example, an expenditure per pupil is often used explicitly and even more often implicitly as a measure of effectiveness for primary and secondary school education. This confuses inputs with outputs.

2. A second inadequate approach, the use of work load measures, is more common. It is to use the observable products of an activity of what it is trying to achieve.

Work load measures are factors such as patient days per year for a hospital, case load for social workers, number of students going through a manpower training program, etc. These work load measures are very tempting to use because they are quantitative and relatively easy to handle.

As indicators of activity, such measures are clearly useful, but they may say little about the effectiveness of the program with which they are associated. They ignore such vital questions as to whether the patients got well, what the social worker did with the case load, or whether the graduates of the training program could hold a job. For a youth vocational training program we need data on the employment that occurs and how long that employment lasts, and information showing the employment would not have occurred if there had been no training program. If there is no demand for the trainees and thus little added employment, putting a large number of youths through the best training in the world is of very little use, if our objective is to decrease unemployment.

Physical standards are also sometimes used. For example, the contracts for most federal buildings, such as hospitals, are let on a fixed price, competitive bid basis, primarily to obtain a low cost per square foot. This has led to prisonlike structures with minimal outside window space and huge rooms inside. Such standards are not useful unless they are related to the program objective and they may lead us into absurdities.[5]

3. A third unfortunate approach is to force all outputs into a common index of worth. In cost-benefit analysis this is done by insisting that each program output be translated into a common unit of measure, dollars in this country, by suppressing those effects for which this cannot be done. In other cases, the scheme is to use a weighting technique of some sort to combine measures that are expressed in different units into a single overall "index of worth." In doing this, the analyst may be forced to make many arbitrary value judgments. Hatry (1970) comments:

> There is no doubt that the job of decision maker would be easier if a single effectiveness measure could appropriately be used. However, I contend that such procedures place the analyst in the position of making a considerable number of value judgments that rightfully should be made in the political decision-making process, and not by the analyst. Such value judgments are buried in the procedures used by the analysts and are seldom revealed to, or understood by, the decision makers.
>
> Such hocus pocus in the long run tends to discredit analysis and distract significantly from what should be its principal role: to present to decision makers alternative ways of achieving objectives, and to estimate and display all the major trade-offs of cost and effectiveness that exist among these alternatives.
>
> The number of very able professionals, in universities, consulting firms, and (though less frequently) in government, who overdo this commensurability business is appalling.

With guidance from the analysts, if the decision-makers are willing to spend the time required, they can work out their own set of weights. They can, for example, even vary the weight experimentally to see what the effect is on the ranking of, e.g., equal cost alternatives. One such procedure is that developed by Miller (1969). It seems better, however, to have the decision-makers spend their time exercising their judgment directly on the impacts rather than on determining a set of weights, say by viewing a "score card" display as discussed in Chapter 6.

---

[5] There are many stories to illustrate the effects of maximizing on proxy measures. A well-known story, usually attributed to the Soviets, concerns the manager of a nail factory, who initially was given a measure of merit for his output in terms of the total *weight* of the output of his establishment. Like a good bureaucrat, he maximized on this explicitly stated objective and turned out only huge railroad spikes. As a result of having a surplus of railroad spikes, his objective function was changed to maximize the total *number* of nails he produced. In a very short time, he was able to switch over to complete production of brads, tacks, and staples.

In choosing a scheme to measure effectiveness, we are looking for a measure that is positively correlated with the objective of the program or policy under consideration. Suppose a federal manpower training program is to be evaluated. If the objective of the program is to increase national output, how should the effectiveness of the program be measured?

One possibility is to measure the increase in national output by the increase in income of the trainee as a result of his presence in the program. This is certainly in the right direction. It requires that wages be total compensation, that is, include fringe benefits; in addition, it requires that there be no displacement of other workers by the trainees.

Assuming that this last assumption is acceptable, how do we actually get the measure? The ideal way to measure the increase in trainee income would be to compare his actual income after training with what it would have been if he had not taken part in the program—an impossible comparison. By following the work history of the trainees after they have completed the program, it is possible to determine their actual income. But how much of the change in the wage income of a trainee should be properly attributed to his training?

One means of getting the required information would be to form a control group by assigning potential trainees on a random basis to either training or to the control group and then following the wage experience of both groups. Social pressures make it hard to carry out experiments of this type but they are being done on a limited scale.[6] In the absence of an experiment, a substitute is to use a group of individuals having similar backgrounds but who have never come in contact with the program as a basis for comparison. This is not fully satisfactory—one reason that the trainee group and the comparison group may differ in systematic yet undetected ways. It does, however, form a far more satisfactory basis for measuring program effectiveness than work load measures. Our belief is that a rough measure, if conceptually correct, is likely to be superior to one based on an erroneous concept no matter how precise the latter can be made.

Taking Account of Distributional Effects

Cost-benefit and cost-effectiveness have seldom tried to take into account who receives the benefits of a decision and who pays the costs. In most social programs the distribution of benefits and costs is crucial to its success.

One way to take into account distributional goals would be to use a system of weights that would reflect the relative values that society places on increasing the well-being of specific groups. But who should decide? Congress has such a system in mind when it enacts legislation but the system changes with the type of legislation. Again, rather than directing the judgement of the decision-makers to the weight assignments problem, it seems more sensible for them to consider the distributional effects directly.

---

[6] See Chapter 15 for a discussion of experiments and other evaluation procedures.

To illustrate the additional considerations that now appear, assume that the objective of the federal manpower program of the previous section is changed from "an increase in the national welfare" to "the increase in the economic welfare of a certain target population." Now transfer payments such as unemployment compensation or scholarship payments to the trainees, while in the program, must be taken into account; previously they were not considered because they did not contribute anything additional to the national output.

For a manpower program targeted on a poverty population, Glennan (1969) formulates the benefits as:

1. The increased earnings (net of taxes) of the target population resulting from participation in the program
2. Plus the net increase in transfer payments to the target population during participation in the program
3. Less decreases in transfer payments to the target population because of higher earnings subsequent to program participation
4. Less losses of earnings from work that have been performed if enrollee had not been in program
5. Less losses of earnings of poor individuals displaced by trainees

and the costs as:

1. The direct costs of the program including subsistence payments
2. Less any decreases in other transfer payments occasioned by the existence of the program
3. Plus losses of income of the nonpoor if they are displaced by the program enrollees
4. Plus any decreases in income to the nonpoor that occur because trainees are temporarily withdrawn from the work force
5. Less long term decreases in transfer payments because of the higher earnings of target population resulting from program
6. Less net external benefits which accrue to the nonpoor and are not reflected in earnings of target population
7. Less the increases in taxes paid by the target population on earnings increments resulting from the program

Because these costs and benefits occur over a considerable period of time, both the cost and benefit stream must be discounted to the present time using an appropriate discount rate. Left out of the above tabulation are the program benefits that are not easily reflected in monetary terms. Glennan suggests: improvement in self-image, improved access to public services because of better knowledge, less alienation from the world of work or from other segments of society, better health, and improved reading and computational skill. A further discussion of quantitative benefit measurement is found in (Dorfman, 1965).

Judgmental methods may also be used to measure effectiveness. The Urban Institute developed a method for the District of Columbia Sanitation Department that can be used to measure the effectiveness of alternative programs for removing litter and other solid wastes from city streets and alleys (Blair & Schwartz 1972). Inspectors carried a set of photographs that pictured streets and alleys in various stages of cleanliness or litter while surveying a neighborhood. They looked up a street or alley, compared what they saw with the sample photographs, and made a judgment as to the condition of what they saw with respect to litter and trash. They recorded this judgment of the condition of the block by means of a tape recorder on a scale of 1–4 where 1 represented the cleanest and 4 the most littered. To see how much difference a new program makes, a record survey can be made and the results of changes in collection activities assessed by considering how the average of the estimates for the streets and alleys in the neighborhood shifted. Because this system produces data on the output of a new program, it is likely to be far better than one that measures something like the number of truckloads collected in the area in question.

In the study in which this scheme was originally conceived, the aim was to improve neighborhood health and safety as well as appearance. Hence it would have been desirable to measure the direct effects of litter and uncollected wastes on health and safety. The ratings were thus a proxy for the health and safety aspects.

Benefit Measurement in Health Programs

As remarked earlier, because ultimate objectives seldom are measured directly, approximate or proxy measures of effectiveness must be used. These rarely approximate all aspects equally well, however. For example, the Department of Health, Education and Welfare supports a number of programs whose objectives are to save lives or prevent disability by controlling motor vehicle injury and specific diseases such as cancer, arthritis, tuberculosis, and syphilis. What measure can we use if we wish to compare these programs in order to allocate funds among them?

One measure could be the number of deaths averted by the program. With this measure, a frequently used criterion is the *cost per death averted* in which the five-year program costs are divided by the number of deaths averted due to the program.

Another measure might be the *economic savings* that result from the program: the direct savings in dollars that would have been spent in the absence of the program on medical care, including physicians' fees, hospital services, drugs, and the indirect savings, such as the earnings saved because the patient did not die and was not incapacitated as a result of illness or injury. To obtain the latter figure requires that the ages of those benefited be determined and applied to the average lifetime earnings for different age groups to obtain the present value of lost lifetime earnings.

Neither of these measures is perfect. The trouble with the first scheme, the cost per death averted, is that no distinction is made regarding the age at which death is averted; also there is no way to compare diseases such as arthritis, which are not primarily fatal, with other diseases. The second scheme does take both the morbidity and the mortality implications of the disease into account. The objection to it is that it discriminates against nonearners and low earners, against the aged, women, whose earnings are relatively low, and the poor.

Fortunately, when these tests actually were applied to various disease-control programs, it turned out that the two methods gave essentially the same indication as to where money should be spent (Grosse, 1971). Hence the two methods did not actually come into dispute. It cannot be expected to happen very often that we can do the benefit aspects of the analysis very crudely and still get results that do not change the ranking.

It is sometimes argued that no one can put a dollar value on human life or on the value of something like a liberal arts education. Nevertheless, whether we attempt to do so consciously or not, the interaction of individuals and social institutions puts price tags on valuable inputs to programs such as human life, or on outputs such as military capability or the value of an investment in education.

In an investigation of ways to increase highway safety, a State Highway Department may, for instance, estimate that separating two way traffic along a particularly dangerous stretch of road by means of a wire barrier may be expected to reduce fatalities due to accidents in which cars cross the median by, say, 10 per year. They may then decide not to spend the $2,000,000 required to build and maintain the barrier for the five-year period until a new divided road is scheduled to be built. One inference, then, is that saving a life is not worth $40,000 of Highway Department funds. That inference, so worded, is likely to make the decisionmakers uncomfortable. The usual answering argument in a case like this is that the money spent elsewhere had a higher payoff.

There are essentially four approaches to valuing the benefits[7] of health programs:

1. Values based on explicit statements by the responsible political representatives or their designees
2. Values inplicit in past decisions
3. Values based on human capital
4. Values based on individual preferences

For the first, a decision-maker responsible for approving health programs may make explicit a set of preferences for lowering mortality and morbidity that can be used as explicit values for comparison with program costs. He may also get assistance from individuals or committees. The objection is that the values are politically, not logically, determined.

Evaluations implicit in past decisions such as court-determined compensations in cases of accidental death could be used. Because decisions of this type have become policy, they have received *de facto* approvals as being reasonably acceptable.

[7] For a good discussion see Acton (1973).

Alternatively, the revealed evaluations implicit in past consumer choices on the means of reducing the probability of death or injury could be used. These choices might include the purchase of safety devices (for example, seat belts), marginal expenditures on health items (perhaps a doctor's examination plus a prescription), or the premium demanded for accepting an elevated risk (for instance, higher wages for extra-hazardous employment).

There are, however, serious problems with evaluations implicit in past decision. A conceptual difficulty is that preferences are considered to be stable. It is assumed that later developments do not change the evaluation significantly. A practical objection is that a bewildering variation in the amounts obtained from implicit evaluations arise. For example, in highway design the implicit valuation of a life saved is usually something like $5000 to $10,000; in the design of an ejection seat for a new fighter-bomber this might be more than a million dollars.

The livelihood or human capital approach to project evaluation is based on changes in the net present value of earnings streams assumed to be due to the project.[8] As used in many government studies, if the expected increase in future income resulting from a program is greater than the cost, the program is assumed to be worthwhile.

Shortcomings of this measure of benefits include the facts that (A) it does not necessarily reflect what people want, (B) it takes no account of externalities, and (C) it operates as if the current distribution of income were the appropriate basis for decision. For example, it discriminates against women even when a homemaker's services are valued at the wages of a domestic worker, rather than zero as is often done.

The fourth approach is based on the community's willingness to pay for a program that will save lives. The underlying assumption is that lifesaving is essentially a consumption good, and therefore is worth whatever people are prepared to pay for it. Because public expenditure decisions necessarily involve the allocation of resources that might otherwise supply different consumer goods, one argument that can be used to justify the incorporation of individual preferences is "no taxation without representation." This argument and the method itself were suggested by Schelling (1968). The willingness-to-pay approach is based on a simple model of individual choice that can be entirely characterized by health status (alive, A, or dead, D) and wealth (W). Assume that, in a given time period, the individual has a probability P of dying but can purchase a change d in that probability. What is the maximum amount x he is willing to pay for decreasing his probability of death from P to P-d? In more technical language, what is the amount x that will make him indifferent between

---

[8] One formula for the livelihood of a person n years old is the following:

$$\sum_{i=n}^{\infty} \frac{P_i(E_i - C_i)}{(1+r)^i}$$

where P_i is the probability of surviving ith year, E_i is earning in ith year, C_i is consumption in ith year, and r is the discount rate.

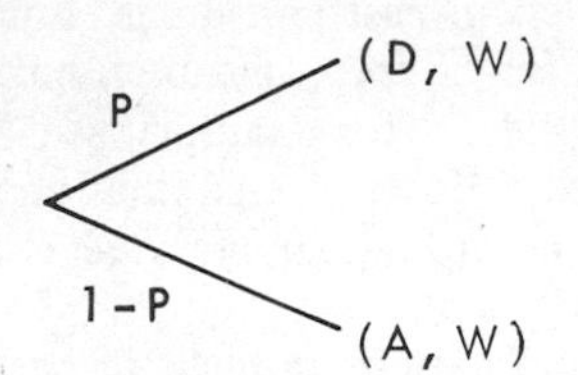

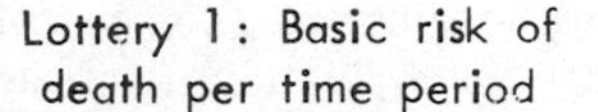

~

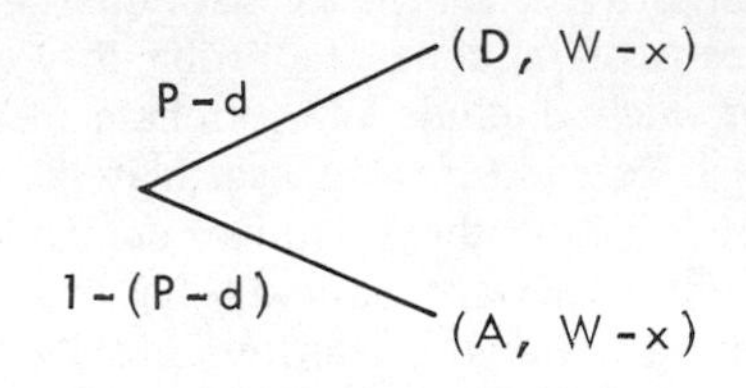

Lottery 2: Modified risk of
death per time period

FIG. 7-1. Lotteries illustrating the calculation of willingness to pay, *x*, to reduce probability of death by *d* (Acton, 1973).

(that is, equally willing to take part in one as in the other) the two lotteries shown in Fig. 7-1 (Acton, 1973)?

The problem is how to get an answer to the question of what the value of x should be that will reflect society's preferences. Factors that should influence the individual in his choice of x include P, d, W, and the number and status of his dependents. Other influences, more from society's point of view, are the willingness to pay of people who know the individual at risk and those who are willing to contribute to lower the general mortality rate.

Problems arise for several reasons. It is necessary to create realistic questions to motivate response, but some of the more realistic situations are very complicated. The form in which probabilities are presented to the subject can make a significant difference in his choice between lotteries. Also people do not always understand probabilities of this type and often give inconsistent answers. There is also the problem that the procedure presents an opportunity for people to misrepresent their preferences strategically. In other words, a person may respond with a "manifest" willingness-to-pay amount that may be different from his "latent" or "true" willingness to pay—but the decision should be made on the basis of the latent amount. All in all, however, manifest preferences are probably a much better guide to latent preferences than are any of the alternatives that do not consider preferences at all.

References

Acton, J. P., *Evaluating public programs to save lives: The case of heart attacks.* R-950-RC, The Rand Corporation, Santa Monica, California, 1973.

Blair, Louis H. & Schwartz, Alfred I., *How clean is our city: A guide for measuring the effectiveness of solid waste collection activities.* The Urban Institute, Washington, D.C., 1972.

Blumstein, Alfred, Cost effectiveness analysis in the allocation of police resources, in M. G. Kendall (ed.), *Cost benefit analysis.* The English Universities Press, London, 1971, 87–98.

Dorfman, Robert, (ed.), *Measuring benefits of government investments.* The Brookings Institution, Washington, D.C., 1965.

Glennan, Thomas K., Jr., *Evaluating federal manpower programs: Notes and observations.* RM-5743-OEO, The Rand Corporation, Santa Monica, California, September 1969.

Grosse, R. N., Cost-benefit analysis in disease control programs, in M. G. Kendall (ed.), *Cost-benefit analysis.* The English Universities Press, Ltd., London, 1971, 17–34.

Hatry, Harry P., Measuring the effectiveness of non-defense public programs. *Operations research,* 1970, **18**, (5), 774.

House Report 1536 Authorizing Defense Procurement and Research and Development, 16 May 1966, Armed Services Committee, U.S. Congress, Washington, D.C.

Miller, J. R., *Assessing alternative transportation systems.* RM-5865-DOT, The Rand Corporation, Santa Monica, California, 1969.

Schelling, Thomas, The life you save may be your own, in Stuart Chase (ed.), *Problems in public expenditure analysis,* The Brookings Institution, Washington, D.C., 1968, 127–176.

Chapter 8

ALTERNATIVES

The purpose of policy analysis is to help someone choose a course of action; therefore it follows that the analyst should devote a considerable effort to uncovering the set of possible actions, or alternatives, that offer some hope of accomplishing what is wanted. Unless he is particularly perceptive or has been specifically trained, a decision-maker almost never conducts a systematic and exhaustive screening of his entire set of possible alternatives. Even when he brings in analysts to help him with his problem, he may bar them from considering certain classes of alternatives, sometimes for no better reason than that "we simply don't do things that way". But no process of evaluation will designate the best alternative if it is not even considered in the analysis. Thus a strong effort should be made to search out and discover further alternatives, or to invent and design them, before a decision to select a particular one is finally made.

Finding the Alternatives

Industry and business understand, at least more than most public agencies, the need for searching out and examining a wide range of alternatives. As witness we find in the private sector far more than in the public, departments devoted to research and development, to planning, to testing and even campaigns to get the rank and file to "submit ideas."

Effort devoted to searching out alternatives is particularly needed in government because the normal processes of planning and budgeting tend to obscure rather than display alternatives as explicit options for consideration by either managers or bodies with control over fund allocation. Public planning and budgeting are largely incremental; each agency expects to get a little more money this year than last and hopes to do a little better with it. A public agency such as a Fire Department has a traditional way of operating, and its budget and personnel requirements for the next year are developed around its program for the current year, and not by taking a fresh look at the problem that that program was designed to meet and *then* considering how other ways and other programs might contribute. For instance, municipal fire departments were established to put out fires and are largely managed with that aim in view. It takes an effort to focus on fire prevention by radical changes in the building code and even more to consider education or recreation or social reform as the path to fire prevention. And still more of an effort is needed to consider such a far-out alternative as turning fire protection over to a private contractor.

But how shall alternatives be invented, or discovered, or identified? How many, which should be looked at in detail, and in what order?

The generation of alternatives is tied up with the definition of our goals; what is a good alternative is largely determined by what we want. It is much more than simple testing of all possible combinations and permutations of a set. The generation of alternatives is, or should be, a creative act. It is easy to think up inefficient or costly ways to an end but it is exceedingly difficult to come up with a truly imaginative one. Genuinely new alternatives are hard to come by simply because it is very difficult for the human mind to think of things someone has not thought of before. The process of model-building is of great help here for it provides a well-defined context and concrete situation to stimulate the imagination.

The process of searching for alternatives also includes a certain amount of evaluation, for in so doing the grossly inferior ones are implicitly screened out by simple tests for dominance or acceptability. Sometimes these tests are based more on similarity to alternatives found acceptable in the past than on estimates of their actual effectiveness. This is simply a reflection of the fact that similarity is often an efficient screening device. Possibly too much so; it is seldom that a radically unfamiliar alternative will appear useful because the screener, with coordination in mind, will tend to eliminate an alternative that does not appear to fit in with other areas of his organization. The familiar alternatives that change only incrementally have at least that virtue of fitting within the organization.

Except for very narrow problems that permit a completely closed mathematical formulation (seldom found except in the textbooks) where algorithms of one sort or another, linear programming, for example, enable us to investigate an infinity of possibilities, it is not feasible to consider all possibilities nor would it be reasonable to try. One can always think of trivial ways to modify an existing alternative to obtain an entirely new variant. But it is also obvious that at some stage it clearly may not be worthwhile to do so; although differences may exist, they may be so insignificant that, within the precision of our model or of our data, their effect on the benefits or the costs cannot be distinguished from previously considered alternatives. Moreover, many will be obviously dominated by others; that is, provide lesser benefits and generate more costs.

To illustrate, consider an industrial machinery firm that is trying to compare alternative methods of delivering parts to its customers (McKean, 1953):

> The company officials or researchers are likely to consider only a partial list of alternatives. Assume for a moment that they compare the results of using railroad shipments with the consequences of owning their own fleet of trucks. Obviously, if they ignore a third method—e.g., the possibility of using air freight—they may not reach the best answer. At first glance, the solution may appear to be simple: just

> make sure that all alternatives which are close substitutes are considered. The difficulty comes about because of the fact that alternatives are not always obvious substitutes. They are often dissimilar in physical appearance and often differ in their specific function. Indeed, to turn immediately to the extreme case, *all* of the various things which a firm or individual buys, or might buy, are alternative objects of expenditure which may contribute to the firm's or individual's general objective, even though their specific functions are as different as day and night. Bookkeepers and overhead cranes–pianos and gumdrops–all are alternatives, substitutable in varying degrees, competing for the purchasing agent's check.
>
> Now it is obviously impossible to consider simultaneously the whole range of alternatives. Moreover, it is surprisingly easy to overlook a crucial course of action which, if considered explicitly, changes drastically the preferred allocation of funds among the various alternatives.

To consider alternatives, as we remarked earlier, one must first find them. To do this to some extent systematically, one must remember that alternatives need not be similar or attain the objective by the same means. Hence the search for alternatives may be thought of as involving two kinds of tasks:

1. Identification of classes of solutions
2. Investigation of solutions in a given class

Of course, there is unlikely to be a hard-and-fast rule to distinguish between classes. Therefore, to combat juvenile delinquency, education, on-the-job training, work camps, recreation centers, big brother programs, and various antipoverty measures are all possibilities and they may or may not be in the same class, depending on one's point of view. But regardless of the lack of a clearcut distinction between classes, it is useful to look at the search for alternatives as a two-stage process.

For the first, one must make a deliberate attempt to stretch one's mind to identify the relevant kinds of solutions. Brainstorming, for instance, might help. Thus, if the objective were to reduce the crime rate, one might first consider alternatives in broad categories such as social measures (preventive education, antipoverty legislation) or police measures (apprehension, punishment) and then turn to what sort of education and how many patrol cars.

There is a need for creativity and imagination, for there are many subtle forces that tend to restrict the range of alternatives likely to be examined, particularly at this level of abstraction. For one thing, accepting a problem as posed is often equivalent to excluding certain alternatives from consideration. There is a tendency, when a problem is first observed in one sector of business or bureaucratic organization, to look for a solution that is completely contained in that sector and under the jurisdiction of the same administrator. An administrator likes, and ordinarily attempts, to solve his problems within the boundaries of his responsibilities. He is likely to pose his problems in such a way as to bar from consideration, or to dismiss as impractical, alternatives that appear far out or that do not fit into the chain of steps by which policy has been made in past years in the field in question. Problems, however, do not respect organizations.

The Bureau of Reclamation, for instance, was established to increase agriculture production by reclaiming desert land in the West for farming. The alternatives it considers all involve damming surface flows to irrigate desert areas with surface water. It does not consider reclaiming land in areas where water is plentiful even though this might bring a far higher return on investment.

Changing the way a problem is formulated or looking at the question from a different point of view can lead one to consider new alternatives. An example is given by Herman Kahn (1960):

> Another example of *asking the right question* is given by analysts who were asked to study the problem of expediting the delivery of truck loads of merchandise to a chain of markets. The stores had found that because of traffic and other delays they could not predict when the trucks would get to the store, making it necessary for them to keep an unloading crew waiting until the truck actually arrived or to keep truck driver and the truck waiting until the unloading crew came on duty. The analysts were asked to improve the chain's ability to predict the arrival time and thus schedule trucks and crews better. The problem as posed turned out to be rather unfruitful, but the researchers suggested that a much better way to do things would be to use trailer-truck combinations and deliver the goods at night, leaving the loaded trailer at the store one night and picking it up again the next evening. It turned out that it was thereby possible to cut transportation costs (because there was less traffic congestion at night) and to make deliveries completely predictable.

To counteract this tendency to restrict the number of alternatives by the way the question is posed, we must remember that the traditional, conventional, or plausible way of carrying out a task is unlikely to be the only way.

A second restricting influence is parochialism—"the single, most important reason for the tremendous miscalculations made in foreseeing and preparing for technical advances or changes in the strategic situation" (Kahn & Mann, 1957). This often takes the form of a "cherished belief" or unconscious adherence to a "party line." An analyst, an agency, or even an organization as large as the United States Air Force can be so caught up in a party line that the full range of alternatives is just not seen.

A party line can have a strong influence on the list of alternatives considered. What can happen is that the participants and successive reviewers become aware that certain assumptions or some of the alternatives being considered are frowned upon by "higher-ups" or by the sponsor. It seems useless, sometimes even hazardous, to argue the merits of such unpopular views strongly, or to work hard to develop what seems like an unacceptable option, and gradually they may be stressed less and less or even forgotten. In its most extreme form, this influence can lead the organization analyst to lose his independence of view.

Next, having blocked out the broad categories, one must engage in a deliberate attempt to exhaust the most attractive and dominant variations in each particular class. Consider, say, the problem of urban mass transportation. Broad vehicle alternatives might include buses, subways, trains, and individual automobiles. For each of these broad alternatives it would be desirable to consider in detail the advantages of different routing, vehicle configurations, operating practices, etc. This is the kind of a task most readily done by computer, which easily can perform the simple repetitive job of changing discrete variables one at a time until all reasonable combinations are considered. When this tactic is put into practice, the procedure usually can be simplified in various ways. One possibility may be to devise a scheme to scan the choices so that only a finite number of dominant possibilities need be considered, as is done when linear or dynamic programming is used.

Even for small-scale problems, the number of alternatives that can be considered in any detail must be reduced until the number left is manageable. To consider the complete range of possible ways to divide a city into councilmantic elections districts and to be "fair" to all the interests involved may far outrun analytic capacity. Fortunately, the vast majority of alternatives will be obviously inferior, and can be left out without harm. The danger is that some alternative superior to the one ultimately uncovered by the analysis might also be left out. Thus, although constraints must be imposed to reduce the number of alternatives to be examined, this should be done by preliminary analysis, not by arbitrary decree. Moreover, such constraints should be flexible, so that they can be weakened or removed if it appears in later cycles that their presence is a controlling factor. On the same basis, any alternatives that now begin to generate support should be examined again, even if they were previously eliminated.

The generation and subsequent exploration of alternatives is not a single operation but an iterative one. The evaluation process may uncover weaknesses and deficiencies in various alternatives. These can be corrected at least in part by modifying the alternative to remove such deficiencies, thus creating new alternatives. Frequently we lack the imagination to do this at the beginning of an analysis; we think of modifications and we invent new alternatives as the analysis proceeds, helping us to learn more about the problem. This is one of the principal payoffs from the iterative character of the analytic process.

The model used to generate the various impacts may call attention to features of alternatives that cause them to perform less well than other alternatives, leading the analysts to seek ways to modify alternatives to improve their impact. Unfortunately, modifications suggested by the model cannot always be successfully translated into modifications of the real world alternatives.

Examining the Alternatives

As a basis for comparison, the alternatives need to be thoroughly examined. This examination should include political and economic feasibility (legal and engineering feasibility also if applicable), costs, including costs from spillovers, benefits, including indirect benefits, if any, and any other impacts that might seem pertinent to the decision being contemplated. The distribution of the costs and benefits is particularly relevant as well as the distributional effects of such externalities as pollution, noise and inconvenience.

The thorough examination of an alternative or a class of alternatives can be a policy analysis all in itself or a series of such analyses. As said earlier, examining every alternative in any detail may be an impossible job. They can, however, be grouped into a finite number of basic types and a representative may be examined. An example would be the *Studies in Public Welfare* prepared for the Subcommittee on Fiscal Policy (1973). Here three represetnatives or variants of the basic approaches are examined in detail with emphasis on interactions with other programs, the maintenance of incentive to work, and on the more equitable distribution of benefits.

It is the examination of the alternatives and the listing of their characteristics and the consequences that result from their choice that give a basis for comparison.

When we begin to compare alternatives, ideas for modification occur–and a modified alternative is equivalent to a new alternative. As rough costs estimates are prepared, someone may realize that certain features of a system or policy increase its cost all out of proportion to their contribution to effectiveness; as a consequence he may suggest a change that will be equally effective but less costly. Similarly, as effectiveness is now explored, new features may be found to increase effectiveness with little added cost. It is in this aspect of analysis–the innovation of ideas and the appreciation of new impacts–that the analysts can make their greatest contribution.

As we have seen, or will soon see if it is not yet clear, in most policy problem situations we are not able to carry out anything like a full optimization. But it can be very helpful to carry the analysis to a stage at which we have explored a large number of alternatives, examined their characteristics, and determined enough of their impacts to reduce the original list to a reasonable number of candidates (by dominance and other arguments). Consider the following discussion by G. H. Fisher (1966):

> Suppose that we are concerned with deciding among alternative proposed water resources projects, and that we have a given budget to spend on such projects in the future. The budget is such that all of the proposed projects cannot be undertaken. We therefore want to choose the "preferred mix." Suppose further that we have an analytical staff and that it comes up with a summary of results of cost-utility analyses of the problem in the format of [Table 8-1].
>
> Assume that in addition to the quantitative data presented in the table, the analytical staff has supplemented the numerical calculations with *qualitative* discussion of some of the more relevant nonquantifiable issues involved in the decision: e.g., political factors, nonquantifiable "spillover" effects, and the like.
>
> Now decision problems regarding alternative water resources projects are usually very complex. The analyst can rarely come up with a preferred solution–particularly in the sense that one mix of alternatives completely dominates all others. I submit, however, that even in such a context, analytical results of the type portrayed above can go a long way toward sharpening the intuition and judgment of the decisionmakers. I think you will all agree that in the above illustrative case, the decisionmakers would be better off if they had the results of the analytic effort than if they did not have such information. Their decision is likely to be a more informed one. That is the goal of analysis.

Readying the Alternatives for Implementation

Even when the alternatives have been compared and a decision made to "go with" a certain one on the basis of, say, dominant cost and effectiveness

TABLE 8-1.
A Comparison of Projects[a]

Analytical Factor	Proposed projects 1	2	3	4	...	n
(1) Present worth[b] ($):						
(a) Discounted @ 2½% (50 yr)						
(b) Discounted @ 5% "						
(c) Discounted @ 8% "						
(2) Possible variability of outcome:						
(a) "Most likely" range of present worth (low–high $)						
(b) Range of present worth outside of which outcome is "very unlikely" to fall						
(3) Effect on personal wealth distribution:						
(a) Number of farms affected						
(b) Average value of land and buildings per farm in the watershed ($)						
(c) Average net benefit per farm owner ($)						
(4) Effect on regional wealth distribution:						
(a) Average increase in per family income in the Basin ($)						
(b) Percentage increase in average income in the Basin due to project						
(5) Internal rate of return of project (%)[c]						

[a]Adapted from G.H. Fisher (1966).
[b]Present value of estimated benefits minus present value of estimated costs.
[c]The rate of discount which reduces present worth to zero.

considerations, there may still remain considerable work to be done investigating that alternative further.

For example, in comparing programs whose objective is to lower fatalities from heart attacks, the decision may have been to select a program of screening for symptoms and pretreatment in preference to improved ambulance service, or mobile coronary care units, or other alternative. This comparison may have been made on the basis of a relatively detailed cost-benefit analysis, but it may not have been detailed enough to answer all questions about implementation. For instance, although the total capability may have been significant to the comparison, this may not be the case for the number of units and their location in the city. There also may be questions about layouts, the ratio of doctors to nurses, the tests to be used in screening, the questions to be asked. For some of these the analyst's help may be needed.

In other words, even after the decision has been made, the selected alternative may have to be further developed or "fine-tuned." Sometimes a preliminary comparison may have reduced the decision to a choice between two alternatives—to make this choice the final two alternatives may have to be developed to a far greater detail than the others.

Concluding Remarks

This has been a short chapter but it deals with possibly the most important aspect of policy analysis—the design of good alternatives. If the alternatives are all good, the decision will be a good one—perhaps not the best, but still good. If some of the alternatives are poor, the issue may be in doubt.

In summary, it is clear that the deliberate generation of a wide range of choices is an essential component of policy analysis. The job of the policy analyst includes policy design, the area in which he sometimes can make his most important contribution. Analysis is easier for strong systems and good policies. It is also easy for very bad ones. A list of better alternatives can make further analysis unnecessary. The "intraocular traumatic test" may take over; the best alternative may just hit everyone, including the decision-makers, right between the eyes!

References

Fisher, G. H., *The analytical basis of systems analysis.* P-3363. The Rand Corporation, Santa Monica, California, May 1966, p. 11–12.

Kahn, Herman, *On thermonuclear war.* Princeton University Press, Princeton, New Jersey, 1960.

Kahn, H. & Mann, I., *Ten common pitfalls.* RM-1937, The Rand Corporation, Santa Monica, California, July 1957.

McKean, Roland N., *Sub-optimization criteria and operations research.* P-386. The Rand Corporation, Santa Monica, California, April 1953.

Subcommittee on Fiscal Policy of the Joint Economic Committee, *Studies in Public Welfare,* Washington, D.C., Government Printing Office, July 2, 1973.

Chapter 9

COSTS

Meaning of Cost

Costs are the negative impacts associated with a decision–the consequences we would like to avoid or to minimize–just as the benefits are the positive values we seek to obtain or to maximize. A cost to one person may, of course, be a benefit to another; it depends on their objectives.

No decision is free of costs whether or not it leads to the expenditure of money. In public policy analysis costs are much more than the expenditure of money–more than the consumption of physical resources, the employment of human labor, and the dissipation of time. Everything we do, every public program generates costs. Costs may be outputs that follow as a consequence of a policy decision as well as inputs designed to implement it.

If the federal government raises the minimum wage, money is required to notify employers and to prosecute violaters but there are also costs which are outputs of the program. Some unskilled workers become unemployed and are forced to go on welfare. Workers at the margin, say, in rural areas of the South where wages have been low, may move North and into the cities where some join the ranks of the unemployed.

Like so many common words, the word cost is used differently in different contexts, differently by different peoples, and usually in vague terms. Numerous interpretations are possible.[1] Bickner (1971, p. 24) remarks.

> To illustrate the various ways in which the word "cost" might be used, consider the "cost" of stopping by the neighborhood bar on the way home from work tomorrow. This might "cost" you (a) an expenditure of several dollars, (b) a chance to watch your favorite newscast and stockmarket report, and (c) a hangover. What would be the "cost" of stopping by the bar?†
>
> ---
>
> †The reader may be in haste to simplify the concept of "cost" by distinguishing "economic cost" from "noneconomic cost." Any of the three cost items mentioned could involve "dollar costs," however, and distinctions between "economic cost" and such things as "psychological cost" or "social cost" or "political cost" will not

---

[1] This chapter borrows heavily from Bickner's ideas. The treatise by G. H. Fisher in which Bickner's work appears contains the best treatment of the cost aspects of systems analysis available.

survive serious scrutiny. The example given does introduce one simplifying distinction, however. It distinguishes costs to *you* from costs to *other people,* such as your wife, the bartender, and local law enforcement agencies. For many government programs, the decisionmaker may be uncertain about whose costs he wants to consider and whose costs to ignore."

Fundamentally, a decision to do one thing implies a decision not to do something else. If we choose to use some of our resources to build a new city hall, then those resources are obviously not available for building a new highway. If we assign our best available engineers to research and development of energy sources, they are not available for work on smog control, or vice versa. An estimate of the cost of any such choice or decision is an estimate of the benefits that would otherwise be obtained. *Economic* costs are benefits lost. A decision to build a school may be a decision not to build a hospital for the resources may no longer be available. It is for this reason that economic costs are often referred to as opportunity costs or alternative costs. It is in the alternatives, that is, in the foregone opportunities that the real meaning of cost always must be found. An estimate of the cost of any decision is then an estimate of the benefits that could otherwise have been obtained if some other decision had been made. After all, the only reason we hesitate to spend our money is because of the alternative things that it could buy.

There are also costs, often called *social* costs, to distinguish them from the more direct and immediate economic costs. These are effects we would like to avoid—pollution of the atmosphere, interference with the movement of people, noise, and other undesirable external effects—that, at first thought, may not seem to be economic costs for they may not appear to be benefits lost. But as Bickner remarks (see previous footnote) the distinction will not survive serious scrutiny. Suppose, for example, some decision leads to "pollution of the atmosphere" or to "noise." Aren't we then foregoing an unpolluted atmosphere? Or quiet? Are not these benefits lost? The distinction is merely a matter of degree—between the good we cannot have and the bad that makes the situations worse. The knowledge of what such costs are, how they are distributed by area, income class, ethnic group, and so forth is extremely important to making the right decision and thus to the success of the analysis done to assist that decision.

There are divergent views on whether "social" costing can be performed with a degree of success sufficient to be really useful. An illustration of how one might go about calculating the costs of automobile pollution is contained in the appendix to this chapter. It is included not only to illustrate social costing but also because it is a good example of first-order or "back-of-the-envelope" analysis.

The obvious costs, e.g., the sum of money a city government may have to appropriate to purchase property and widen a residential street to make it an arterial highway, are the easy costs to determine. There are always other costs—the neighborhood may deteriorate, for instance—just as there may be benefits other than increased traffic flow—businesses may move in and tax revenues increase. The problems mostly lie elsewhere.

Why Costs Are Considered Easier to Measure Than Benefits

Because economic costs are simply benefits lost it would seem that they would be considered equally difficult or equally easy to measure. Yet we often see statements (I have made them myself), that cost is easier to measure than benefits or effectiveness. There are several reasons for this common assertion. One is the confusion that often exists between dollar expenditures and total costs. Government officials and budgeters tend to think of the costs of a program in terms of the inputs to the program: resources, labor, and time. These costs can easily be transferred into dollar expenditures (or at least more easily than other costs).

Another reason is the frequent practice of dismissing all those costs that cannot be measured conveniently in dollars as noncost considerations or as qualitative factors. That is, arbitrarily or unconsciously, we modify our concept of cost to mean only those costs that can be readily estimated or evaluated in dollars. If we do not forget about the hard-to-evaluate costs completely, we attempt to take them into account by subtractions from the benefits as negative benefits of a sort. Essentially what we customarily do is to transfer any costs that are difficult to measure in dollars over to the benefit or effectiveness side of the cost-benefit or cost-effectiveness equation. Such costs then become negative benefits and are left for the benefit analysis rather than the cost analysis to evaluate. If there happen to be any benefits readily measured in dollars they are likely to be transferred over to the cost side of the equation as cost savings or credits against costs. If we make such transfers then there can be no doubt that costs are going to turn out to be easier to measure than benefits.

Another argument as to why benefits are harder to measure than costs is that they are likely to be displaced in time. But then so are many of the costs. For example, consider the federal minimum wage hike we mentioned earlier. The effects on cost due to workers becoming unemployed are likely to be just as displaced in time as any benefits that may arise from the law.

For further discussion of the problems with benefits and costs see (Mishan, 1971) and (Prest & Turvey, 1965).

To be certain that all costs are taken into account, it may be well to break them down in the following way.

1. Dollar expenditures
2. Costs that can be measured in dollars
3. Costs that can be quantified
4. Costs that are nonquantifiable

Identifying and Determining Costs

Suppose we are trying to decide whether or not to procure a particular site for a new public park in a mountain area on the outskirts of the city. Consider the following things that we might do and the relationship among them: (1) We might estimate and list the resources required for the proposed new park. This would include the land, the materials and labor to put in the access roads, to landscape, and to build the recreation facilities; the time of park employees, the transportation facilities; and so on, plus miscellaneous items such as the legal staff that would be required to carry out the condemnation in order to acquire the property. (2) We could identify and describe some of the alternate uses of these resources. For example, we could determine what could be done with recreation areas that were already city property with better planning or better facilities. Also, we might consider the value of a recreational building located in the center of the city rather than a mountain park on the outskirts, which might present some problems of access for the low-income people who we hope might use it. We might also consider the resources used alternatively, that is, not just to produce other recreational areas or capabilities but for training, education, or various other activities. (3) If we were ambitious enough, we could attempt, not only to identify the alternate uses of the required resources but to estimate the value of these alternatives. In other words, we could attempt to evaluate the benefits that would be given up if we were to go through with this proposed park project. (4) We could estimate the total dollar expenditure that will be entailed by the procurement, development, and maintenance of the proposed capability, including tax loss.

In short we can try to estimate (1) the resources required; (2) the alternative uses of these resources; (3) the value of the resources in their alternative uses; (4) the value of these resources in dollars. In light of our definition of costs, of course, we should examine all four aspects to the extent that it is possible.

The basic meaning of cost analysis as we have stressed before, lies in estimating the value of the alternatives; this is the third procedure above. The other procedures are useful only as they contribute to this task; the first two are necessary to it. Before we can evaluate the alternative uses of resources, we must first identify those alternative uses; this is the second procedure. The possible alternative uses, of course, depend both upon the ingenuity of the analyst and the scope of authority possessed by the decision-maker. But before we can identify the alternative uses of given resources we have to determine what these resources are. This, the first procedure, is frequently the easiest of the four steps, at least in philosophical sense. But the required list may be quite long.

Money As a Measure of Costs

It is not obvious that dollars can reflect the real cost of even the resources required to implement a program, let alone the full opportunity cost. If we examine the planning of future programs by a government department or by a public body, however, it becomes obvious that money costs are very pertinent. Public bodies always face a budgetary constraint—usually they do not face a significant constraint on other inputs for they can almost always buy more of them if they have the wherewithal to pay the price. To implement a course of action, a government body gives up freedom of action, of course, but it also sacrifices money. If it is within the budget, one item can be substituted for another by paying the price of the one instead of paying the price of the other.

This is not to say that money costs perfectly represent resources sacrificed in carrying out government programs. The prices of goods to be bought in the future are uncertain. A given course of action, in itself, may drive up the price of particular items and it is not possible to predict these effects with complete accuracy. The cost of some items will change as technology advances or if the program is so large that the marginal cost is affected. Nonetheless, imperfect as it is, the money cost of a program usually serves as a reasonable proxy for the sacrifices that would be required of the organization to do something [see also Mishan (1971)].

But even if dollars reflect resource and opportunity cost fairly well from the viewpoint of a given government program, do they do so from the standpoint of the national interest? If the economy is a reasonably competitive one, the answer is surely yes. The reason is that market prices in a competitive economy reflect not only the approximate rate at which one government organization can exchange one item for another but also the approximate rate at which the whole economy can substitute one article or material for another.

In our political–economic system, it is customary to allocate dollars to various government departments rather than to allocate resources, such as manpower, materials, or transportation facilities directly. This procedure gives the departments more flexibility and permits them to choose those resources that can accomplish their mission at a lesser cost—that is, at the sacrifice of other possible benefits.

In order for this system to work, of course, it is necessary that dollar values be assigned to the various resources available to the department, and that these dollar prices reflect the value of the benefits that could otherwise be produced with these resources. This is the function that our free competitive market accomplishes for us. Individual consumers or other government departments that can use these resources to derive benefits bid for them with their dollars. The resultant prices are then the value to them of these resources. Money helps us measure costs because, and only because, and only to the extent that, they help us identify and evaluate alternatives.

Costs are associated with both inputs and outputs. If we decide to build a freeway from point A to point B to supplement surface streets, there are input costs (resources) to be considered—material, labor, equipment rental, etc. There are costs that follow as a consequence also. When completed, the new freeway will attract traffic, eventually so much that the motorists who use the road impose costs on themselves—in delays and in direct costs, chiefly fuel—which can, of course, be translated into dollars. There are also other outputs that become costs but, unfortunately, they are not borne only by the users or by the public entity that built the freeway. Until recently, such costs were seldom made the concern of government analyses. Among such external costs or spillovers are the costs suffered by the people whose access is cut off by the new limited access road, by people in the neighborhood who are annoyed by fumes and by merchants who are no longer visited.

Resource analysis is the name given to the procedures which work out the cost of the inputs or of the resources required to implement a program. We do not have a special name[2] for the process of working out the other costs.

What Costs Are Relevant?

The relevant costs lie in the future, not in the past. Suppose, for example, that we are estimating the cost of developing a new public park from land acquired several years ago for the purposes of building a high school that was never constructed. In this work, the cost of purchasing that land is no longer a relevant factor. Costs that have already been incurred are costs resulting from past decisions. They are not costs of any conceivable current or future decision. When we decide to use land we already have for a park, the alternative uses of that land are the relevant costs, and not what we could have done with the money that we spent to buy that particular land. Past alternatives no longer matter since the opportunities to forego those alternatives no longer exist.

Past costs are sunk costs, and the analyst must be careful not to include any sunk costs in his cost estimates. Sunk costs do not represent alternatives that we can do anything about and hence they are no longer real costs. The costs that are important are future costs or incremental costs. What matters in deciding whether or not to develop city land into a park is not the costs already incurred by the city but the additional or incremental money necessary for completion of the project plus the surrender of the opportunity to use the land for some other city purpose.

Another question is how far into the future must the cost analyst look. The only possible answer to that question is to look as far as he can. We must remember, however, that because money in hand today is ordinarily worth more than the same sum a long time off, costs far in the future may appear to become vanishingly small. But there are other considerations that may have to be taken into account. One consideration is costs to whom? This generation or the next? Another is the time preferences of people. Still others are whether the investment opportunities that can make dollars today will be worth more the following year.

All costs are relevant to some decision or other. The responsibility of the cost analyst, however, is not simply to add up any and all costs indiscriminately but to identify and measure that particular collection of costs that are contingent on the specific decision or choice under consideration. To do that he must distinguish the relevant from the irrelevant ones. Any costs that will be incurred no matter what choice is made, any costs that must be borne regardless of the decision at hand, are not costs of that particular choice or decision. Relevant costs are those costs that depend on the choice made, given the choices available.

---

[2] The name "external impact analysis" is sometimes applied to the process of working out the social costs.

External Costs and Internal Costs

Costs may be relevant but they may not concern us. For example, costs falling upon hostile nations may not concern us in the same way as costs falling upon our own population. External costs are those costs of a program or decision that fall outside the boundaries of the decisionmaker's interest or beyond the scope of his organization. Whether a given cost is internal or external thus depends on where in the decisionmaking hierarchy the decisionmaker happens to be and how comprehensive his concern.

Fairly standard examples of external costs include the adverse affects on flora and fauna caused by cutting down trees in a forest or the increase in the mosquito population by the creation of artificial lakes and other ecological repercussions that ultimately affect the welfare of the people. The offense given by the erection of a building which interferes with television reception is a cost to those who must buy a special antenna; the builder, however, is likely to consider it to be external. The number of external costs in the real world are virtually unlimited.

Externalities occur when not all the costs of production are imposed on the producer. Air pollution from automobiles is a typical example of an external cost that government action could make internal. Externalities, of course, may have positive as well as negative effects. Negative external costs such as air pollution have positive counterparts such as a lake created by dams built jointly for irrigation and power that may also furnish recreation. An example of an externality that is a benefit occurs in the case of vaccination for contagious diseases. Consumption of polio vaccine not only benefits the consumer, who pays for the medical care, it benefits those who might catch the disease from him if he were not immunized. To offset the positive externalities associated with medical care there are also negative externalities; for example, those imposed on nonsmokers and nondrinkers by the consumption of alcohol and tobacco.

Because decisions must be made as to what costs to internalize and what costs to treat as external to the analysis, there is no obvious, no clearly unique set of cósts that can be assigned to a government policy or program. The analyst must give meaning to the cost of an alternative by deciding to what degree the various costs are of concern to the contemplated decision and how they should be weighed and evaluated.

These decisions have to be made with great care. What is today "outside" the decision-maker's "boundary of interest" or organizational scope may suddenly become "internalized." A change in the law, public opinion, or the political pressure of a special interest group may bring this about. For example, the clamor from the environmentalists forced public officials to consider the impact of airports, highways and other construction on the ecology of the area of interest. It is therefore the responsibility of the analyst to point out the more important external effects even though they may be "out of bounds" at the moment.

Marginal Costs

To guide choice, in analyzing the cost one must focus his attention on marginal costs although total costs are the ultimate concern.

Marginal costs are those costs incurred as we make marginal changes in a program. If, for example, we add one more safety inspection during the process of constructing a bridge, this will add a marginal increment to the cost of making such inspections. At the same time, it should add some marginal improvement in the probability that the completed bridge will not collapse.

Because there may be hundreds of alternative ways to increase the effectiveness of a program, we need to estimate the marginal cost of these competing alternatives so that we can allocate our resources to the best of them. We need to know the marginal ratio of benefits (increases in effectiveness) to cost for each of these numerous alternatives. Only then can we be sure of getting the greatest effectiveness with the resources available or of achieving a given level of effectiveness for the least cost.

Effectiveness is maximized, or costs are minimized, only when the marginal ratios of benefits to costs are equal for all the competing alternative ways available to make marginal improvements in a program. This is why the analyst and the decision-maker must be continually concerned with marginal costs. We must be sure that our dollars, our manpower, or our facilities are used in ways that give us the greatest marginal increase in benefits. Unless we equate cost-effectiveness or cost-benefit ratios at the margin we can be sure we are not getting the greatest effectiveness from the resources committed to a program.

Cost As Output

We have, I hope, reconciled the two seemingly different concepts of cost: cost as the value of the alternatives that we cannot have and cost as the undesirable consequences of a policy or a program. The first, the economic concept of cost, which includes the second, is made up of the benefits lost; the second, which includes the first, emphasizes the so-called social costs—pollution and other environmental impacts, for instance. In either case, costs are consequences of our decisions.

It is clear that the direct consequences from any particular use of resources can include both beneficial and noxious elements. The latter should be included among the costs of a decision, but, as we remarked earlier, they are often considered negative benefits rather than costs, in keeping with our tendency to transfer all difficult measurement problems over to the benefits or effectiveness side of our analysis. Just as costs of a decision are benefits lost, so also are the benefits of a decision costs avoided. This may not be of great significance; in complex situations the main accomplishment of analysis may be to assess the major consequences or impacts—both positive and negative—of the alternative choices. We don't necessarily have to display these under the headings of "costs" and "benefits" for, as we noted earlier, we may be forced to leave that determination to the decisionmakers.

Costs are consequences; therefore we must note that only decisions or choices have costs. In our sense of the word, programs do not have costs. The decision to set them up or carry them out has costs. Also we cannot estimate the consequences or the costs of any decision without presupposing some alternative decision or choice.

How can we estimate the consequences of a decision? We can only do this by comparing alternative states of the world. We must estimate what will happen if we make a decision and also what will happen if we do not or if we make some other decision.

Costs depend upon the decision-maker as well as upon the decision. The higher up a decision-maker is on the ladder of authority the more resources, and the more kinds of resources, he is likely to have. Thus the costs to accomplish a particular task are likely to be lower because there are more and more alternative sets of resources from which he can choose. An example might help to clarify this. Suppose a city Department of Sanitation needs additional area for garbage fill and has a plot of land available. Suppose, in addition, that the city itself has a number of plots of land that could be used for the same or other purposes such as parks, etc. The decision to supply a fill area, a specific task, is likely to result in lower costs if made at the city administration level than if made at the department level as one of the city plots may be more suitable for the desired use than the department plot and the department has no alternative. Looking at the decision in another way, the cost of using the department plot, a specific resource, for the fill area is likely to be higher than the cost of using one of the other plots since the department plot may have a higher value in an alternative use.

This dependence upon the decision-maker brings us to the question: What decision-making level should we consider? It might seem that the analysis should focus on the highest conceivable decision-making level. This may be logical theoretically but we are likely to find it exceedingly difficult to define and difficult to make cost estimates that are relevant for so undefined a decision-maker. Even a cost analysis undertaken for a congressional committee must take for granted most of the numerous constraints on decision-making and the numerous boundaries on alternatives that are incorporated into federal laws or into the contemporary framework of the economy and the social order.

We might also assume that the decisionmaker and the analysis have no limit on ingenuity. Consequently, with cost defined as the best alternative use of resources, the cost of a program depends also on imagination and ingenuity. Bickner (1971, p. 47) explains:

> Again, it might seem that the cost analysis should assume, for purposes of estimating costs, that the decisionmaker has no limit of ingenuity. This procedure may also be laudable, but it is no more practicable than the procedure of assuming unlimited authority. There is always certain to be a better use of resources than the use which

we select—if only we were ingenious enough to find it. *If we define cost as the best possible alternative use of resources, the cost of any practicable program will exceed its benefits.* The cost analyst will have to content himself with those very limited alternative uses that the decisionmaker, or the cost analyst, can readily discover. Helping the decisionmakers discover and evaluate alternatives is the function of cost analysis. It is also the function, incidentally, of our competitive free-market economy.

Discounting

Time is valuable and yet the value of time is sometimes forgotten, particularly whenever someone compares dollar expenditures for this year with those of next year and the year after.

Such dollar values are not equal. Resources on hand today are usually worth more than the identical resources deliverable tomorrow. Consequently, the dollars with which we can buy resources today are worth more than those same dollars if they are not available until tomorrow. Therefore, before we can add dollars spent or received in different periods, we must discount future dollars, for they are worth less than current dollars.

The procedure for discounting is simple, although the choice of a particular discount rate may be very difficult to justify. Discount is calculated exactly the way the banks do it. If the discount rate is 6%, the bank will give us $1.06 in a year for a dollar today. The cost of a program which does not require that the money be laid out until some time in the future must be discounted, if it is to be compared with an alternative where the money is to be paid today. By discounting the value of future dollars we can reduce all future dollar expenditures to their present value equivalent. We can then compare programs for which the expenditures come at different times.

Although future dollars are worth less than currently available dollars, the question is how much less. To establish that time has a value is easy. To say specifically how much, however, is difficult.

The proper choice of a discount rate depends just as cost depends, on the alternatives. In turn, these alternatives depend upon the decision-maker's authority and interest. The appropriate discount rate for use in comparing future dollars with today's dollars depends upon the alternative opportunities available for exchanging one for the other.

Other Comments on Costs

INFLEXIBILITY

The option to change one's mind free of cost does not always exist; inflexibility in a system must be considered a cost—a design or program that does not allow for the possibility of change must be assessed a cost over one that does.

RISK

Risk is associated with the concept of cost; the probability of unfavorable outcome as opposed to a desired one. Because risk is undesirable, we demand compensation for accepting it. It thus represents a cost.

EXPERTISE

The cost of expertise will very likely be more than the cost of the expert's salary as the opportunity value for the use of his knowledge can greatly exceed the market value of his salary.

OPPORTUNITY COSTS

For decisions involving public money we should be looking for the best alternative use to which the given resources can be put. In other words, if we did not do a particular project, could the money be spent for a project that would yield even greater benefits? There are, of course, practical constraints. Money must be spent legally, for instance, and the original sponsors of a study are unlikely to countenance a lesser benefit to themselves than to others.

IMPLEMENTATION

Another cost that should be explicitly considered is the cost of implementation, including the cost of organizational changes, if any, that need to be made if a new program B is to supplant ongoing program A. Almost always there are problems with organizations and with some individuals that arise from the use of analysis. It is more than resistance to those changes in practice that will ordinarily be required as a consequence of finding a better way to do something. For the use of analysis may itself affect the implementing organization in serious ways—for instance, by changing the power relationships, by impairing incentives, or by showing a tendency to increase centralization. One aspect of analysis should thus include explicit discussion and evaluation of the implementation and organizational costs (and possibly benefits) associated with each alternative.

OVERHEAD

When several projects are run by the same organization or agency, certain costs are often not identified to a particular project or program but considered to occur as common costs for the group as a whole, with a share allocated to each individual project. These costs are called overhead or indirect costs. They are usually fixed costs that do not fluctuate very much with the level of effort expended for a particular project. The allocation must be made somewhat arbitrarily. The usual procedure is to assign a burden charge to the directly related cost. This burden charge is a fixed percentage of the direct. Consequently, these burden charges are allocated as if they fluctuated in proportion to the fluctuations of the costs that are directly associated with a project.

There are some obvious disadvantages to this approach of burdening the direct costs with an indirect component. The indirect rates, of course, are computed on the basis of what happened in the past. If the level of activity in one project changes while the others remain fixed, leading to an increase in the costs charged as overhead, the result will be a misallocation of the new indirect costs to the other projects. This scheme will not provide for a correct allocation to the increased project since all the costs resulting from the decision to change the project should be allocated to the project.

INPUTS FROM OTHER GOVERNMENT AGENCIES

All of the costs of a program do not occur within the agency administering the program; sometimes one agency may impose a major cost upon another. For example, delivery by mail of a "free" government pamphlet still involves cost, even though the agency benefiting from the service does not buy stamps.

A particularly good example is pointed out by Hovey (1968) in the potential reciprocal imposition of costs on each other by the Army Corps of Engineers and the interstate highway program. Interstate highway bridges must be built over navigable rivers with sufficient clearance to permit river traffic to pass. The extra height or other measures required to ensure clearance are quite expensive for the Bureau of Public Roads or the various state highway departments to provide. They are costs imposed by the Army Corps of Engineers, in its role as the protector of inland navigation, on the various highway agencies which are ultimately paid for by those who buy gasoline. On the other side of the coin, if the highway planners would build causeways or obstruct navigation on waterways with low bridges they would be imposing extra burdens on those who use the rivers for transportation and on agencies such as the Corps of Engineers whose responsibility is to keep rivers navigable. Moreover, they would not have to consider the costs they impose in their road-building budget.

Governments can also make its citizens pay for public benefits without going through the process of appropriation and justification. For example, a school board may decide to change the rules under which it transports students to school by increasing the distance that they must live from the school before they are transported. Also a city government may appropriate private land and reimburse the owners who may then move their business elsewhere without compensating the workers who now may have to travel much farther to work.

An illustration of how the apparent cost of a program may be substantially lower than its full cost is given by Hovey (1968).

Consider a federal program of grants-in-aid to major cities aimed at eliminating, say, at least one-half of the rat population in those cities. If the program were a typical 50% grant program covering such approaches as poisoning, increased sanitation efforts by the city, enactment of ordinances to require private citizens and businesses to follow more sanitary procedures, etc., the apparent federal costs would be the annual appropriation to the federal rat control agency for (A) operating and administrating expenses and (B) grants to the participants. The full costs of the decision to initiate such a program could, however, involve far more. Some of the additional costs might be federal, such as the costs of office space in existing federal buildings (until recently not budgeted for), the increase in the review and administrative costs in agencies such as the Bureau of the Budget and the Public Health Service, which result from the program. There will also be some future expenditures which will be required to keep the impact of some programs from being lost by reversion to their earlier ineffective status when the grant money stops coming. There would also be added costs to private individuals and businesses because more stringent sanitary procedures would require them to keep their premises cleaner, and so forth. Finally there would be the 50% to be paid by the state and local governments.

The above example demonstrates that the federal budget would very likely understate the costs of the program. In this case the question: What would Congress have to appropriate to eliminate half of the rats in our Nation's cities? is by no means the same as the question: What would it cost to eliminate half of the rats in our Nation's cities? One basic element in cost analysis is to make sure that the answer to a question of this type is not substituted for the answer to a question of the second type.

Accuracy

As we shall see in Chapter 11 where we talk about quantification, all measurements are approximations. Hence, if one is to avoid getting lost in a series of detailed refinements, one must appreciate the degree of approximation which is appropriate to the particular problem at hand. Accuracy is costly. The more accurate we are in our measurements, the more costly the work will be and the more time will be required. In public programs the highest degree of accuracy attainable is ordinarily unnecessary in most decision situations.

Cost measurements are approximations. Although it is traditional in accounting practice to express measurements of dollars in two decimal places, that is to cents, long-range cost estimates used for examining alternate public programs cannot be stated with such accuracy and do not require such accuracy. (After all one does not need to know the temperature outside to one-tenth of a degree in order to decide whether or not to wear an overcoat.) To decide between various public policy alternatives, it is usually sufficient to know only that one program costs something like twice as much or three times as much as a competing program without knowing within hundreds of dollars what either of them will actually cost. Also, a rough approximation available today is much more valuable than a more accurate cost figure that might be obtained a year or so from now.

References

Bickner, R. E., "Concepts of Economic Cost" in G. H. Fisher (ed.), *Cost considerations in systems analysis.* American Elsevier, New York, 1971, 24–63.

Coates, Joseph F., letter, *sppsg Newsletter* (now *Public Science*) June–July 1972, 3, 31–34.
English, J. Morley, *Concepts of systems resource requirements in cost effectiveness; the economic evaluation of engineered systems.* Wiley, New York, 1968.
Hovey, Harold A., *The planning-programming-budgeting approach to government decision-making.* Praeger, New York, 1968.
Massey, H. G., Novick, David & Peterson, R. E., *Cost measurement: Tools and methodology for cost effectiveness analysis,* P-4762. The Rand Corporation, Santa Monica, California, February 1972.
Mishan, E. J., *Cost benefit analysis.* Praeger, New York, 1971.
Prest, A. R. & Turvey, R., Cost benefit analysis: A survey. *The economic journal,* 1965, Vol. **15**, (300) 638–735.

Appendix

The Costs of Automobile Pollution: A First-Order Analysis[3]

Joseph F. Coates

The following is a schematic outline of the procedure followed in an exercise at the Engineering Foundation Conference at Deerfield Academy in August 1971 in which a dozen of us—as an exercise—blocked out and carried through an analysis of the social cost of automobile pollution.

The results were not intended to be definitive, but only to provide an opportunity for mutual education in how a complex social cost problem might be approached. Consequently, no effort will be made to present the detailed quantitative analysis, but only to rough-out the steps which established the conceptual framework.

STEP 1. DEFINE THE PROBLEM

It was essential to define the scope of the problem as crisply and sharply as possible to reduce the intellectual noise and to focus the analysis. Consequently, we took the problem to be: "Calculate the cost to society of pollution generated by the automobile where the automobile is taken to be private vehicles in the United States and pollution is taken as all those adverse effects generated by all things given off by an automobile or associated with its ultimate disposal." This definition thereby excludes problems associated with automobile accidents and traffic congestion.

---

[3] From Coates [1972]. In his original letter the author remarks:

It should be clear that this is at best a first-order analysis, although I think you will see from the number and texture of results that it has some intrinsic interest.

STEP 2. A FRAMEWORK FOR EFFECT

To provide some hooks on which to hang concepts, we worked out rough categorizations of where sources of pollution may originate. After some discussion (about five minutes), we focused on the following categories:

	Sites of Effects			
Source	People	Things	Processes	Natural Environment
Air				
Water–liquid				
Solid				
Aesthetics				
Noise				
Miscellaneous				

STEP 3. IDENTIFYING EFFECTS

This next step took about thirty minutes. It involved only the attempt to identify adverse effects associated with pollution, on the various categories mentioned above. No attempt was made to define their importance. Rather, we strove for completeness, comprehensiveness, and the rough assignment of category. The specific approach was to take a category, say "air" (exhausts to), and just discuss among the team what adverse effects may result: eye irritation, cancer emphysema, intoxication, e.g., from lead. That list generation was beat around until we had fairly well exhausted that row of the matrix against each column head. It was repeated for the next row. We systematically went through all sorts of things we thought might be given off by automobiles, the gaseous exhalations, lube drippings, parts dropped off, etc. Just working this over and noting all this down on a blackboard allowed us to identify some 95 adverse effects from automobile pollution.

STEP 4. INTELLECTUAL SHAKE-DOWN

Having defined a list of effects, we then went back over it, combed it out a bit, reorganized some of the effects to better fit the categories, expanded some, and collapsed others. That was a tightening-up, ordering phase, taking about 15 minutes. The miscellaneous category dropped out. It was empty.

STEP 5. TOOLS FOR SOCIAL COSTING

Recognizing that many of the effects were going to be medical or involve recurring quantitative factors we felt that in making costs estimates, it would be useful to agree on some numbers ratios, and proportions beforehand. For example, we took the population of the country as 200 million, but the principal focus of exposure as major metropolitan areas, and estimated that population. We knew we were going to need some value of a life. After discussion, we set a tentative figure of $250,000 for a man in his middle years and scaled this back in value if we were talking about the elderly or the young. The value we used for old people in the terminal years of their lives was $250,000 divided by three times the average working life (3x40) per year of reduced longevity. Many disabilities reduce life expectancy. This value costs out the terminal years, discounted for reduced productivity. Obviously, this was not an elegant or unquestionable value, but a useful one. We then attempted to agree on a rough ratio between morbidity and mortality and decided that for every mortality there was probably an average 50-fold morbidity.

By discussing these tools separately, they could be isolated emotionally from their specific applications later. The group was surprised, considering their divergent views about the plausibility of social costing, at the easy agreement on the numbers for the tentative analysis. It was recognized that an error of 50% or even 100% in many cases was not going to be a strong driver of the analysis, for our purposes.

STEP 6. THE STRATEGY FOR THE ANALYSIS

Having looked over the range of effects it seemed that many of them could be dealt with by direct calculation of the cost. In other cases, it was not clear how one would deal with the cost directly. We agreed, in those cases, to try the "at-least method." That basically says if you cannot cost-out the effect itself, you can cost-out what it would take to remove, eliminate, or alleviate it. Then one can decide if it is worth "at least" that much to deal with the problem. This, incidentally, is a technique used in the Bureau of Mines Circular 8414.

Having defined our strategy to use the direct calculations where feasible or the "at-least method" when we had to, we then jumped into the main event and went down the list, item-by-item, calculating the individual social cost.

It was a little sticky at first until the group got into the swing of it. It went straightforwardly. Again, the group was surprised and pleased with the relatively ready consensus on the numbers and calculations.

To illustrate the "at-least method," we were unable to calculate the direct cost of the aesthetic insults from automobile scrapyards, but decided that one way to deal with that problem would be to hide them behind high but attractive fenses. We first estimated the number of scrapyards in the United States, then their average size. From that we calculated their total circumference and with that estimated what it would cost to fence them in with a fifteen-foot high fence.

STEP 7. CODIFICATION OF RESULTS

We were then able to reorganize the data, add things up and calculate the table in which the various pollution effects were laid out against the points of impact of man, the natural environment (flora and fauna), aesthetics, etc. From that we developed a textured view of the cost of pollution. The calculated aggregate cost of pollution, when analyzed, came to $100 per vehicle or 11 cents per gallon of gasoline, $10.4 billion/year.

The Social Cost of Automobile Pollution[a]

	People	Things	Processes	Natural Environment
Air	$6,955,000,000	$481,000,000	X	$500,000,000
Water	70,000,000	210,000	X	100,000,000
Solid	122,500,000	X	X	X
Noise	1,250,000,000	X	$16,000,000	X
Aesthetics	880,000,000	X	X	X

[a]Total 10.34×10^9 dollars/year; 90×10^9 gallons gasoline/year; 11¢/gallon.

Chapter 10

MODELS

What Is a Model?

The heart of any policy analysis is the creation of a clear, precise, manageable process designed to produce information about the consequences of a proposed action. This process is provided by one or more models, devices that range from no more than an image of the situation in the mind of the analyst to an elaborate simulation involving men and computing machines. Models are fundamental to policy analysis. Although they cannot predict the consequences of decision with the assurance of scientific models, models tell us what the possibilities are, based on various assumptions about the factors of concern, and thus produce information that helps us to understand the situation more clearly. In fact, analysis of a problem might be defined as the search for a solution with the aid of one or more models.

Decision-making itself is a process of sequential model-using. Decision-makers constantly shift from noncalculated or unconsciously calculated decisions to explicitly calculated decisions, that is to say, from calculations based on primitive intuitive or tacit models to occasional calculations based on sophisticated models. Decisions are often made intuitively without explicit models or even without the decision-maker being aware he has a model in mind; but a model exists even though considerable effort might be required to make it explicit. There is no such thing as not using a model in analyzing a decision.

As an example of the use of implicit models, taken from Hovey (1968), consider the following arguments that might be offered by a principal for repainting his school:

1. The current paint is dirty and flaking; if uncorrected, this situation will create an atmosphere of dirt and disorderliness which will cause student performance to worsen.

2. The current paint is dirty and flaking; if uncorrected, this situation will create an atmosphere of dirt and disorderliness which will cause teacher performance to worsen.

3. The current paint is flaking; if uncorrected, this situation will cause cracking and structural deterioration which would cause higher costs later in order to keep the school operating.

4. The current paint is in good condition, but a change of colors every two years is necessary for good teacher and student performance.

5. The current paint is in good condition, but it is "sound practice" to repaint on a two-year cycle.

The first four of the above statements are each based on a different model that is implicit to the person making the argument, of a relationship between painting and education or educational operations; the fifth models the practice of painting itself. The school board that receives these arguments might review a recommendation based on (3) differently from one based on (1) or (2); (3) is based on a model that draws technical conclusions about the relationship between painting and structures and is thus probably outside the principal's expertise; (1) and (2) relate painting to student and teacher performance, an area more within the principal's ken but for which conclusions may rest on highly fragmentary evidence.

In the more elaborate analyses, formal, explicit models are ordinarily built as part of the investigation or, in a few rare cases, borrowed from other model builders. This need not be the case; a model does not necessarily have to be elaborate in structure or even explicitly set out in full detail, but to argue for any action convincingly requires at least some sort of explicit model to show others how we determined what to expect.

Again, a model is central to every analysis. Its presence may not always be obvious but it is there nevertheless. Its role in policy analysis or in any analysis of choice is, as we said, to provide a way to produce information about the outcomes that follow a choice of alternatives. Ideally, we would like the model to do more, to actually predict or forecast the outcomes with a high degree of reliability but in policy analysis we must usually settle for much less.

To fulfill its role, a model may take a wide variety of forms. What are some of these forms? A map, an organization chart, a set of mathematical equations, a wind tunnel, a series of random numbers—all of these can serve the role of model. In a beauty contest, to predict the contestants who survive to later rounds, the model[1] may be as simple as three measurements—in the United States the magic numbers are 36-24-36. To investigate an air-defense plan the model may consist of a group of men and machines acting as if they were an air-defense combat information center.

All of these models have certain aspects in common. Each is an idealization, an abstraction of some part of the real world and each is an incomplete representation of the real thing—an imitation of reality.

Olaf Helmer (1966) amplifies this as follows.

> A characteristic feature in the construction of a model is abstraction; certain elements of the situation may be deliberately omitted because they are judged irrelevant, and the resulting simplification in the description of the situation may be

---

[1] The example is from R. D. Specht (1968). Several sections of this chapter follow Specht's ideas very closely.

helpful in analyzing and understanding it. In addition to abstraction, model-building sometimes involves a conceptual transference. Instead of describing the situation directly, it may be the case that

> 'each element of the real situation is simulated by a mathematical or physical object, and its relevant properties and relations to other elements are mirrored by corresponding simulative properties and relations. For example, any geographical map may be considered a physical model of some sector of the world; the planetary system can be simulated mathematically by a set of masspoints moving according to Kepler's laws; a city's traffic system may be simulated by setting up a minature model of its road net, traffic signals, and vehicles.'

A model involving such transference, in addition to abstraction, is called a "simulation model."

When an operations analyst constructs a model, simulative or not, he usually does so in order to determine the most appropriate action to take in the face of a given situation. His function is, after all, to give operational advice to a decision-maker. Often, he may find himself at the frontier of the state of the art, and he may have to rely heavily on whatever expert judgment may be available, rather than on a solid (nonexistent) theory. His model is therefore apt to be ad hoc, tentative (that is, subject to modification and improvement), future-directed, and policy oriented. Frequently, the reliability of such a model may leave much to be desired; yet its justification should derive from the fact that recommended actions based on it have a good chance of being more appropriate than actions selected without use of the model.[2]

A model, then, is a substitute for reality—a representation of reality that is, hopefully, adequate for the problem at hand. It is made up of factors relevant to a particular situation and to the relations among them. We ask questions of the model and from the answers we hope to get some clues to guide us in dealing with the part of the real world to which the model corresponds.

A model depends on more than the situation being modeled, however. It must also depend on the question being asked; that is, the purpose for which it is to be used. If I want to ride a bicycle from point A to point B, the model that I need to tell where I might find a lunchroom if I take the left fork instead of the right—a simple road map—may be entirely different from the model I need if I want to route a freeway between the same two points.

We must not object if the model does not look like the "real thing" or if it does not represent all aspects of reality. It seldom accomplishes the first and it cannot do the second. The important thing is whether or not the outputs from the model, that is, the answers it gives to our questions, are reasonably appropriate and valid.

We would like to be able to test the predictions based on our model and determine the correctness and relevance of these predictions for real-world decisions. In some circumstances we may actually make this sort of a test; for example, we may be able to validate the model above that tells us how to ride a bicycle from point A to point B by tracing the route in a car. Unfortunately,

---

[2] Olaf Helmer and Nicholas Rescher, "On the Epistemology of the Inexact Sciences," *Management Science,* 6 (1959), 48.

there are many cases in which such tests are impossible or too expensive. It is almost never possible when major policy issues are involved. For example, we hope we may never find out whether our models of all-out nuclear warfare are accurate.

The models in policy analysis, then, are the means of producing information about the consequences of adopting an alternative. Sometimes this may be done with a single model. Usually, however, it is more convenient to consider the over-all process as made up of a number of lesser models or submodels—a cost model, for example, and an effectiveness model. Note that I said "producing information about the consequences," which is a rather weak phrase. In rare instances we can predict with certainty or assign high confidence to the results that come out of our model. But the world is usually too full of uncertainty for this sort of thing. For one thing the system or situation we model is always imbedded in a more general situation or larger system that we have taken as given—but the "givens" may change.

To recapitulate our earlier remarks, models in policy analysis are used in somewhat the following manner. Given the context, goals, and the alternatives, including those that the analyst may have invented during the course of his work, the analyst designs one model or a sequence of models that reflect the context or state of the world of which the alternatives are a part and predicts the consequence of each choice. The analyst then determines in the light of his various models what consequences to expect if an alternative is selected; for example, how effective the alternative is likely to be in attaining the objective, the resources it may require from the economy, and how it might impact on the environment.

An Example of a Mathematical Model

To illustrate the construction and use of an analytic model we will describe a model developed at the University of California (Jewell, 1963) for the prevention of forest fires.

> It is generally agreed that a strong initial attack upon a fire will prevent most, if not all, fires from causing substantial damages, both in values burned and in suppression costs. The questions of "how much" and "how soon" manpower, equipment, and other suppression forces should be committed to the initial attack do not have a clear-cut answer. The extreme alternative of sending "everything you have as soon as possible" is, from economic and strategic points of view, just as undesirable as the other alternative of sending too few men; the optimal strategy lies somewhere in between. . . .
>
> The costs incurred on a given fire may be broken down into the following categories:
>
> 1. *Fixed Costs:* that are associated with the training and maintenance of a fire-suppression force, and various prevention and detection systems.
>
> 2. *Emergency Costs:* that are standby and other "emergency state" costs of the complete system, irrespective of how large the fire is, or how many forces are mobilized.

3. *Suppression Mobilization Costs:* that are related to the number and type of forces employed in suppressing the fire. This cost is primarily transportation expense, portal-to-portal wages and other "one-shot" logistic support costs.

4. *Hourly Suppression Costs:* that are proportional to the total force-hours spent in suppressing the fire. . . .

5. *Cost of Values Burned:* that not only include the potential market value of any timber destroyed, but the losses to future production, and the necessity for reseeding and restoration. Somewhat harder to determine, but just as important, are the losses to wildlife and wildlife habitat, damage to watershed values, and recreational loss. . . .

If we use x to denote the number of initial attack forces of a certain type sent to a given fire A, the ultimate area of burn, and T the total time to control the fire from the moment of initial attack, then, if linear costs are assumed, we have a total cost:

$$C(x,A,T) = C_F + C_T T + C_S x + C_H x T + C_B A \quad (1)$$

where the constants are related to the cost categories above.

To illustrate the possibilities in intial attack planning, let us consider the simple problem illustrated in Fig. 4.

A fire front of length L moves at a velocity V_f through homogeneous fuel under constant weather conditions. The attack plan is to construct a fire break at length L and of width W at an appropriate distance downwind from the fire; after thus stopping the head of the fire, the weaker flanks may be easily mopped-up. The chosen width W is a function of the wind, the flame velocity, and type of fuel.

Let us suppose that it is planned to send x men to construct the fire break. If we assume that each man can clear and scrape α square feet/hour in this type of fuel, then x men will construct $\alpha x/W$ feet of fire break per hour, assuming proportionality. Efficient location of the break (neglecting any possible topographic advantage) will be such that the break will be completed just as the fire reached it. (Any fixed safety factor or evacuation time may be effectively included by delaying the start of suppression.)

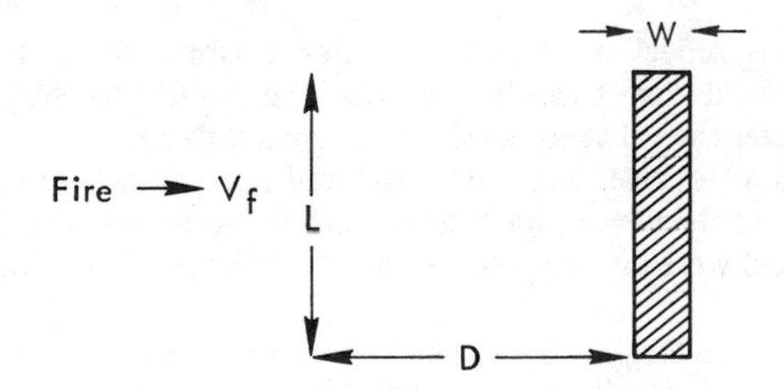

FIG. 4. A simple fire-spread model.

Thus, the time to construct the break and control the fire is just:

$$T = LW/\alpha x \text{ hr} \quad (2)$$

and the area burned during this time is

$$A = L^2 W(V_f/\alpha x) \text{ sq feet} \quad (3)$$

The total cost [equation (1)] then becomes:

$$C(x) = C_0 + C_S x + C_B L^2 W V_f/\alpha x \text{ dollars,} \quad (4)$$

where C_0 is a fixed cost associated with the fire

$$C_0 = C_F + C_H LW/\alpha \text{ dollars,} \tag{5}$$

and C_B' is an equivalent cost of burn:

$$C_B' = C_B + C_T/LV_f \text{ dollars/square foot.} \tag{6}$$

There is an optimal fire-suppression force, x^*, which will minimize the total costs of suppression and damage,

$$x^* = L \sqrt{\;} V_f (C_B'/C_S)(W/\alpha) \text{ men.} \tag{7}$$

The fire is to be controlled in a time

$$T^* = \sqrt{\;} (1/V_f)(C_S/C_B')(W/\alpha) \text{ hours,} \tag{8}$$

resulting in a minimal total cost of:

$$C(x^*) = C_F + C_H LW/\alpha + 2L \sqrt{\;} V_f C_S C_B' (W/\alpha \text{ dollars.} \tag{9}$$

Despite the simplicity of this model, there are several instructive results. First, we notice that the least-cost solution balances not only the marginal costs of suppression and damage, but also the total (variable) costs as well. The number of forces per unit length of fire break is independent of the size of the fire, but proportional to the square root of the fire-spread velocity; the faster a fire is spreading the quicker it should be controlled. Cost factors enter as a square root of their ratio, and so on. . . .

The value of x^*, which gives the optimal force size, is determined by differentiating Eq. (4) with respect to x, setting the derivative equal to zero and solving for x.

Crude as this model is, it is useful. For example, from the form of the expression for total cost it can be seen that the penalty for sending too small a force is far greater than that for sending too great a force.

The analysis, of course, was not considered to be anywhere near complete at this point. In fact it was hardly begun. Various extensions and elaborations of this simple model were investigated. As listed by (Jewell, 1963) these were:

1. The effect of varying assumptions about fire geometry and rate of spread.
2. Investigations of the efficiency of various fire-fighting tactics, and their possible analytic representations.
3. Incorporation of elements of risk, particularly in the rate of spread of fire, and in imprecise estimates of suppression effort.
4. Consideration of alternate types of suppression forces.
5. The problems of reinforcement when it is discovered that initial attack forces are inadequate.
6. Interactions among allocations of suppression forces in dynamic multiple-fire-situations.

In order to make progress in investigating the initial attack on forest fires, the model we have just sketched had to both omit some of the relevant factors and make some simplifying assumptions and idealizations. For example, the effect of topography and of natural barriers to fire spread had to be neglected; cost has been assumed to depend literally upon force size, etc. The model is therefore incomplete, but then this is a characteristic of every model. The world is simply too complicated to be included in any quantitative expression, or even to be included conceptually. But the model does have the characteristic elements of every model: it is an abstraction of the real fire situation and of fire-fighting forces; it picks out factors that are assumed to be relevant; it makes explicit the relationship between these relevant factors; and it predicts certain outcomes.

The "Role" of the Model

In constructing a model for a given problem situation, the actions taken are: (1) to single out certain elements as being relevant to the problem under consideration, (2) to make explicit the significant relationships among these elements, and (3) to formulate hypotheses regarding the nature of these relationships. It is these functional relationships that are most often conveniently expressed in mathematical form.

Systems analysis and operations research, for example, have functioned well for industry and defense, areas that lack the benefit of a comprehensive theory for guiding action, by relying on the systematic utilization of a large body of only partly articulated and largely intuitive judgment by experts in the relevant fields. The standard operations research procedure for such utilization of experts is that of constructing with their help an appropriate model of the situation. Systems analysis, operations research, and related techniques are now being called on in public policy areas precisely for the same reasons they were called on in industry and defense—a satisfactory theoretical foundation for guiding action is lacking. Consequently, we must rely again on model building.

Without exception, insofar as I know, writers on operations research and related disciplines recognize the importance of model-building even though they may put their emphasis on model-using. They agree that the major element in tackling a problem is the construction and use of a model, tailored both to the situation under study and to the question being asked.

The "distinctive approach," as the Operational Research Society of Great Britain put it in their definition of operational research, "is to develop a scientific model of the system, incorporating measurements of factors such as chance and risk, with which to predict and compare the outcomes of alternative decisions, strategies or controls." Now I am not certain how the word "scientific" should be interpreted beyond a plea for maintaining the traditions and practices of scientific research, but I do know how "scientific model" is interpreted.

The most utilized models, often the only type even considered by the usual practitioner, are scientific in a narrow sense; that is, they attempt to be completely mathematical, to express through a set of mathematical equations, or a computer program, the advantages and disadvantages of alternative actions in terms of parameters the different values of which discriminate between the various options or alternatives. Other parameters enable us to investigate changes in the contingency or context. By operating with such models, either analytically or numerically, we attempt to predict the more important consequences of alternative choices.

Now it is true that, in ideal but seldom-realized circumstances, the model in a systems or policy analysis provides a way of forecasting the outcomes that follow the choice of alternative actions and, even of indicating a preference among them and do so with confidence. Circumstances, however, are seldom that ideal and we must usually settle for a good deal less. But when the situation being modeled is an existing one, or is structured like an existing one, and we know or can learn a good deal about the variables and their relationships involved, a mathematical model may provide a close approximation to what would actually happen. But even under ideal conditions, the representation is seldom so faithful that one is willing to accept the solution from the model as the solution of the problem.

Even though a mathematical formulation with which one can optimize is the most sought after model, there are other routes to estimate the consequences that follow a choice of alternatives. In fact, when the situation is fuzzy, they may be more satisfactory. These routes are based on a more direct use of judgment and intuition.

ROLE IN FOCUSING JUDGMENT

Reliance on expert judgment and intuition is crucial to every decision. This reliance permeates every aspect of public policy analysis—in limiting the extent of the inquiry, in deciding which hypotheses are likely to be more fruitful, in designing the model, in determining what the facts are, and in interpreting the results. A great virtue of model building is that it provides a systematic, explicit, and efficient way to focus the required judgment and intuition.

We should not look at models simply as devices to provide an analytic route from hypotheses to predictions about the real world. So narrow a view ignores an important process: in using, and particularly in building models, analysts and the experts they confer with are compelled to use a precise terminology, to develop their ideas and to exercise their judgment and intuition in a well-defined context and in proper relation to each other, and to explore the reasons for their disagreement. In addition, a model provides feedback to guide the participants in refining their earlier judgments. This point is important; by "exercising" the model and testing for sensitivity, the model can provide information that may lead the builders to alter their original judgments. It is these features associated with building and using a model that are important to its role in supplying a route from hypotheses to prediction—not merely how faithfully it simulates the real world or whether it provides a formal or quantitative scheme for optimization.

The realization that intuition is important is certainly not new; intuitive judgment has been used to predict outcome from the time man first began to think. The use of systematic ways to exploit intuition and judgment is new, however. Operational gaming, that is to say, exercises in which estimates of outcomes under various initial conditions are obtained from the actions of the participants who interact by playing roles that simulate individuals, or factions in a society, or even such things as sectors in an economy, is one such scheme. It is clearly a step away from the traditional model. And, as we shall see in later chapters, we can usefully go even further afield using scenarios, for instance, or even a series of questionnaires to provide a framework that serves the same prediction function but looks still less like the traditional mathematical model.

Classification of Models

In policy analysis we use the term model in a very broad sense to apply both to a description of some part of the real world, which has explanatory and predictive properties, or to a process using people, with the same properties. In science models are used to build theory. This modeling is a crucial step in the chain of hypothesis formulation, model building, prediction, and validation. A model that is proven valid attains the status of a theory. The net result is a better understanding of some part of the real world. In policy analysis, however, models are used to gain insight into the value of various alternatives on which a decision-maker might take action, and thus to provide a better and more explicit basis for choice than the direct application of judgment and experience without the use of an explicit model. The net result is usually a better basis for action. Our focus, of course, is on this latter use of models. We are only slightly interested in modeling as a theory-building tool, but are strongly interested in models as a tool in the problem solving process.

Because there are so many kinds of models, it is convenient to have some scheme of classification, some set of characteristics according to which we can group them.

One way to classify models is to separate them into two types: qualitative or (strictly) quantitative. A quantitative model[3] in this sense is a mathematical model defined by a precise set of assumptions expressible in terms of a well-defined set of mathematical relationships. These might be equations or other analytic expressions or instructions for a computer. The behavior of such a model is determined completely by the assumptions, and the conclusions are derived as logical consequences of those assumptions without recourse to judgment or intuition about the real world process or problem being modeled.[4]

Frequently, a conclusion will be drawn by a decision-maker or analyst using an analytic model that does not follow as a logical consequence of the assumptions defining the model. For example, it may follow from some intuitive stroke

[3] The usual practice is to use the term quantitative much more loosely.

[4] The set of assumptions for such a model is almost always larger than that which may have been specifically enumerated by the analyst. For instance, certain mathematical assumptions are always included, but possibly implicitly; those of probability theory and decision theory, for example.

of insight sparked by the model or from the addition of some consideration that may not have been taken into account in the model. When this happens the model being used becomes partly implicit and ceases to be a strictly quantitative model, at least in relation to that conclusion.

There are very many real problems and processes, however, which cannot be described in the precise mathematical terms required for a quantitative model in this strict sense. For these problems, qualitative or partially quantitative models must be used. A qualitative model typically is based on less precise assumptions than are required for a quantitative model and its behavior may be described by a combination of deductions from these assumptions and by further subjective judgment about the process or problem being modeled. Because so very few public policy models are quantitative in this strict sense, this classification is not very useful.

There are many other ways in which to characterize our knowledge of models. For example, we could classify models according to the following scheme:

Purpose: Training, research, decision, etc.
Field of application: Transportation, inventory, education, health, etc.
Level: Office management, national policy
Time character: Static or dynamic
Form: Two-sided or one-sided, conflict or not
Analytic development: Degree to which mathematics is used
Complexity: Detailed or aggregated
Formalization: Degree to which the interactions have been planned for and their results predetermined.

A fairly standard but not very useful classification for policy analysis separates models into *iconic, analog,* and *symbolic.* Iconic models are images; they look something like what they simulate but differ in scale—maps, architectural drawings, scale "models" of airplanes built for testing in a wind tunnel are examples. Analogs are models that parallel the operation to be studied, e.g., by using fluid flow to represent traffic flow. Symbolic models use numbers, letters, or other symbols to represent the variables and parameters in the situation and their relationships; a mathematical model is thus a symbolic model.

We will use a classification that is a modification of one formulated by Robert D. Specht[5] that seems better suited to policy analysis.

In our modification, the five categories are:

1. Mathematical Models
2. Computer Simulations
3. Operational Games

[5] Specht (1968) classified models into five categories in order of increasing analyticity. Each category was broken into two subcategories according to whether or not an active opponent was involved and conflict (or cooperation) an essential part of the model.

4. Verbal Models
5. Physical Models

These categories overlap to a considerable extent; the title refers to the main means or elements used in the model to determine the consequences of the various alternatives considered. Thus, for example, an operational game, say an exercise in which human players act the part of land-use planners, may employ both computer simulations and mathematical models as submodels to handle certain aspects.

A few words about some of the categories. Let us begin with the most mathematical.

1. MATHEMATICAL MODELS

Here are found the models that use techniques such as linear and dynamic programming, queueing theory, network theory, and so on. The forest fire model of the previous section is also of this type. A computer may be used with these models, but as an aid or calculator, not as a simulator, and after the mathematician has finished most of his work. It is characteristic of these models that they deal not with specificity, but with generality; not with a single play or instance of a situation (game or otherwise), but with all possible plays or instances of the situation.

To handle conflict analytically, such as a strike or labor dispute, we have the models of a game theory. Here the analyst is concerned not with playing a game like Monopoly or some much more sophisticated version—games that involve people and actual play are described under Operational Games as well as in Chapter 13—but with the manipulation of mathematical relationships. In these conflict cases, the decision-maker must be concerned not only with his own decisions, as in the case of the fire-fighting example of the previous section, but also with the decisions of an opponent who may not, like nature, be neutral. You may remember the cartoon that showed a high-level conference in Washington, with the speaker saying, "The way I see it, Russia thinks we think they think we're not willing to go to war." Game theory cannot solve problems such as that one, but it does furnish a framework in which one can think more clearly about the difficult problems involving a second decision-maker who can cooperate with or hinder your operations. Conflict situations occur in political contests, strikes, or bidding contracts and so on, although most applications have been military.

2. COMPUTER SIMULATIONS

In the forest fire example of the previous section, the relevant factors were few in number and the relations among them were simple enough for us to trace out the interactions with pencil, paper, and a little mathematics. We arrived analytically at the relation that specified, for any value of the parameters, the optimum number of men to be deployed. Moreover, this was a general solution;

that is, the optimal force size was determined for any value of velocity, length of front, width of break, and so on. In more complicated problems, the relevant factors are too numerous or their interrelations to complex to be handled analytically. Instead, we must write out instructions for an electronic computer, and the model thus appears as a computer program.

In the usual case, this computer program would be a "simulation." In operations research parlance the term "simulation" is applied to the process of representing numerically, but without using formal analytical techniques, the essential features of a system or organization and analyzing its behavior by operating with the representation on a case-by-case basis, in fact, a pseudo-experiment. Simulation is also used broadly to describe various physical or analog devices such as a flight trainer, or, as we mentioned before, a group of people and machines acting as if they were an air defense control center.

The line between a simulation and a mathematical model is not clear cut. The representation of a transportation network by means of a linear programming model would not be termed a simulation, however, because the programming model would be a mathematical model. But if the network were represented by a miniature physical analog involving roads, traffic signals, vehicles, etc., or by random numbers on a computer, it would be called a simulation. If the representation or model simulates a conflict or if working with it includes any aspects of playing a game, or even if it merely involves people to make decisions during the running of the experiment, the simulation is often called a "game." If the model of the real situation is stochastic and the analysis is carried out by sampling experiments, the simulation is often termed a Monte Carlo simulation.

An example of a large computer simulation model, designed to supply policy assistance, that has attracted considerable attention and much criticism[6] is that of Jay W. Forrester in *World Dynamics* (1971) and D. H. Meadows *et al.* (1972) in *The Limits to Growth.*

Categories 1 and 2 together constitute the class of "analytic models" or, using the term more loosely, quantitative models. They are the subject of Chapter 11.

3. OPERATIONAL GAMES

Humans are, of course involved in all models—as designers, users, and as the suppliers of data. But in the models of this category humans are the essential element. Here people are used to act out the role of real-world decision-makers, or to simulate a policymaking institution. The war game, the business game, and the military and political crisis exercise are in this category.

Both people and computers may be part of this sort of model. An example would be an air defense simulation in which computers are used to present realistic surveillance information about enemy invaders and the people play roles simulating the behavior of those who must order the proper steps to counter an enemy attack.

---

[6] For criticism, see H. S. Cole *et al.* (1973).

Categories 2 and 3 together include the class of simulation models, sometimes called "gaming models." We will say more about these later.

4. VERBAL MODELS

As we have seen, the model-builder decides what factors are relevant to his study, determines the relations among them, and traces out their interactions and implications. This activity, more or less, traces what anyone does when thinking about a problem. The model-builder merely does these things explicitly and, where possible, quantitatively—his assumptions laid out on the table for any man to inspect and criticize.

If a model has no quantitative content it goes, perforce, in this category. The most common study, of course, is one that combines verbal and analytical models; it is a mixture of 4 with 1, 2, and/or 3.

The scenario, whether used alone or in conjunction with other models, is usually cast in the form of a verbal model. Scenarios are discussed later in the chapter on Extraquantitative Methods. The simplest form of verbal model is the analogy.

All of us use models of various aspects of the real world, although in most cases we do not make them explicit and, indeed, would probably have great difficulty in laying them out for others—or even ourselves—to see. And most of our decisions must be made on the basis of these implicit models.

Whereas the verbal models often remain unstated and implicit, there are exceptions. For example, in a verbal model built by Anthony Downs (1964), a bureaucracy is defined as an organization that has the following four characteristics:

> 1. It is large; that is, the highest-ranking members know less than half of all the members personally. This means that bureaus face substantial administrative problems.
> 2. A majority of its members are full-time workers who depend upon their employment in the organization for most of their income. That is, the bureau members are not dilettantes but are seriously committed to their jobs. Also, the bureau must compete for their services in the labor market.
> 3. Hiring, promotion, and retention of personnel are at least theoretically based upon some type of assessment of the way in which they have performed or can be expected to perform their organizational roles (that is, rather than on some characteristics such as religion, race, or social class or periodic election).
> 4. The major portion of its output is not directly or indirectly evaluated in any markets external to the organization by means of voluntary tit-for-tat transactions.

Some typical examples (Specht, 1968) of bureaus covered by the theory are the Roman Catholic Church (except for the Pope, who is elected), the University of California, the Soviet central planning agency, the U.S. State Department, the New York Port Authority, and the Chinese People's Liberation Army. The model or theory has been designed to make practical predictions about the likely behavior of real-world bureaus. It does this by generating specific propositions linking certain elements of the internal structure of bureaucracies with certain aspects of their functions and their external environments. An example (Downs, 1966) would be "... *in any large, multi-level bureau, a very significant portion of all the activity being carried out is completely unrelated to the formal bureau goals or even to the goals of its topmost officials.*" or the "... Law of Ever Expanding Control: *The quantity and detail of reporting required by monitoring bureaus tends to rise steadily over time, regardless of the amount or nature of the activity being monitored.*"[7]

5. PHYSICAL MODELS

Physical models are not very important in public policy analysis. An example might be a physical model of something like a housing development or a new city. Another might be a PERT network–a graph (a collection of points, or nodes, connected by lines or arcs) that represents a set of decisions in a complex development program. A more established model of this type, from which extremely reliable predictions can be made, might be an engineer's diagram of the forces in the structure of a bridge. Such a model, a series of straight lines on a sheet of paper, is a simplification that is known to work. An engineer thus does not need to experiment by building bridges to see how many stand up. But the predictions from even models such as this are sometimes in error.

Some Characteristics of Models

As we remarked earlier, in the more elaborate policy analyses, say one contracted for by a federal agency such as HEW and involving several man years of effort, formal explicit models are often required. These are likely to be computer simulations or mathematical models heavily dependent on computer calculations. Sometimes such models are even contracted for directly; there are times when model builders have built a model and then attempted to peddle it to an agency with a problem.

We would like our model to be both valid (so that a high degree of confidence exists that the inferences obtained from it will, in fact, occur in the real world) and relevant[8] (so that it throws light on the problem of interest to the decision-maker).

There are two ways to go about constructing a model that is valid and relevant. One is to design the model with validity uppermost, attempting to capture the real world by simulating the situation to prefection. Using this approach, for example, there have been attempts to model the behavior of a city

[7] This "Law" is strikingly similar in phrasing, and to some extent in content, to Parkinson's First Law. Our indebtedness to Parkinson's general approach to formulating principles he has formulated is indeed immense. See C. Northcote Parkinson, *Parkinson's Law and Other Studies in Administration,* pp. 2–13.

[8] The terminology follows Ansoff and Hayes (1973). Their comments on corporate decision-making have wide application to decision-making in the public sector.

so realistically that the model reproduces faithfully what goes on–growth, population movement, industrial development, and so on. When such models have been asked to help with questions of interest, however, the results have been poor (Brewer, 1973; Wildavsky, 1973). A second approach is to construct the model with relevance in mind, i.e., with the question uppermost. This latter approach is almost always better for unless the model builder has the question in mind he has no guide to tell him what to leave out–and the real world is too complicated to be encompassed in any model.

The decision-maker (and the analyst) should also be interested in another aspect of the model: cost. Building, developing, and applying a model, including the gathering of data, is not a cost-free exercise and the decision-maker would like to be sure that the improvement that is to result from the model will exceed the expense of model building.

Another set of criteria of main interest to analysts describes the quality of the model. Thus a model is *nontrivial* if it permits inferences to be drawn that are not obvious or readily perceivable by direct observation; it is *powerful* if it offers a large number of such nontrivial inferences; and it is *elegant* if a minimum of carefully selected analytic tools are used to produce a model of great power. Criteria such as these are a major concern of model builders whose scientific reputation, chance of promotion, and feeling of self-fulfillment are all highly correlated with the quality of the models they build.

The decision-maker and the analyst can have different points of view or interest with respect to models. The decision-maker's objective is to produce results; although the analyst seeks to help his client, as a model builder he typically seeks model quality. It is not ordinarily the analyst's responsibility if a particularly pressing public problem is not receiving the attention that it should. On the other hand, the decision-maker is definitely responsible. He may be penalized for failing to focus attention on a critical problem and he receives little reward for commissioning an elegant model for solving a problem of secondary importance.

These differences in outlook can lead to differences in preferences for the kinds of models that are used. The model builder's natural inclination is to construct models that will give the greatest latitude for the application of his skills and knowledge and that will produce, as a consequence, the greatest recognition and reward. This appears to be found in highly nontrivial, powerful, and elegant models–mostly large computer simulations. He may, in fact, seek to work only on problems that offer opportunities for models of high quality. In practical problems, these considerations then often become the criteria for choosing how to tackle the problem and what kind of model to build.

In the effort to make a model more realistic, common practice is to gather more information and add it to the model, but the costs of overcomplication and excessive size may negate the benefits of increased realism. The ability of the analyst to understand and to manipulate his model decreases rapidly as the analytic size of his formulation increases. As a guide he must keep the question he seeks to answer clearly in mind.

Rather than model quality in the above sense and comprehensiveness, the decision-maker wants relevance and validity. This usually requires a close working relationship between the decision-maker and the model-builder. One recent help in this direction is the advent of real-time computer technology, which has made it possible to involve a busy manager in model building. It has also made it possible to explore many more alternatives and to do this very much more rapidly than we have been able to do in the past.

For reasons of cost, inaccessibility, or risk, the analyst cannot always test his models satisfactorily. He must frequently be satisfied with a pseudo-experiment or simulation. Often the best that can be done is to determine answers to questions such as the following:

1. Can the model describe correctly and clearly the known facts and situations?
2. When the principal parameters involved are varied, do the results remain consistent?
3. Can the model handle special cases in which there is some indication of what the outcome might be?
4. Can it assign causes to known effects?

The Need for Nonquantitative Models in Public Policy Analysis

In industrial and military applications of analysis the problem is likely to deal with an enterprise largely designed and directed by man—a manufacturing process, a weapon system, a railroad network, for example. The enterprise will be characterized by a set of operations involving men, machines, and materials. It has a goal or goals (sometimes these may be hard to state precisely) against which the value of the output can be measured, and a policy, a set of rules, to guide the operations. Now as we said earlier, the key to successful analysis is the creation of an operational model: a clear, precise, manageable description of the processes involved. Ideally, this model should have two attributes—it should be numerical, this is, quantitative, and it should be succinct. To be succinct, the description must abstract from the host of features of the enterprise those that are relevant to the problem being investigated and only those.

Now it often requires a great deal of effort and ingenuity to "find" such a description. Yet, when one is investigating something designed by man, the creation of a useful model seldom requires the same sort of profound digging that is required to determine something like the role of hormones in regulating body functions.

A major reason for the success that systems analysis and operations research have encountered in military and industrial applications is, as expressed by Normal Dalkey (1967), traceable to the fact that they have been dealing with artifacts—that is, the subjects of study have been designed and built by man. For instance, a transportation system is something that is more or less created with a purpose in mind and has a structure that follows the laws of engineering and economics. Goals can be stated, authority is clear-cut, cooperative, interested in improvement, and the underlying principles governing seemingly haphazard behavior can be discovered and modeled.[9] A savage, wending his way through a forest, may trace out a path that defies description. But a highway is constructed with careful forethought and is likely to have geometric simplicity. A telephone system, growing under the rising demand for communications, is structured according to the laws of economics and communication engineering.

In contrast, an attack on problems of air pollution, urban renewal, vocational rehabilitation, or criminal justice involves investigating a "system" that has grown without conscious design, the goals of which may be obscure and conflicting, and doing this in a situation in which authority may be diffuse, overlapping, and have different sets of goals. Data collection is difficult; the act of investigation may even bias the data. Linear programming, queueing theory, statistics, and the computer are still enormously useful here, but this is usually true for segments of the problem and not for the central question. It is not surprising, therefore, that attempts to build quantitative models with which one can optimize in the conventional sense tend to fail. Fortunately, this does not mean that we must give up systems analysis entirely, for, as we have seen, there are substitutes for the traditional quantitative model, but they must be sought. There are possibilities like Delphi and gaming (which we consider in a later chapter on nonquantitative analysis) that can be completely nonquantitative. Others are still partially quantitative, leaving, say, whatever optimization is done to the decision-maker.

Miscellaneous Remarks on Models

Building and using models is fraught with its own types of pitfalls. One of these has to do with the question of determining the total composite of variables that are relevant to the situation for which policy is being formulated. A second has to do with accurately determining the functional relationships among these variables, particularly those for which no data are available.

There are, however, even greater pitfalls in the intuitive and mental models that many policymakers and planners employ. These models may often invoke the wrong variables; so may explicit models. In addition, the policy-makers and planners who use such mental models often assume incorrect interactions between the factors that they regard as components of the system being considered. Analysts do also. But as long as these models are implicit and cannot be observed by others these dangers not only exist but are much greater than they need to be.

When decision problems arising in the real world are analyzed, the distinction between the problems and the models used to analyze them are often blurred or suppressed. In the case of textbook examples, where the problem has a clear and

---

[9] This does not mean there are no difficulties. For instance, one of the driving considerations in future transportation systems is a behavioral question of demand. This has proven to be a difficult area to deal with.

well-defined structure usually identical with that of the model, this may be of little consequence. In problems of public or civil order, however, the distinction between the problem itself and a model used to help solve it is important. Real problems tend to be complex and extremely fuzzy; a model on the other hand, is a simplified abstraction or analog of that problem or portions of it, used to understand or gain insight into important aspects of the problem itself.

Models, in fact, are often deliberately oversimplified. For example, queueing theory usually assumes that the input rate and the holding times are not affected by the queue length. This is rarely true. People simply do not join a queue of excessive length.

It should be pointed out in this connection that frequently the use of numerical parameters and thus of a mathematical model is an excellent device for conceptual clarification.

If we simplify the model too much, we may not be able to use it to explain or to understand what happens in the real world. If we make it too complicated, it may no longer be an aid to understanding. The dilemma is thus how much to simplify. A model must be both simple enough to allow the analyst to think with its aid and realistic enough to produce reasonably valid predictions. Increased complexity goes with increased realism.

Those critics of analysis who object that its models deal with an artificial and oversimplified version of the real world often have in mind an even more artificial, more highly simplified model of the real world, although their view is implicit rather than explicit. For example, they sometimes appear to think that the arms race may be summarized simply by referring to budgets or by counting warheads, that weapons are either invulnerable or vulnerable, first strike or second strike, that postures are characterized either by superiority or stability, that both the wartime objectives of United States and Soviet Union are simple, and that many things can be assumed as certain to lead to escalation or all-out war or spread of arms. It is our contention that the explicit model that at least attempts to consider all the major factors in the problem (or at least those relevant to the issue at hand) and to allow for the uncertainties is far more likely to lead to sensible results.

Sometimes a decision-maker with a policy problem feels that he or his analyst does not have the knowledge or facilities to build the required model. He may then commission someone not associated with the problem to prepare the model or he may take the very risky course of allowing some entrepreneur to sell them a "general purpose" model.

When someone else builds a model or modifies a model already built for another purpose, it is necessary for him to be very clear as to why the model is wanted, what is wanted out of it specifically, and how it will contribute to solving the decision-maker's problem. This is necessary, when any model is prepared but it must be particularly stressed when someone not associated with the preliminaries does the work. Attention to the formulation stage cannot be overemphasized; we need at least a crude idea of what the answer might look like in order to have some idea of how to approximate it and what type of model to use.

A model is a device for assembling the information and hypotheses scattered throughout the "community" pertinent to the question being considered in such a way that all components are put into proper relationship. The people who have the problem can help a great deal but often they can do no more than to identify the important factors that have to be considered, possibly with some qualitative associations. The analyst has to search out the data and establish the relationships. Iteration–formulation and reformulation of the problem–is important here.

Before one undertakes the construction of an elaborate computer model or considers modifying an existing model, certain questions, at a minimum ought to be answered. Brewer (1972) suggests the following list:

1. Are the variables related to the policy question accessible and measurable at acceptable costs?
2. Is there a good data bank containing these variables already? Is it accessible? Are the variables in a flexible, usable format?
3. Is there sufficient, reliable theory concerning the question to enable the construction of a model having good representation? Or, are the phenomena encompassed by the question well-understood?
4. Will it be necessary to consider alternative theoretical possibilities? If so, are there data or are the prospects for obtaining them acceptable, relative to their costs?"

If these questions can be answered satisfactorily, the attempt to construct a relevant and valid computer model may be successful. If not, an alternative approach with a simpler model may be more productive.

References

Ansoff, H. I. & Hayes, Robert L., Role of Models in Corporate Decision Making, in Ross, Miceal (ed.), *Operations research '72*, American Elsevier, New York, 1973, 131–162.

Brewer, Garry D., *Policy analysis by computer simulation: The need for appraisal,* P-4893. The Rand Corporation, Santa Monica, California, August 1972, p. 18.

Brewer, Garry D., *Politicians, Bureaucrats, and the Consultant: A critique of urban problem solving,* Basic Books, Inc., New York, 1973.

Cole, H. S. D., et al, *Models of Doom: A Critique to the Limits of Growth,* Universe Books, New York, 1973.

Dalkey, Norman C., *Operations research,* P-3705. The Rand Corporation, Santa Monica, California, October 1967.

Downs, A., *Inside bureaucracy,* P-2963, The Rand Corporation, Santa Monica, California, August 1964. (Published commercially in A. Downs, *Inside Bureaucracy,* Little, Brown, Boston, 1967.)

——, *Bureaucratic Structure and Decisionmaking,* The Rand Corporation, RM-4646-PR, March 1966.

Forrester, Jay W., *World Dynamics,* Wright-Allen Press, Cambridge, Massachusetts, 1971.

Helmer, Olaf, *Social Technology,* Basic Books, Inc., New York, 1966.
Hovey, H. A., *The Planning-Programming Budgeting Approach to Government Decision-Making,* Frederick A. Praeger, New York, 1968.
Jewell, William S., "Forest Fire Problems; A Progress Report, *Operations Research 11* 1963, pp. 678–692.
Meadows, D. H., et al, *The Limits to Growth,* Universe Books, New York, 1972.
Specht, R. D., "The Nature of Models," in E. S. Quade and W. I. Boucher, Eds., *Systems Analysis and Policy Planning: Applications in Defense,* American Elsevier, New York, 1968, pp. 221–227.

Chapter 11

QUANTITATIVE METHODS

What Is Quantitative Analysis?

A policy analysis would be completely quantitative if the problem under investigation were described by a mathematical model so faithfully that a decision could be made purely on the basis of the results that are obtained from it. There may be such problems but they are not found in the domain of public policy analysis. We shall therefore use the term "quantitative analysis" loosely, as is ordinarily done, to characterize any analysis that is centered about a mathematical model or a computer simulation.

When critics speak about PPBS as a "technocratic utopia where judgment is a machine product," (*New York Times,* 1967), or of people who have created "the illusion that they are capable of relating cost to military effectiveness by scientific analysis" (Rickover, 1970), they refer to the type of activity that I am here calling "quantitative analysis."

Quantification is the ideal toward which we strive in the belief that quantification improves communication and understanding thus leading to better decision-making. Prediction in physics and engineering, where it is based on quantified models, is, on the whole, more accurate and reliable than in those disciplines such as political science in which quantified models are in their infancy. Thus, in the behavioral sciences, sociology, psychology, and so on, the tendency is to regard quantification as a means for becoming more scientific and rigorous. But an answer obtained by quantitative methods is not necessarily better than one obtained by qualitative arguments, for in order to quantify a model, too many aspects of the problem it seeks to illuminate may have to be suppressed.

Quantitative analysis is sometimes identified with scientific analysis (even though much that is termed quantitative analysis is clearly not scientific or even objective). Their purposes are different, but they have much in common; for one thing they are both limited in their application. Although there have been a number of attempts, the limits of scientific analysis have not been adequately defined. There is hardly an area today in which science has been applied successfully that was not considered at some time in the past to be beyond its reach. Quantitative analysis is undergoing a similar expansion. Hence, we cannot expect to define a clear boundary that can be used to separate problems that can

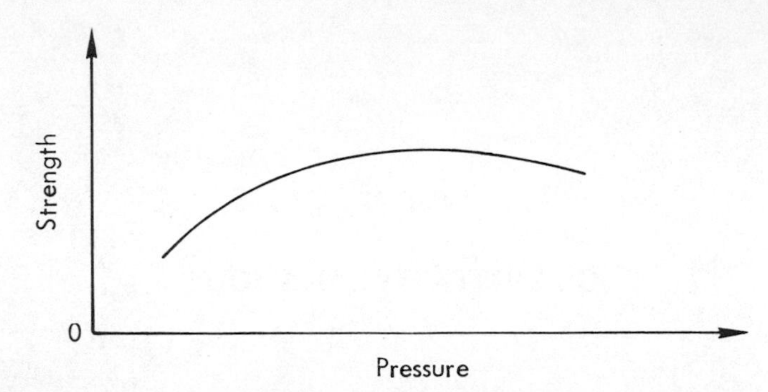

FIG. 11-1. Simple maximization

be handled by quantitative analysis from those that must be treated by methods that are qualitative or only partially quantitative. We can, at most, say something about what we can quantify today. Moreover, the use of quantitative analysis is not only limited externally, in the sense that there are problems that we do not know how to tackle quantitatively, but it is also limited internally, in the depth to which we can make an analysis purely quantitative.

Some Limitations of Quantitative Analysis

It is not merely a matter of time until we can attack every factor quantitatively. It is true that many parameters that could not be quantified a few years ago can be quantified today, but the process is an unending one.

Let me illustrate what I mean by means of an example.[1]

Suppose we have a process for making a glue joint and wish to improve the strength characteristics. As a first approach, we can consider variations in clamping pressure. This should produce a curve of strength versus pressure (Fig. 11-1).

This process gives rise to a further question. What temperature? To answer, suppose we generate a series of curves (Fig. 11-2) of strength versus pressure for varying temperature.

Fine, but how long should it cure? We can run more tests but we then have more answers than we can illustrate on a simple chart. Possibly a set of several overlays might be prepared. But there are still more questions. How much glue? What viscosity? How long should it set before joining? How smoothly should the surface be prepared? It is clear that we might as well give up on using conventional two-dimensional charts to illustrate all these relationships.

There are things we can do; strategies we can adopt. For example, we can use a sequential strategy. After the first curve is computed, take the pressure for maximum strength and using this value, vary the temperature to find maximum

---

[1] I am indebted to Dr. W. C. Randels, Lockheed Missiles and Space Corporation, Sunnyvale, California, for this example.

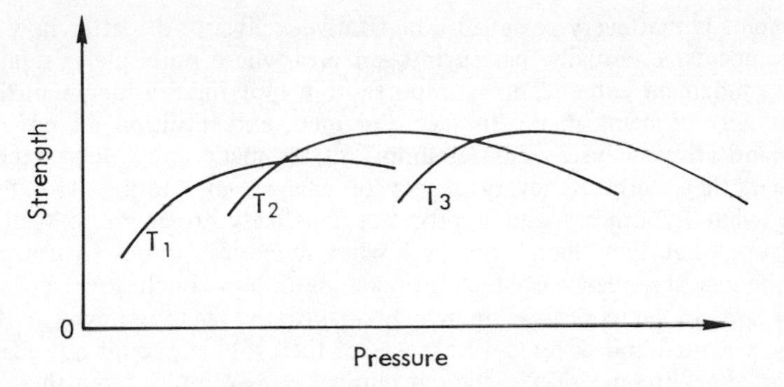

FIG. 11-2. Sequential maximization

strength, and then using these values vary the curing time, etc. It is clear that each step represents an improvement upon the previously existing situation. Many strategies may be used, but the choice is subjective, based on the analyst's experience and his judgment about the smoothness of the functions involved.

Now suppose that to attain maximum strength a three-week cure is required, a period that involves tying up a good deal of equipment for the period. This may make the cost prohibitive; hence, one must take cost into consideration. This might yield the relationship shown in Fig. 11-3. But to make a choice of the point at which to operate we need to know how badly we need strength. In other words, what is the value of strength? But the value of strength is specific to a particular use, to a particular "scenario." If the uses are many and are to vary with future events, we must again depend on judgment.

Sometimes rules of thumb can be used that have general applicability. For example, when the cost–strength curve has a sharp bend (as in Fig. 11-3), one frequently operates close to the "knee" of the curve. But just where and how sharp is a matter of judgment.

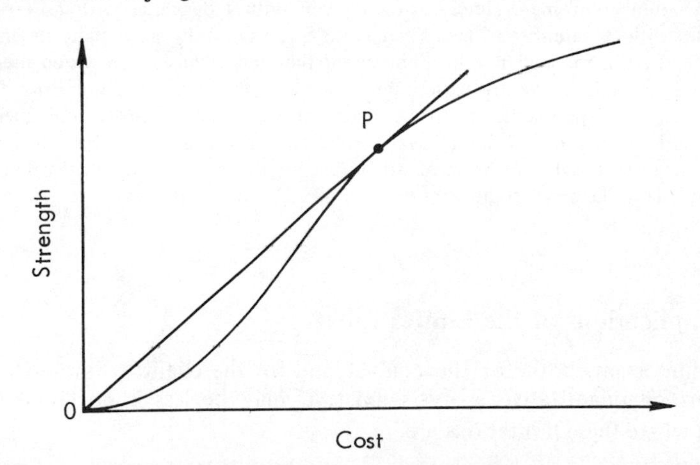

FIG. 11-3. Finding an operating point

The point is that every so-called quantitative analysis, no matter how innocuous it appears, eventually passes into an area where pure analysis fails and subjective judgment enters. This is important; in applying this judgment the real decisions may be being made. In fact, judgment and intuition do not merely enter quantitative analyses when assumptions are made and when conclusions are drawn; they permeate every aspect of analysis in limiting its extent, in deciding what hypotheses and approaches are likely to be more fruitful, in determining what the "facts" are and what numerical values to use, and in finding the logical sequence of steps from assumption to conclusions.

There are, of course, areas in which quantitative analysis, even if it were perfectly executed and covered all aspects of the problem, could not guarantee to predict the outcome of a particular choice. For example, game theory may tell us how to behave in a given competitive situation to obtain, say, the greatest expected return, but it will do so only in a probablistic sense, and will not determine the outcome for a particular instance.

As pointed out to me by the following example,[2] it is not quantification per se that is difficult to achieve, but rather useful quantification.

> Consider an arctic station. Suppose, unrealistically, that the output of the station is clear and simple, a certain number of observations on a specified phenomenon and that the objective of the expedition is to maximize the number of these observations for a fixed budget. Since the cost of fuel is serious, would it be best to have the temperature at 68° so everyone could work comfortably or would it be better to keep the temperature at 50° so that everyone would be miserable in sweaters and gloves but with facilities for two more men on the expedition? You might say that there is a serious problem in quantifying misery. But there are a million and one easy ways; we can measure misery as a function of temperature, for instance, by the amount per day that will induce a free workman to tolerate it. It then becomes a relatively easy matter to say how the number of men depends on the temperature within the fixed budget. What is difficult is to know output as a function of misery and the number of men. Merely knowing that output decreases with misery and increases with the number of men (if that be true!) is wholly inadequate in finding a solution. Also, we may not have measured the kind of misery on which the performance depends. Extensive experimental work would reveal the unknown functions, but the experiments might be really prohibitively expensive and guessing unavoidable, in practice. What I have learned from this example is that difficulties that you at first sight might have attributed to the need for quantification, are actually due to the presence of what you would call real uncertainty.

Some Implications of the Limitations

It is important, both for the analyst and for the client, to understand that limitations on quantitative analysis exist; it may be less important to know what and where these limitations are.

---

[2] By the late Leonard J. Savage in a letter criticizing my stress on quantification (June 1961).

Once the decision-maker understands that, for complex questions, computers and mathematical models cannot treat all aspects of any problem, let alone all problems (and that those they do treat can be treated very badly if the model builders have poor judgment), he should realize that it takes more than error-free mathematical methods to reach "correct" decisions. Also, once he learns that he can never expect an answer from analysis that does not at least in part depend on someone's subjective judgment, he should see little advantage in trying to shunt to an analyst his responsibility for the judgments he can better make himself.

Moreover, once the analyst admits to himself that he is not going to obtain a full answer by purely quantitative and objective analysis, he can help with a clearer conscience with what he has. It does a decision-maker very little good to be told that in two years the systems analysts will have a finished model and will then be able to give an answer. Often, before he is willing to give any answer, an analyst will plead for more time, more testing, more debugging. Yet, although he may bemoan the fact that the decision is not going to be made on the basis of the best analysis he is capable of carrying out, he knows he is never going to get an answer that is completely free of judgment. He should thus more agreeably use his judgment to provide interim help. Whereas we realize that almost all analysis is partial, we should also take into account that partial analysis, well done, is likely to be far better than no analysis at all.

For both, a belief that analysis can lead to complete answers is likely to lead to overrun and late contracts and to a great deal of marginal work. Quantitative and scientific answers are always relative; they hold only under certain conditions. Constraints of time and staff introduce some of these conditions; the problem introduces others. Both the analyst and the decision-maker really know of the existence of limitations; it is just that the analyst tends to forget certain kinds and the decision-maker forgets others.

Because of the growing use of quantitative analysis to assist management in commerce, industry, and banking, there is a fear on the part of some decision-makers that computers eventually will take over their functions.

It is true that the solution to a large number of repetitive and routine problems—even some not so routine—can be reduced to decision rules applied by clerks or computers. Thus, for example, we can have a computer program to simulate the behavior of an investment trust officer, one that "picked such marvelously uninspired and safe portfolios for its customers that by now it may have become a member of the Union League Club" (Shubik, 1970). Consequently, decision-making can be relieved of some problems—but only a few, and most of these will not be relegated completely to the hands of a computer but partially to less skilled and subordinate personnel who input the problem to the computer and interpret its answers.

A decision-maker who is not aware of the limitations may call on analysts for help even when the latter have nothing to contribute beyond their judgment and intuition; because these analysts are ordinarily smart and well educated their judgment is frequently very good. Herein lies a danger—not in the use of judgment but in the failure to emphasize the difference in results and recommendations based on judgment alone. An analyst's judgment is more likely to be in error than his research, and the use of his unaided judgment is likely to reflect on quantitative analysis unless the processes are kept distinct in the user's mind.

As indicated in the previous section, it is not merely a matter of time until we can attack every question quantitatively or even approach it with analysis. To use Herman Kahn's illustration, in 1910 no type or amount of analysis on the impact of the motor car on the future of American life would have been likely to uncover its impact on the dating habits of American youth, nor, for that matter, would expert judgment and intuition have been of much help. The perspective of a group of teenagers might have worked, however. Moreover, of the major issues facing the nation today, it is hard to think of any that can be resolved purely by what we know now. Analysis, should thus be viewed more as a method for investigating problems than for solving them.

So long as the nature and limitations of quantitative analysis are known and kept firmly in mind by the decision-maker, there is little cause for concern that the analysis will lead him astray. But for the analyst dangers exist. One is the failure, in the pressure to quantify, to recognize the unsuitability of quantitative analysis for the problem at hand. Another is that attention will be focused on efficiency rather than on goals and on how to determine as preferable one pattern of results over others. James Schlesinger (1963) put it well:

> Operational research is a new field, and in the early years practitioners of new arts tend to regard them as panaceas. . . .
>
> . . . . The ordinary practitioner of a discipline must be expected to accept the formulas of discipline uncritically and to apply the methods mechanically. In quantitative analysis, a long run threat to optimal decision-making exists in the pressure for quantification and in the failure to recognize not merely the qualitative but the inherently subjective nature of the decisions that must be made. While estimates of exchange ratios can be made more or less accurately on the side of costs, they cannot be on the side of objectives. Precise utility maps exist only in the textbooks. In quantitative analysis, energy is diverted to the search for numbers, and, what is worse, a decision may be distorted because of the disproportionate weight given to those aspects of a problem that are quantifiable.

On the other hand, even in broad problems, say those involving the choice of objectives as well as the allocation of resources, quantitative analysis can provide relevant information on performance and costs. Such information, even though not complete is almost always useful. Yet, in spite of its virtues, it is remarkable that quantitative analysis has gotten as far as it has in policy-making and in government because politics is geared to winning the voters by promising the benefits of programs before their quantitative aspects, the costs and the means, are determined.

Mathematics and Policy Analysis

To the uninitiated, mathematics may appear to offer a real hope for solving policy problems. Mathematical methods have been extremely successful in dealing with many complex problems in the physical world, particularly in engineering; hence, it is quite natural to hope that by extension they might perform equally well, or at least reasonably well, in dealing with the questions that policy analysis tries to answer. Certainly mathematical methods in business, in industry, and in the military have had success. It is not by mere analogy that mathematics might be expected to be pertinent to policy analysis. In fact, it is obvious from both the context and from problems that we have mentioned, that policy analysis is concerned with the relationship between a large number of parameters and quantities; also, the numbers are increasing so that more rather than less mathematics is bound to be employed.

The term quantitative analysis conjures up visions of computers and numerical calculations. Neither may be involved and a mathematical model itself may be nonquantitative–one based on symbolic logic, for instance. Thus (Garvey, 1966):

> ... To say that a subject will be analyzed mathematically is *not* necessarily to say that all, or even that any, of the elements in that problem can be quantified. It is not even to say that numbers or measurements will be used in developing a solution. Mathematics can be applied in various ways and at several different levels of generality. It is often possible to define the important variables of a problem with sufficient clarity that they can be expressed in mathematical terms and variables to be expressed as mathematical relationships. It is then possible, without even once "quantifying" the variables, to follow through the logical implications of one's assumptions. Used in this general way, mathematics becomes a tool of logic rather than a means of numerical analysis.

The last point made by Garvey is an important one. It is amplified by Margolis (1970):

> ... When we come to practical applications of mathematics–that is, when we come to any use of mathematics in connection with practical decisions–there is no longer any argument. Applied mathematics is applied logic. The underlying chains of reasoning are exceedingly complex, and for that reason cannot be grasped without the use of a marvelously efficient symbolic language which is unfamiliar to those who have not studied mathematics. But it is logic, nothing more nevertheless. *If* the premises are true, and *if* no errors of logic have been made in handling the argument, then the consequences will be true. Exactly the same statement applies to all arguments, whether they are made by a scientist or mathematician in a technical paper, or a lawyer in his brief, or by a politician making a stump speech.
>
> Now the language of mathematics is fantastically powerful. It allows us accurately and often easily to deduce the consequences of assumptions which no human mind could deduce if the chain of logical reasoning had to be written out with conventional words. But the validity of the conclusions still depends entirely on the validity of the assumptions. The people who work with computers have a down to earth slogan for this: "GIGO," meaning Garbage In, Garbage Out.
>
> From this, it should be clear why no one of any sophistication thinks, "you can't argue with mathematics." No one wants to argue with mathematics. We want to argue with the assumptions to which the mathematics (i.e., chains of logical reasoning) are applied. To say that calculations should be looked on as mathematical arguments does not say you should refuse to believe the consequences of a nonmathematical argument, that you should believe the calculation only insofar as you believe the underlying assumptions and have confidence in the technical correctness

of the mathematics. For a great many situations, neither is in dispute. You accept the results of the calculation not because a mathematical argument is inevitably more reliable than a non-mathematical one, but because there is no serious counter-argument. That situation very often prevails on narrowly technical questions. It rarely prevails on decisions which are substantially political, and in which differences in judgment and values and the assessment of uncertainty leave plenty of room for counter-arguments. The great merit of quantitative arguments is not that they make possible to deduce correct policy, but that *if used well* they help to eliminate from contention those subissues on which there is no serious question about the assumptions, and focus our judgment on those parts of the issue where judgment is critical.

If one were to make even a casual survey of the literature of operations research and systems analysis–which, insofar as their quantitative tools are concerned, are identical with those of policy analysis–one could easily get the impression that success in those fields[3] depends upon a thorough knowledge of certain rather special mathematical techniques. In fact, in their rather short life as recognized disciplines, systems analysis and particularly operations research have so firmly adopted certain mathematical approaches–linear programming, computer simulation, and game theory, to list just the three most often associated with operations research–that together with the more traditional quantitative methods–probability theory, statistical inference, econometrics, and so on–these seem almost to sum to the complete activity.

The mathematical tools of analysis have been well mastered–so much so that when the study is a poor one or in error, it is rarely because the analyst lacked sufficient sophistication in his analytic tools or blundered in his mathematics or in his use of a computer. On the contrary, the trouble is almost invariably elsewhere.

Most frequently, the difficulty lies in problem formulation, e.g., in setting up the hypotheses to be tested more to fit the set of tools on hand than to illuminate the decision to be made. Also, questions about what issues are most relevant, what relationships are most significant, and how conservative or optimistic the estimates and calculations ought to be occur here, and are most influenced by partisan considerations. Interpretation of the results obtained from a quantitative model when other considerations must be taken into account is a second contributor.

The difficulties in problem solving range from the philosophical or conceptual to the analytic or computational. There is no clear-cut separation, however. The mathematical techniques of operations research mentioned above are designed to overcome the difficulties at the analytic or computational end of the scale. And although these difficulties are dwarfed by those at the other end, it does not follow that they are not troublesome or insignificant. For this reason, the recipient as well as the practitioner of policy analysis may find it profitable to learn something about the mathematical techniques that have proved extremely useful in dealing with so large and important a class of problems. Moreover, even

---

[3] Systems engineering, systems design, and management science might also be included here.

though an understanding of fundamental concepts is much more important than analytic techniques—partly because more elementary methods ordinarily will serve, although far less efficiently—new technique frequently leads to new understanding.

The mathematical aids associated with policy analysis range from tools such as a computer or a table of random numbers, to broad techniques such as dynamic programming, Monte Carlo, or queueing theory. We will not attempt to teach or to say very much about these aids beyond limiting ourselves to a description of their limitations and a few words about Monte Carlo and computers in order to be able to make the subject of computer simulation understandable. Almost any operations research textbook would be a good source for further study.

Computers

The high-speed digital computer is sometimes equated with modern decision-making. There exists a belief that all that is needed to solve the most difficult problems is a bigger, faster, computing machine, which is sure to come along. On the contrary, a computer alone does not solve problems of interest to the public decision-makers; all that it can do is execute a series of instructions laid out by some analyst or mathematician that may lead to a solution. The computer is just a tool; it cannot do anything with problems it is not told to do. Solutions obtained from computers are only as good as the people who define the problem, state the objective, and choose the criteria.

We do not, however, want to deny the importance of computers as aids to analysis. They make feasible the application of powerful ideas and techniques that could not have even been considered before their advent. Indeed, the new techniques we have just mentioned, e.g., linear programming and Monte Carlo, are powerful aids in certain kinds of problems only because good computers now exist to put them into operation. But if time were not a factor, it is questionable whether we could do things with a computer that we could not do without one.

One turns to the mathematician not only for his skill in organizing computation, but also for his ability to avoid it. With ingenuity, he may be able to substitute simplified for intricate mathematical manipulations and narrow down an unworkably large number of cases (for instance, by dominance arguments) to a number that may still be large, but is manageable.

Computers, of course, have limitations as well as virtues. Three of these limitations are particularly significant to the analysis of problems of public policy. For one, certain intangibles of great importance in planning a public policy decision are extremely difficult to handle by computer—morale, resistance to change, the stability of political alliances, the personalities of the actors—these are crucial. Even the most elaborate model adapted for high-speed computers is unable to take these factors into consideration with proper emphasis and subtlety; we simply do not know how to represent them numerically. The second limitation is that practical considerations force many simplifications on a computer representation, requiring the use of aggregate variables and the omission of many details. A third limitation is that sometimes several years and many people are required to formulate programs and debug an elaborate computer model. Thus, the large computer model is likely not only to be expensive and not ready to perform until after the time for decision is passed, but, more importantly, to be inflexible. The assumptions on which a model is based inevitably call for successive corrections as one learns through its application and as changes in factors that affect public policy occur. Thus the learning process may be frequently hindered rather than enhanced by the use of an elaborate computer model. Furthermore, it is in the nature of the process that only selected stages of a computer computation are readily visible to an observer, while most of the intermediate steps remain hidden in the "black box" of the machine. Hence, the direct influence of the variables upon one another, the knowledge of which is crucial to any intuitive reappraisal of a given conclusion, must generally be inferred indirectly.

The concept of a "black box" brings to mind another point. If the analysis is intricate and involves sophisticated mathematical techniques, the decision-maker, a man who is seldom an expert in the field of analysis, may find it extremely difficult to apply his judgment to the logic of the analysis. He may then have to depend more than he likes on his confidence in the competence of the analyst. For this reason, the calculations most useful in helping a decision-maker with important decisions are likely to be of a simple nature—of the sort that go little beyond the "back of the envelope."

The analyst, however, may have needed both the computer and the mathematical sophistication to have found out how to put his arguments in understandable form.

Simulation

A solution may be extracted from a mathematical model either by mathematical analysis or by conducting experiments on it (i.e., by the process of simulation). In some cases, the mathematical analysis can even be conducted without any knowledge of the values of the variables (i.e., abstractly or symbolically), but in most cases the values of the variables must be known either concretely or numerically.

If certain types of functions are involved in a mathematical model and if the parameters are not too numerous, then mathematical tools such linear programming, differential equations, and so on provide powerful means for finding a solution to the problem represented by the model. In other cases, the computational process is often so involved that it is not feasible to find a solution by analytic needs. In these cases we may be able to find a solution by conducting model experiments in which we select values of the uncontrolled variables with the relative frequencies dictated by their probability distribution. This allows us to compute the corresponding values of other variables. Sometimes such experiments can be performed entirely within a computer.

Simulation, except for the use of physical models, has a modest history, particularly in the policy area. As wargaming it stretches back possibly a hundred years or more in the military area, but it was not until high-speed digital computers became widely available that simulation became a tool of operations research and then of policy analysis.

The basic idea is simple: faced with a phenomenon that is too complex, or too dangerous, or too expensive to study directly, a natural way to proceed is to study a simulation, a model, something that resembles the phenomena of interest in its essential features, but is more manageable, less expensive, less dangerous—in short, easier to study. The model can be iconic and the resemblance physical, as in the case of a model ship in a test basin, or it can be an iron wire and colored ball representation of a chemical molecule (the scaling is not necessarily down) or it may involve men playing roles, or men and computers, or even computers alone. In any case we attempt to learn something about the phenomenon through experimentation with the model.

Generally speaking, a simulation tends to have a greater structural similarity to the phenomenon being studied than occurs with the more traditional mathematical model. There is a tendency for a simulation to exhibit a one-to-one correspondence with the elements of the process under examination. In a traffic flow study, for example, the elements of the simulation might be individual automobiles, and a freeway might be represented in geometric fashion; a more traditional analysis might make use of a hydrodynamic analogy, where the elements are traffic density and speed. Again, a mathematical analysis is likely to be formulated as a set of equations. Although equations describing interactions may occur in a simulation, the overall model is more likely to be formulated as a set of procedures for carrying out a sequence of computations—a flow diagram. The goal of a mathematical analysis is usually the computation of a specific number or a few numbers—the equations are "solved," whereas with a simulation, the outcome of the computation is likely to be many series of numbers—such series representing a synthetic history of a process. Finally, a mathematical analysis will be used where a theory does not exist, and "runs" of the model are surrogates for experimentation. Barton (1970) is a good elementary reference.

In some cases, the role of the humans involved in a system may not be understood well enough for explicit representation in a model. Then a simulation may be formulated involving the humans in a role-playing capacity—as in a laboratory study of a business organization where one individual represents a division of a company. Such simulations, often called operational games, make direct use of the judgment and intuition of human participants; this subject is discussed in Chapter 12. Alternatively, the resemblance can be abstract and the representation can be made completely by computer program. It is this form of simulation we discuss in this section—the computer or abstract simulation.

An Example of a Simulation

In order to show how simulation is carried out and, incidentally, to make the distinction between an analytic model and a simulation clearer, an example is needed. Consider the following hypothetical situation.

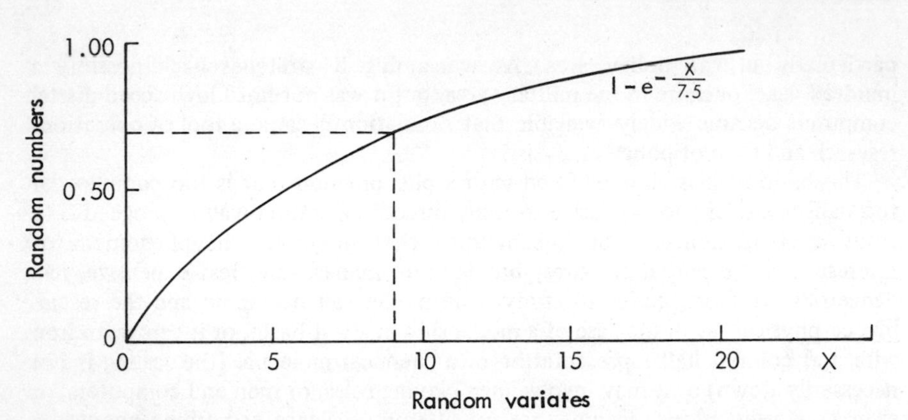

FIG. 11-4. Conversion of random number to random variate

A county health department has been offering inoculations on request without charge. These inoculations require three minutes time. People who take advantage of the offer walk into the clinic off the street at random on the average of eight per hour. The threat of an epidemic of another sort forces the department to decide to give a second shot, that is, two altogether to each visitor. The time required for the two shots together will be 6 minutes. The question is: Do they need to add another physician or can the one man currently on duty handle the two shots in the time he had available for one?

Although random arrival fits this sort of operation very well, the assumption that the arrivals are random has been made deliberately, for under that assumption it is possible to solve the problem analytically as well as by simulation. By random arrivals we mean that the arrivals are equally likely to occur at any point in time. Put another way, the assumption states that the chance of the next arrival's occurrence is independent of the time that has past since the last arrival. More precisely, if ΔT is a sufficiently small amount of time and λ is the mean rate of arrival (i.e., the unit of time divided by the mean time between arrivals), the probability of an arrival in the interval T to $T + \Delta T$ is $\lambda \Delta T$ and is independent of the time T. Under these assumptions, the time between arrivals has an exponential distribution.[4] To conduct our simulation we will generate a series of times separated by intervals of random length with mean time between arrivals of $7\frac{1}{2}$ min or $\frac{1}{8}$ hr. At the end of each such interval we assume someone arrives at the clinic ready for inoculation.

There are numerous ways of generating numbers to represent the lengths of these random intervals. We shall use a procedure that is relatively simple in principle, although in many cases it may be very difficult in practice. What we do is to plot the relevant distribution in cumulative form as in Fig. 11-4. We

[4] The distribution of arrivals generated by this assumption is called Poisson, because it may be shown that the probability of n arrivals in any finite interval of time T is $e^{-\lambda T} \lambda T^n / n!$ This is the Poisson distribution with parameter λT. For an explanation of the mathematics involved and discussion of distributions of this type, the reader is referred to almost any text on operations research, e.g., Ackoff and Sesuni (1968).

divide the vertical scale of probability into equal parts. A random number is selected from a uniform distribution, then located on this scale and projected horizontally over to the function. Then one projects down to the abscissa and reads the corresponding random variate. The random number may be selected from a printed table or it may be generated by a device designed for that purpose, say a computer program, or by almost any chance device.

Since our cumulative probability distribution is $1 - e^{-(1/7.5)T}$ the relationship between the random number and T can be found analytically by solving for T. The result we get will be $T = -7.5 \ln (1-N)$. Using a table of natural logarithms and a table of the uniform random numbers between 0 and 1 we can then construct a series of intervals which represent the intervals between arrivals at the clinic.

When an individual arrives at the clinic, if there is no one being inoculated, he enters, is inoculated, and finishes in six minutes. However, if he arrives when someone is there, he must wait until those before him have been inoculated before he can get his shots. His waiting time is obtained by subtracting his time of arrival from the time of departure of the previous client. A table showing these arrivals, departures, and waiting times has been constructed from which we can trace the behavior of the clinic for a typical day (cf. Table 11-1). By adding up the waiting times and dividing by the number of clients, we can determine the average waiting time.

This simulation by Monte Carlo experiment gives us a total waiting time of 697 min for the 60 arrivals or an average of 11.7 min/patient.

The experiment does indicate, however, that there would occasionally be five or six people waiting, some with delays of 30 min or more. In practice, this would mean that someone arriving and seeing several others waiting might postpone his visit. To allow for this possibility we might carry out another simulation making an assumption that the probability that an arrival would join the queue depends on the number in the queue. This would require the determination of another set of random variates.

This problem can be solved analytically as well as by simulation. A mathematical model has been developed that gives the waiting time as a function of the arrival rate and the service time, provided the arrival rate is random and follows the exponential distribution.

This means that waiting time, excluding the time in service, for random input and constant service time, can be obtained from the work of D. G. Kendall (1948). The formula or mathematical model, for this time, usually denoted $\bar{t}_w$, is

$$\bar{t}_w = 1/\lambda \left\{ \lambda/\mu + (\lambda/\mu)^2/[2(1-\lambda)] \right\} - \bar{t}_s$$

where $\bar{t}_s$ is the time in service, λ is the mean arrival rate, and μ is the mean service rate. This model and other queuing models with derivations are given in most operations research texts, for example, Churchman *et al.* (1957).

In our example, $\bar{t}_s = 6$ min, λ is 8/hr or 1/7.5/min, and $\mu = 10$/hr or 1/6 min. Thus, $\lambda/\mu = 0.8$ and we have the following calculation.

TABLE 11-1.
Simulation of Service at a Clinic

N = Arrival no.	$1-R$: Random no. in (0,1) × 100	$-\ell n\,(1-R)$	Length of interval between arrivals (min)	Time of arrival from $t = 0$ (min)	Time finished from $t = 0$ (min)	Time waiting for service	No. waiting
1	23	1.470	11.0	11.0	17.0	0	0
	98	.020	0.2	11.2	23.0	5.8	1
	80	.223	1.7	12.9	29.0	10.1	2
	44	.821	6.2	19.1	35.0	9.9	3
5	24	1.427	10.7	29.8	41.0	6.2	1
	39	.942	7.1	36.9	47.0	4.1	1
	26	1.347	10.1	47.0	53.0	0	0
	90	.105	0.8	47.8	59.0	5.2	1
	97	.030	0.2	48.0	65.0	11.0	2
10	22	1.514	11.4	59.4	71.0	5.6	1
	87	.139	1.0	60.4	77.0	10.6	2
	85	.163	1.2	61.6	83.0	15.4	3
	25	1.386	10.4	72.0	89.0	11.0	2
	18	1.715	12.9	84.9	95.0	4.1	1
15	71	.342	2.6	87.5	101.0	7.5	2
	59	.528	4.0	91.5	107.0	9.5	2
	37	.994	7.5	99.0	113.0	8.0	2
	83	.186	1.4	100.4	119.0	12.6	3
	16	1.833	13.7	114.1	125.0	4.9	1

20	59	.528	3.9	118.0	131.0	7.0	2
21	91	.0943	0.7	118.7	137.0	12.3	3
	55	.598	4.5	123.2	143.0	13.8	3
	84	.174	1.3	124.5	149.0	18.5	4
	37	.994	7.5	132.0	155.0	17.0	3
	89	.116	.09	132.9	161.0	22.1	4
26	41	.892	6.7	139.6	167.0	21.4	4
	64	.446	3.3	142.9	173.0	24.1	5
	34	1.079	8.1	151.0	179.0	22.0	4
	30	1.204	9.0	160.0	185.0	19.0	4
	72	.328	2.5	162.5	191.0	22.5	5
31	89	.116	0.9	163.4	197.0	27.6	5
	45	.798	6.0	169.4	203.0	27.6	5
	91	.0943	0.7	170.1	209.0	32.9	6
	03	3.507	26.3	196.4	215.0	12.6	3
	07	2.656	19.9	216.3	222.3	0	0
36	17	1.772	13.3	229.6	235.6	0	0
	16	1.833	13.7	243.3	249.3	0	0
	26	1.347	10.1	253.4	259.4	0	0
	60	.415	3.1	256.5	265.4	2.9	1
40	18	1.715	12.9	269.4	275.4	0	0

TABLE 11-1 (continued)

N = Arrival no.	$1-R$: Random no. in (0,1) × 100	$-\ln(1-R)$	Length of interval between arrivals (min)	Time of arrival from $t = 0$ (min)	Time finished from $t = 0$ (min)	Time waiting for service	No. waiting
41	33	1.109	8.3	277.7	283.7	0	0
	85	.163	1.2	278.9	289.7	4.8	1
	02	3.912	29.3	308.2	314.2	0	0
	80	.223	1.7	309.9	320.2	4.3	1
	48	.734	5.5	315.4	326.2	4.8	1
46	23	1.470	11.0	326.4	332.4	0	0
	71	.342	2.6	329.0	338.4	3.4	1
	79	.236	1.8	330.8	344.4	7.6	2
	86	.151	1.1	331.9	350.4	12.5	3
	52	.654	4.9	336.8	356.4	13.6	3
51	44	.821	6.2	343.0	362.4	13.4	3
	66	.415	3.1	346.1	368.4	16.3	3
	31	1.171	8.8	354.9	374.4	13.5	3
	98	.020	0.2	355.1	380.4	19.3	4
	56	.580	4.3	359.4	386.4	21.0	4
56	47	.755	5.7	365.1	392.4	21.3	4
	90	.105	0.8	365.9	398.4	26.5	5
	98	.020	0.2	366.1	404.4	32.3	6
	28	1.273	9.5	375.6	410.4	28.8	5
60	03	3.507	26.3	401.9	416.4	8.5	2

$$\bar{t}_w = 7.5\ \ 0.8 + 0.64/2(1 - 0.8)\ \ -6 = 7.5(0.8+1.6) - 6$$

$$= \ 12 \text{ minutes}$$

Questions as to whether or not we can depend on a simulation for drawing policy inferences frequently arise.

Very commonly, computer simulation models are employed when it is impossible, too expensive, or too inconvenient to experiment with an actual process or system. Thus, in addition to the problem of verification–ensuring that the simulation model behaves according to intentions of the experimenter–the analyst is faced with the problem of determining within some reasonable degree of accuracy whether the model correctly portrays what actually happens. This is not easy to do; for further discussion see Van Horn (1969) and Brewer and Hall (1972).

Of course, we can do simple checks. For instance, in our example we might check values for the intervals between arrivals against the exponential distribution they are supposed to simulate. This is shown in Fig. 11-5. It is clear that we have a preponderance of shorter intervals since their mean length is 6.7 min instead of the 7.5 min of the distribution they are supposed to simulate.

An example of simulation used to study and reduce fire engine response time is presented in Chapter 19.

Advantages and Drawbacks

One of the principal advantages of simulation models for studying complex enterprises is that they allow the construction of precise manageable manipulable descriptions of processes for which analytic models do not exist. Moreover, the description of intricate processes can be built up out of elementary activities.

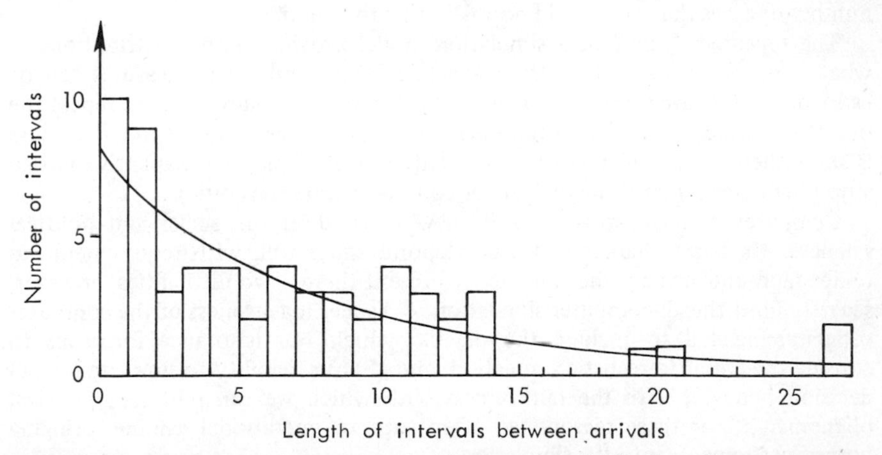

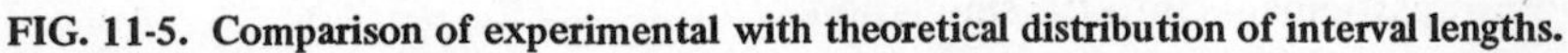
FIG. 11-5. Comparison of experimental with theoretical distribution of interval lengths.

Traffic in a city can be investigated in terms of simple concrete events, such as cars turning left at an intersection, trucks obstructing a traffic lane by double parking to unload, and the like. Traffic movement can be described in terms of simple rules such as lane changes and turns are made according to a given probability distribution, a vehicle attempting to park forces following cars to stop until it has completed its parking maneuver, no vehicle drives backward, and so on.

Because the description depends on elementary events a large measure of concreteness and realism is present. It is possible to include specific operational constraints that are hard to take account of in more abstract computations. The elementary activities simulated are usually well understood and data can be obtained in a direct way. Thus, for example, it is possible to construct a model that will simulate, say, a new procedure for jury selection, although we may have no experience to go on, because most of the activities are likely to be similar to activities that are conducted, at present, in jury selection to which current data are relevant. It is possible to include specific operational constraints that sometimes render worthless more abstract models.

The use of high-speed computers offers many advantages. A computer is an ideal device for performing the massive bookkeeping required to deal with large numbers of elementary events. Once the fairly time-consuming task of programming the computer is accomplished, it is relatively easy to replicate cases so that synthetic experimentation can be carried out on a large scale.

Despite these advantages, there are drawbacks. Simulation is usually a slow and cumbersome way of conducting a study. Whether or not a computer is involved, the construction of a program, the set of instructions for carrying out the computations, is usually time-consuming, and lucky is the analyst who does not have to revise his program many times before useful results can be obtained. For large simulations where a computer is essential, running time for production cases may be a major consideration. This problem is exasperated by the large number of cases that one would normally like to examine.

The repeated "runs" of a simulation model provide sample distributions of what happens at each stage from which averages and other measures can be estimated. This brings up the practical question of how many times to repeat the operation, which is similar to the question of how large a sample to take. There is some theory, but judgment is important. For example, it is expensive not to stop the operation until the budget for computer time runs out.

Computer simulation is a fairly new method in the social and political sciences. Its basic characteristics correspond quite well with requirements to understand and manage the complexity present there. Two difficulties, however, severely limit the all-computer simulation: (1) sheer massiveness of the computer programs needed to include the details, which our ignorance forces us to consider essential to realistic simulation; and (2) our inability to simulate human decision behavior with the faithfulness with which we can represent physical phenomena. For these reasons we often turn to operational gaming, bringing human participants into the simulation.

For reasons of cost, inaccessibility, or risk, the analyst very often cannot test his models satisfactorily in a social context. For such simulations, Brewer and Hall (1972) have prepared an extensive list of questions for guidance. These are repeated below.

> General questions that might be asked *after the fact* of model construction include the following:
>
> 1. Is the distortion between the model outputs and what the policymaker is looking for so large that the model is rejected out of hand? Can this be reduced? At a reasonable cost in time, effort, and money?
> 2. Are the model's output and input generally intelligible; are they in a form that is familiar to the policymaker?
> 3. Does the model offend "common sense"?
> 4. Are elements of the identified question excluded in the interests of generalization or precision?
> 5. Is the model static and descriptive in the interest of simplification?
> 6. Is it possible to include submodels or to change individual behavioral relationships that appear to have a bearing on the policy question without destroying the processing of the logic of the model without significantly increasing its operating costs?
> 7. Are relevant variables, as determined empirically and by virtue of sensitivity testing, omitted in the interest of precision or expense?
> 8. Is the model able to predict, through reconstruction, the time series upon which it is formulated? Has it been able to predict time series from the reference context developed subsequent to the model's formulation?
> 9. If there are known structural changes in the empirical context, are provisions made in the model to capture these? I.e., if there are increasing numbers of disaggregations, changing parameters, or precipitating discontinuous events in the context, are these taken into account? Or, are these events ignored or assumed anyway?
> 10. Are the policy interpretations of various model entities, structures, and recommendations, consonant with the ethical–moral and professional standards of the policymakers and the affected population?

Brewer and Hall go on to say: "To the extent that these questions are not satisfactorily answered, a given model's proper application must be determined, for it is certainly not suitable as a policymaking device.

The ease with which a simulation can be put together makes it tempting to employ the technique where insufficient data exist to justify such a model. Because of the apparent concreteness and detail, a specious air of realism can be imparted to a study, thus masking the incomplete information on which the model is based. The tendency to think of a simulation—especially if coded for a high-speed computer—as a black box that accepts inputs and generates outputs is one of the perils of creating a model that is too extensive to allow examination of the detailed computations. A good deal of the difficulty arises from the fact that the theory is incomplete, for simulation is still mainly an art. This situation is not too serious when conclusions reached by simulation studies can be verified by experience. This is possible in studies involving industrial processes, logistics operation, etc. In these cases the simulation is itself a theory of the process and is subject to observational check. The situation is decidedly serious in the case of long-range planning studies and "social engineering" such as the simulation of urban growth patterns or the problems of environmental protection where verification is indirect or lacking or long delayed.

On the whole, simulation applied to the problems of society is in its infancy and the results to date have been most unsatisfactory (Brewer, 1973; Wildavsky, 1973).

References

Ackoff, Russell L. & Sasieni, Maurice W., *Fundamentals of operations research.* Wiley, Inc., New York, London, 1968.

Barton, Richard F., *A primer on simulation and gaming,* Prentice-Hall, Englewood Cliffs, New Jersey, 1970.

Brewer, G. D., *Politicians, bureaucrats, and the consultant: A critique of urban problem solving.* Basic Books, New York, 1973.

Brewer, G. D. & Hall, O. P. Jr., *Policy analysis by computer simulation: The need for appraisal,* P-4893. The Rand Corporation, Santa Monica, California, August 1972.

Churchman, C. West, Ackoff, R. L. & Arnoff, E. L., *Introduction to operations research,* Wiley, New York, 1957.

Garvey, Gerald, Capt., USAF, Analysis, tool for decision-making. *Air university review,* Jan.–Feb. 1966, **17** (2), 7–14.

Kendall, D. G., On the role of variable generation time in the development of a stochastic birth process. *Biometrika,* December 1948, **35**, 316.

Margolis, Howard, A note on the SS9/safeguard debate, IDA N-541, July 1970.

New York Times Editorial, Stacked analysis. 27 August 1967, p. 14E.

Rickover, Vice Admiral H. G., Cost-Effectiveness Studies. In *Planning programming budgeting,* Senator Henry M. Jackson, Subcommittee on National Security and International Operations, Government Printing Office, 1970, 599–613.

Schlesinger, James R., Quantitative analysis and national security. *World politics,* January 1963, **XV**, (2), 295–315.

Shubik, Martin, A curmudgeon's guide to microeconomics. *Journal of economic literature,* June 1970, **8**, (2), 419.

Van Horn, Richard, Validation. In T. H. Naylor (ed.), *The design of computer experiments,* Durham, North Carolina, Duke Univ. Press, 1969.

Wildavsky, A., *Consumer Report (Book Review), Science,* 28 December 1973, **182**, (4119), 1335.

Chapter 12

WHEN QUANTITATIVE MODELS ARE INADEQUATE

In public-policy affairs, no matter what problem the analyst investigates, there will always be aspects for which quantitative techniques are clearly inapplicable or inadequate. Sometimes (but rarely) this is of little consequence; the aspects that can be treated quantitatively dominate and the advice to decision-makers can be based on purely analytic models. Problems dealing with routine operations such as traffic and supply may sometimes be of this sort. In other cases, a quantitative model, which may not include all aspects of the problem, may provide the analyst or decision-maker with insight sufficient for him to modify the results in the light of his additional knowledge about aspects that could not be incorporated in the model. There is, however, also a large class of problems for which analytic models have not been developed or even conceived of that realistically can take into account the dominant organizational, political, and social factors.

Problems associated with welfare, urban development, and education are frequently of this latter type. Here the nature of politics and the environment for public decision-making can make things difficult for quantitative analysis, which may not even be perceived as relevant. The emphasis of quantitative analysis tends to be on items that are of little concern to practicing politicians. Consider, for example, how the gains and costs of a public housing project might appear to an alderman in a city such as Chicago (McKean, 1967). The gains most important from the viewpoint of the nation (or even of the city) and certainly from the point of view of the quantitative analyst–those involved in the criteria used in cost-benefit analysis–may not even appear on his list. For the alderman, the loss or gain of support from constituents or from other aldermen, the avoidance of possible racial strife and troublesome issues in his ward, the possible loss of support from highway officials who might want the site for freeway access, the loss, say, to his personal insurance business from the buildings displaced, etc., may loom much larger. In general, the criteria for good policy used by each participant in the political process are likely to differ greatly from the criteria used in cost-benefit or other quantitative analysis based on economics. Quantitative analysis is unnatural for most politicians; their questions about a project are more likely to be: Can I live with it? Will the public buy it? Does it have political sex appeal? These questions are not the easiest to deal with by means of quantitative analysis.

This chapter surveys various methods for tackling problems and those aspects of problems that are ill-suited to the quantitative approaches discussed in the previous chapters. These methods are characterized by a much more direct and explicit use of judgment and intuition than was previously discussed. This chapter reviews the role of judgment and intuition in analysis (particularly as exercised by experts), considering ways of identifying appropriate experts and of utilizing their expertise individually or in concert. The scope of the ways examined range from an appeal to individual judgment and intuition to fairly elaborate schemes for bringing to bear the opinions of many people in systematic fashion. Most of the procedures treated in this chapter were mentioned in chapter 10. Here we will amplify our earlier remarks, discussing two approaches for using expert judgment in detail: scenario writing and group judgment technology or the Delphi technique. A third important approach, operational gaming or simulation using human participants is treated in detail in Chapter 13.

Two additional points should be made. First, the methods of this chapter are not without many quantitative aspects. Experts may base their judgment on quantitative models—Delphi computes its consensus and operational games may make extensive use of computers. Second, if competing analysts take different approaches to a problem, the recommendations from the more quantitative are not necessarily the more reliable. The presence or absence of quantification in a study is usually much less important than are a number of other attributes. The breadth of relevant influences taken into consideration in analyzing the policy process, the use of the most trustworthy, the strongest, and most recent evidence that is relevant to the topic at hand, and the use of concepts and tools that are likely to illuminate the aspects of importance to the decision-maker all may be more significant. Yet, although it has become fashionable to deprecate quantification in analysis (as reaction against its excesses), one must remember that decision-makers seldom take action until the discussion produces numerical estimates.

Experts and the Role of Judgment

As emphasized in the previous chapter, policy analysis is critically dependent upon the use of judgment. The standard preferred framework for employing this judgment—some form of mathematical model—has become so through the remarkable success of techniques, such as operations research, which use these models in areas ranging from aeronautics to welfare. But for an analysis to provide sound advice it is not essential that the model on which it is based be expressed in quantitative terms; it is inevitable, however, that the model rely on expert judgment and intuition. Furthermore, the more expert the judgment and intuition, and the more systematically and efficiently they are applied, the greater the likelihood that the advice will be sound.

Let me again illustrate the necessity for reliance on expert judgment. Suppose we were charged with the design of a new system for interurban transportation. We would approach this task by means of a series of models. For automobiles, models would be required to provide a way to relate travel demand and cost to the cost of highway construction, parking, congestion, accidents, travel time, and other factors. Similar models, possibly some of the same models, would be required to predict demand and costs for the various alternatives to motor vehicle transportation–railroads, aircraft, helicopters. Insofar as possible we would try to make these models quantitative, that is, to express the relationships by means of a set of mathematical equations or by a computer program. But we would not build these models without the advice (based on the exercise of a great deal of judgment) from specialists in transportation and many other disciplines. Even if an analyst were a transportation specialist himself, he could not know all that he would need to know about what routes to consider, about future population growth, about demand, about noise, about air pollution, about what approximations to use, and about what to include and what to leave out.

Furthermore, we may have to rely on nonexpert judgment for there may be areas we have to take into account in which no true expertise exists. For example, with regard to route selection, there may be no experts to tell us how convenience or inconvenience to the users, relocation problems of business and private owners, and aesthetic factors affect public support. For instance, in California, a few years ago a great uproar arose over the planned routing of a freeway through a coastal wilderness area. According to their calculations, that uproar must have seemed to the highway planners to be all out of proportion to the damage that might be done to the environment. Roads through similar areas had been built without recent comment, and no formula based on experience could have predicted this reaction. Minimizing the costs per freeway mile may seem to be a good rule for the public at large to use in deciding how to route the freeway, but what most people have regarded in the past to be benefits–say, the bringing of more people and business into the area–may now be regarded by others as costs.

Value judgments are an integral part of policy analysis. In speaking of national security, Alain Enthoven (1964), then an Assistant Secretary of Defense, noted that analysts can be honest in the sense of not deliberately biasing their analysis but they cannot make it objective in the sense of being independent of values. Value judgments must be made throughout the analysis, by both analyst and his client. It is part of the task of the analyst to make his own value judgments explicit and to see that the policy-maker has the opportunity to review them and to posit his own in the light of as much relevant information as possible.

Value judgments on the part of the analyst, however, introduce additional dangers, for how can the analyst divorce himself from–or even be aware of–the influence on his judgment resulting from his organizational and personal commitments, tacit knowledge and beliefs, political leanings, etc.?

In order for an analysis to provide sound advice, it is essential, of course, that the information and data on which it is based be sound. To some extent this information and data can be found in published reports and books or can be gathered by the analyst but to a large extent it must be found in the minds of experts. Moreover, expert judgment and intuition must be used to interpret it.

Reliance on expert judgment is essential to all analysis. As Olaf Helmer (1966, page 11) puts it

> While model-building is an extremely systematic expedient to promote the understanding and control of our environment, reliance on the use of expert judgment, though often unsystematic, is more than an expedient: it is an absolute necessity. Expert opinion must be called on whenever it becomes necessary to choose among several alternative courses of action in the absence of an accepted body of theoretical knowledge that would clearly single out one course as the preferred alternative. It should be noted here, incidentally, that an inability to determine a preferred alternative on theoretical grounds may have one or both of two essentially distinct reasons: there may either be a factual uncertainty as to the real consequences of the proposed courses of action, or, even if the consequences are relatively predictable, there may be a moral uncertainty as to which of the consequent states of the world would be preferable. The latter kind of doubt often arises even when there is a clear-cut basic ethical code, because the multiple moral implications of a complex change in the environment may not be directly assessable in terms of the basic code. The following examples illustrate these points: the design of a vehicle for landing on Mars is subject to the first-mentioned, factual uncertainty; the question of whether or not the workweek should be shortened below forty hours involves real uncertainty as to which corresponding state of the world would be preferable; in weighing the pros and cons of permitting birth control, the Ecumenical Council must be gravely disturbed over the conflicting moral consequences of such a policy that are implied by their seemingly unambiguous basic moral maxims.

The handling of a complex problem usually calls for the judgment of several kinds of experts who may be regarded as belonging to two general categories, specialists and generalists. The specialists provide substantive information and prediction whereas the generalists offer problem formulation, model structuring, or preference evaluation among the predicted alternatives. The analyst himself presumably belongs to the latter classification and may sometimes belong to the first as well. Thus he is looking primarily for people in the first category to provide him with substantive information and prediction.

Certain practical questions arise with respect to experts. How are they to be selected? In other words, how can we tell who is an expert or the most expert? A second question is: Can we do anything to help an expert in his work? A third question arises when we have a group of experts and we need to find some way to pool their judgment.

The first question leads to another: Is there any way or method for determining or rating experts on the basis of past performance? One obvious way is to base an expert's performance on his reliability: an individual's degree of reliability would be the relative frequency of instances in which he ascribed a greater personal probability to the alternative that was eventually correct than to the other choices open to him. On this basis, the more often he proved himself correct, the greater would be his authority as an expert. This measure has its use but (Helmer and Rescher, 1959) state

. . . . it must yet be taken with a grain of salt, for there are circumstances where even a layman's degree of reliability, as defined above, can be very close to 1. For instance, in a region of very constant weather, a layman can prognosticate the weather quite successfully by always predicting the same weather for the next day as for the current one. Similarly, a quack who hands out bread pills and reassures his patients of recovery "in due time" may prove right more often than not and yet have no legitimate claim to being classified as a medical expert. Thus what matters is not so much an expert's absolute degree of reliability but his relative degree of reliability, that is his reliability as compared to that of the average person. But even this may not be enough. In the case of the medical diagnostician . . . the layman may have no information that might give him a clue as to which of diseases A and B is the more probable, while anyone with a certain amount of rudimentary medical knowledge may know that disease A generally occurs much more frequently than disease B; yet his prediction of A rather than B on this basis alone would not qualify him as a reliable diagnostician. Thus a more subtle assessment of the qualifications of an expert may require his comparison with the average person having some degree of general background knowledge in his field of specialization. One method of scoring experts somewhat more subtly than just by their reliability is in terms of their "accuracy": the *degree of accuracy* of an expert's predictions is the correlation between his personal probability p and his correctness in the class of those hypotheses to which he ascribed the probability p. Thus of a highly accurate predictor we expect that of those hypotheses to which he ascribes, say, a probability of 70%, approximately 70% will eventually turn out to be confirmed. Accuracy in this sense, by the way, does not guarantee reliability, but accuracy in addition to reliability may be sufficient to distinguish the real expert from the specious one.

Unfortunately, this may not be very helpful in fields other than weather or medicine. It may be hard to find a class of similar issues or questions for which we can determine whether an expert is good or not so good. There are, of course, other standards for measuring an expert's qualification such as years of professional experience, number of publications, academic rank, etc. These are objective standards, but their value is not above doubt. Unfortunately they may be all we have and in many cases the analyst must rely on his judgment as to who is an expert and who is not.

In broad problems, the range of expertise required is not likely to be provided by a single individual. Almost inevitably a variety of expert advisors must be consulted, some individually, others as a group.

In working with individual experts, it is important to insist that each one make explicit the logic behind his opinions or judgments. Only when reasoning is explicit can someone else, whose information and perspective may be different, use the work of a first expert to modify his own opinion. Often it is the analyst himself who must serve as a bridge between experts, going from one to another to get them to explain the limits each has set on the problem and to find counterarguments for better schemes. This effort sometimes even leads the analyst deliberately to propose naive or outrageous schemes to shock the experts into reaction, for experts often tend to cling to the conventional wisdom of their specialties. Established experts tend to be very conservative and for new ideas it may be more sensible to go to someone on the periphery of the field.

An expert specialist can be effectively aided in his performance if he has ready access to relevant information that he may know exists but which he does not have available. The greatest aid to an expert's performance, however, is to place him in a situation in which he may interact with other experts in the same field, or in different fields related to the same problem. Model building can furnish the interaction. Models do this by providing (1) a context that makes clear to the expert the nature of the problem and his relationship to it, (2) a means of communicating his judgments unambiguously to the others involved, and (3) feedback that enables him to revise his earlier judgments.

UTILIZING GROUPS OF EXPERTS

The traditional, and in many respects the simplest, method of using a group of experts to arrive at a judgment is to conduct a round-table discussion among them, having them eventually agree on a group position. This procedure is open to a number of objections. In particular, the resulting judgment is very likely to be a compromise between divergent views arrived at often under the influence of certain psychological factors, such as specious persuasion by the member with the greatest supposed authority or merely the loudest voice, a reluctance by certain of the participants to abandon positions that they might have taken publicly, and the bandwagon effect of a majority opinion. Unfortunately, committees often fail to make their assumptions and reasonings explicit, sometimes not even recording the opinions of dissenters. Experimenters have, in fact, shown that whereas committee consensus may be an effective way to get a group position, the judgments reached in this way tend to be inferior to group judgments reached in other ways. There is even some question as to whether it is not better in the long run to select one favored expert and to accept his sole judgment. What is needed is a way to pool the judgments that avoids the psychological drawbacks of a round-table discussion, and thus provides a setting in which the pros and cons of an issue can be examined systematically and dispassionately.

A form of committee activity that meets many of the objections above is the continuing "decision seminar" (Lasswell, 1960). The critical requirement for a decision seminar is a nucleus or team who are determined to work together on some problem or problem area over a number of years, meeting with high frequency. There may be changes in the nucleus, but not too many, and other participants, consultants or collaborators on specialized tasks, may come and go. The heart of the seminar, however, is a core of 12 to 15 dedicated specialists, each of whom contributes his particular set of tools and point of view to the problem, all willing to devote a considerable fraction of their time over a period of years (Brewer, 1972).

No matter how they approach a problem, communication among the experts remains important. To emphasize this further, we once more quote Olaf Helmer (1966, pages 17–18):

> When dealing with a multi-faceted problem with the aid of a variety of experts of different backgrounds, perhaps the most important requirement in the interest of an efficient use of these experts is to provide an effective means of communication

> among them. Since each of the participating experts is likely to have his own specialized terminology, a conceptual alignment and a real agreement as to the identity of the problem may not be easy to achieve, and it becomes almost imperative to construct a common frame of reference in order to promote a unified collaborative effort.

In Helmer's view, as we have noted earlier, the ideal way to enforce a common usage and an understanding of concepts among experts is to present the problem to them in terms of an analytic model or, even better, to have them participate in formulating the model as well. When this cannot be done readily (but not excluding all cases when it can!), other schemes can substitute with varying effectiveness. These range from simple organizational devices such as the contextual map to elaborate pseudo experiments or operational games.

The contextual map, for instance, ("matrix" might have been a more descriptive term to use than "map") was first described in the literature of operations research in terms of an application to an anthropological experiment conducted in Peru in 1955. To quote John Kennedy (1956), a participant,

> As an interdisciplinary team of planners faced with the complexities of a large interacting cultural system and its own problems of internal communication, we needed a method for systematically utilizing the special talents and experiences of the planners despite the frustrations of having to establish a common vocabulary, an agreed-upon ideology, a set of reasonable goals, a common context for symbols, and ways of translating ideas into actions. Our solution was to design and make up a "map room," whose walls contained a large matrix with time (in years) on the ordinate and the "variables" the group was interested in along the abscissa. This matrix was the "contextual map."

Since that experiment (and probably earlier also, as the idea is so obvious), contextual maps have been used to display goals, predictions, and achievements, thus furnishing each member of the team at all times with an up-to-date exhibit of the project's status–its accomplishments and remaining tasks. The detailed displays used by many firms and agencies to project plans and budgetary requirements over a period of years are a form of contextual map.

The Goeller "scorecard" (Goeller, 1974) mentioned earlier is another example of the use of a visual display to help experts, in this case the decision-makers themselves, to integrate a host of considerations. Used, as it was originally, with a set of models behind it, the analysts could change the shading of entries on the scorecard (attributes of alternative future transportation systems) in response to alternative sets of assumptions (about regional futures in the California corridor). This was important because it enabled the decision-makers to exercise their intuition and to sharpen their basis for making judgments by asking "what if" types of questions. Also in its original use, the display was particularly helpful because the alternatives had been "normalized" so that they had equal budget costs.

In addition to the contextual map and the scorecard there are three other schemes for using expertise fairly common in policy analysis. Two of these, scenario writing and the Delphi procedures, are discussed in some detail in this chapter. Simulation of the sort in which the experts participate as simulators, often called operational gaming, is a third procedure that has not been much used as yet for policy analysis, will be mentioned just briefly here but is the subject of the following chapter.

Scenarios and Scenario Writing

A scenario is a description or prediction of the conditions under which a system or policy that is to be analyzed, designed, or evaluated is assumed to perform. It consists of an outline of a sequence of hypothetical events, usually including a forecast of what will happen to the environment during some period of time. It is primarily a communications device, used most often to provide the setting for a game or analysis.

A complete policy analysis may involve many scenarios, some stipulating typical tasks, typical conditions, and typical constraints under which the policy being investigated may have to operate, and other scenarios stipulating unique, unlikely, and even extreme conditions under which analysis may operate. Scenario writing is both a way of using expertise individually, as scenarios can be written by one person, and also a way of using experts collectively, for scenarios can and often are written by groups of people. A scenario also can be generated out of a computational model or through the sequence of plays in an operational game.

Although in the opinion of some OR (operation's research) people, scenario writing should not be considered a form of model-building activity, I cannot agree. It is certainly a closely related activity. Olaf Helmer (1966, page 10) describes it as follows.

> . . . Scenario-writing involves a constructive use of the imagination. It aims at describing some aspect of the future; but, instead of building up a picture of unrestrained fiction or even of constructing a utopian invention that the author considers highly desirable, an operations-analytical scenario starts with the present state of the world and shows how, step by step, a future state might evolve in a plausible fashion out of the present one. Thus, though the purpose of such a scenario is not to predict the future, it nevertheless sets out to demonstrate the possibility of a certain future state of affairs by exhibiting a reasonable chain of events that might lead to it.
>
> Thus, scenario-writing has been applied, in particular, to the exploration of potential military or diplomatic crises, but it is a technique that promises to have useful applications in a wider field. By providing a sample of future contingencies, a set of scenarios may serve to warn of dangers ahead, it may afford an insight into the sensitivity with which future trends depend on factors under our control, and thereby it may enhance our awareness of available policy options.
>
> The connection with model-building is at least threefold. First, the process of writing a scenario may be looked on as a primitive, one-man mode of simulation, inasmuch as the author forces himself to go through the thought experiment of examining a plausible developmental chain of events. Second, a useful heuristic device in setting up a formal model concerning a future situation may be to begin by writing several scenarios leading up to it. These can be of great help in discerning the decisive relationships among the elements of the situation and in eliminating negligible irrelevancies. Third, once a simulation model has been constructed, whether a

gaming model or not and whether a computing model or not, the records of repeated runs of the model constitute a systematic source of scenarios and thus afford a methodical sampling of contingencies.

What should a scenario look like? In some cases it may be presented in computer language. In other cases the most useful scenario may resemble an historic essay, rich in detail with the purpose of not only conveying the tangible features of a situation, but its tone and mood. The character of a scenario–its language, numbers of words, or other symbols, the kinds of details it presents–cannot be settled in principle, but only with reference to the specific research task at hand.[1]

Scenarios, provided they are well executed, can be extremely useful. By providing an insight into possible futures they can make us aware of the potential consequences of particular policies. They can help the analyst working on other aspects of the problem to discern the important relationships among the elements of a situation and to eliminate irrelevancies. Thus, they can lead to formal models.

Scenarios have uses other than those associated with setting the scene for analysis.[2] Seyom Brown in Quade and Boucher (1968, page 305) remarks

> There is yet a grander task for scenario writers. The trouble is that there are too few fools willing to rush in where most angels, having been burned, fear to tread. Essentially, this task involves viewing the scenario as a device for altering not only the systems analyst's model, but such criteria as those of costs and effectiveness held by the top decisionmakers. Where such an approach is meaningful, the scenario writer (or scenario-writing group) operates autonomously, not directly coupled to any existing systems analysis project. He conceives it to be his purpose to serve as a kind of advance sentinel, to be able to alert the decisionmaker, the military planner, the systems analyst, to state-of-the-world changes and specific situations which may require the application of military force in ways that are not now planned for, or which may alter prevailing expectations of what planned-for applications of military force can accomplish. In a word, the scenarist at his grandest (and most insufferable) is an iconoclast, a model breaker, a questioner of assumptions, and–in rare instances–a fashioner of new criteria.

As an example of scenarios used in this way see Herman Kahn (1965).

As remarked earlier, a scenario is primarily a communication device. The dramatic appeal of a well-written analytic scenario (which may be in the form of a novel for that matter) that stretches the bounds of plausibility by logical and imaginative use of factual data has proved very useful in broadening the number of contingencies taken seriously in military and defense as well as in business and industry planning by illustrating forcibly the advantages or pitfalls of various proposals or of a new capability. Most of us need concrete illustration of how new concepts or ideas may fit in with the other complexities common to policy analysis; and scenarios used responsibly can contribute to better understanding

---

[1] DeWeerd (1967) and (1973) are good references on the use of scenarios.

[2] For example, to study organizational behavior (Jones, 1967).

of the arguments on both sides of the controversies that always arise over novel suggestions.

The case for some new proposals rests almost entirely on judgments of whether or not conditions for their use seem likely to come about. An advocate of a change in policy or a system is expected to be able to describe in concrete terms at least one plausible and self-consistent set of circumstances in which the proposed change seems worth its cost or other necessary compromise. The discovery that no plausible scenario can be prepared leading to a certain predicted situation, although not a proof, represents evidence of a sort that the situation is unlikely to come about. An attempt to write a plausible scenario or a set of them can be very instructive to anyone who has a serious proposal on either side of a controversy. In showing a deficiency a single dramatic failure is often sufficient, but in proposing a new policy, technique, or system the scenario requirement is quite different.

As is true for participants in gaming, the main benefit of a scenario may be in the self-education of its writer. In the process of writing down the details of a scenario, the educational value to the writer depends on the efforts he puts into making his assumptions both plausible and reasonable and in applying realistic planning factors to calculate the implications.

The use of scenarios is open to a number of pitfalls. It is easy to bias a scenario and thus the analysis by the scenario that sets the context. This doesn't have to be done deliberately–a carelessly prepared scenario can get a game started on the wrong track.

Operational Gaming

In addition to scenarios and visual displays, other "partially quantitative" techniques may be employed to overcome difficulties that cannot be handled by the use of computers and mathematical models. These techniques may surrender some of the features of the traditional model, but still provide the basic requirements for using expertise efficiently: context, communication, and feedback.

Operational or manual gaming, i.e., simulation with human participants–exercises in which the participants interact by playing roles that simulate the actions of individuals, or factions in a society, or even such things as sectors in an economy–is a first step away from computers and mathematical models. Its predictive quality is very clearly a function of the intuitive insight provided by the participants. Because it allows the participants to introduce their judgment at every stage, a game provides an opportunity to consider intangibles often seen as beyond the reach of analysis. Both the experts who control the game and the players can let their decisions be influenced by their appraisal of the effects of the simulated environment. For example, the player can take into account how the success or failure of an economic action may depend upon assumptions about the willingness of a population to accept such diverse actions as a change in diet or a modification of the political structure to accommodate a new power bloc. In any mathematical formulation or computer simulation, factors of this type must be anticipated and decisions about how they are to influence other factors must be made in advance; in a game such decisions can be made seriatim, and in context, as the need arises.

But gaming—despite the fact that its use sacrifices both the capability to replicate exactly and to optimize—is still, like the traditional model, a simulation. It may sometimes be an advantage to sacrifice this too and make a more direct use of expert judgment; we will consider how this might be done in the next section, postponing the further treatment of operational gaming to Chapter 13.

Group Judgment and the Delphi Technique

Delphi[3] is an iterative procedure for eliciting and refining the opinions of a group of people by means of a series of individual interrogations. Usually this is done by questionnaires (a "framework" that replaces the usual representative model). In practice, the group would consist of experts or especially knowledgeable individuals, possibly including some of the responsible decision-makers. The idea is to improve the panel or committee approach in arriving at a forecast or estimate by subjecting the views of the individual participants to each other's criticism in ways that avoid the psychological drawbacks associated with unstructured face-to-face confrontation. The Delphi approach is characterized by three simple ideas: anonymity, iteration and controlled feedback, and statistical group response.

ANONYMITY

The participants are queried and they respond by means of a formal mode of communication. Originally this was by a written questionnaire but recently, with increasing frequency, by on-line computer console. In determining an estimate or prediction, the responses are not matched with the respondents, and even the identity[4] of the participants may be concealed from each other until the end of the exercise.

ITERATION AND CONTROLLED FEEDBACK

Discussion is replaced by an exchange of information controlled by a steering group or exercise manager. After each questionnaire, the information, or part of it, generated in the previous stages, is fed back to the participants in order that they may use it to revise their earlier answers. In this way "noise"—irrelevant or redundant material—can be reduced.

---

[3] For a more extensive discussion see Dalkey (1969) and Chapter 2 of Dalkey (1972).

[4] Otherwise, it may be possible for some respondents to match the response to the respondent. There may even be very rare instances in which there would be reasons to maintain anonymity after the exercise is over.

STATISTICAL GROUP RESPONSE

Although the group opinion tends to converge with feedback, the normal outcome is a spread of opinion even after several iterations. Rather than making an attempt to force unanimity, a statistical index, usually the median, is used to represent the group response. This way of defining the group judgment reduces pressure for conformity and ensures that the opinion of every member plays a role in determining the final response.

Let me now be more specific. Consider the common situation of having to arrive at an answer to the question of how large a particular number N should be. (For example, N might be the estimated cost of a measure, or a value representing its overall benefit.) We might proceed as follows.[5] (1) We would ask each expert independently to give an estimate of N, and then arrange the responses in order of magnitude, and determine the quartiles, Q_1, M, and Q_3 so that the four intervals formed on the N-line by these three points each contain one-quarter of the estimates. If we had 11 participants, the N-line might look like this.

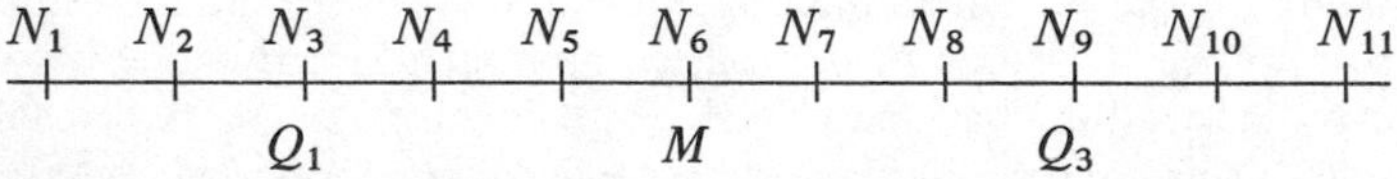

N_1 N_2 N_3 N_4 N_5 N_6 N_7 N_8 N_9 N_{10} N_{11}

Q_1 M Q_3

(2) We would communicate the values of Q_1, M, and Q_3 to each respondent and ask him to reconsider his previous estimate. Also, if his new estimate lies outside the range (Q_1, Q_3), we might ask him to state briefly the reason why, in his opinion, the answer should be lower (or higher) than that of the 75 percent majority opinion expressed in the first round. (3) The results of this second round (which, as a rule, will be less dispersed than the first) would be communicated to the respondents in summary form, including the new quartiles and median. In addition, the reasons for raising or lowering the values, elicited in round 2 and suitably collated and edited, would also be given to the respondents (usually preserving anonymity). We would then ask the experts to consider these new estimates, giving the arguments the weight they deserve, and, in the light of the new information, to revise their previous estimates. (4) Unless additional rounds seem advisable, the median of these round 3 responses may then be taken as representing the group position as to what N should be.

Out of the thousand or so Delphi studies that have been conducted since Gordon and Helmer (1964) almost all have dealt with forecasting, and a preponderance have followed the same format. In this format, a long list of miscellaneous potential events, largely suggested by the respondents, are rated with respect to their probability or date of occurrence and on some scale of desirability or impact on the sponsoring organization. In general, this format is not helpful to a decision-maker with a specific problem before him.

[5] There are, of course, many possible variations; the one presented here was used in Gordon and Helmer (1964).

For illustrations of the potentialities of Delphi for providing policy advice, see Dalkey (1972) and Turoff (1970).

Moreover, there may be other serious difficulties in using Delphi for policy advice. The results from any Delphi process, of course, depend on two critical factors: the makeup of the panel and the directors who implement the process. Here, in a situation in which a decision-maker has to rely on the advice of experts, he may be well advised to select his experts so as to represent different schools of thought on the subjects in question. By and large, the criteria used for selecting experts used by Delphi practitioners are still the standard academic criteria of publications, reputation, recommendations by peers, and the like. Hence, the choice of "policy advisors" and "experts in the area" for the panel may bias the outcome. And this bias is likely to be toward conservatism in dealing with a situation where the only hope for improvement may lie in innovation. Moreover, policy analysis is shot through with judgments about values and goals that one may not like to entrust to a panel, let alone to the group of analysts conducting the exercise.

The procedures used in policy applications need much more experimental work before the superiority of a Delphi approach over ordinary committee action is well established. Experiments here are hard to carry out and we may have to wait on experience.

Major credit for development of Delphi must be given to work done at The Rand Corporation by Olaf Helmer and Norman Dalkey. The ideas presented here originated in their papers.[6] Although still experimental, Delphi has been used, among other tasks, to determine promising educational innovations (Adelson, 1967), to select possible technological developments that appeared to offer unique advantage for an industrial organization (North, 1968), to provide short-range forecasts of business indices (Campbell, 1966), to investigate the capability of doctors to estimate the incidence of diseases in normal populations (Williamson, 1970), and to determine the market for two-way information service in homes (Baran, 1971). Except in the applications for which direct comparison with the forecasts from a control group can be made, the superiority of such exercises over more traditional methods has been hard to demonstrate.

The potential usefulness of the Delphi approach is much wider than the above applications indicate. Extensive use has been made by industrial and urban planners, research managers, and policy-makers (in the United States government and elsewhere) of Delphi procedures for technological forecasting, corporate planning, organizational decision-making, and policy evaluation. Suggested applications range from the drafting of diplomatic notes and long-range political

---

[6] In addition to Dalkey (1969, 1972, 1973), see Helmer (1959, 1966), N. Dalkey and O. Helmer, An experimental application of the Delphi method to the use of experts, *Management Science, 9,* 1963, p. 458, N. Dalkey, *Delphi,* The Rand Corporation, P-3704, October 1967, N. Dalkey, An experimental study of group opinion, *Futures,* September 1969, pp. 408–426, and N. Dalkey, Analysis from a group opinion study, *Futures,* December 1969, pp. 541–551. The earliest paper is A. Kaplan, A. L. Skogstad, and M. A. Girshick, The prediction of social and technological events, *Public Opinion Quarterly,* Spring, 1950, pp. 93–110.

forecasting to determining what products to market. Unfortunately, many of the applications being made or considered are marginal at the moment, in the sense that greater effectiveness of Delphi procedures over more conventional techniques (when such techniques are available) has yet to be demonstrated.

Much remains to be learned about Delphi and the use of expertise. For example, we would like to know how much of the convergence that takes place is induced by the process itself rather than by elimination of the basic causes of disagreement. Placing the onus of justifying their responses on the respondents clearly tends to have the effect of causing those without strong convictions to move their estimates closer to the median, for those who originally felt they had a good argument for a "deviationist" opinion may tend to give up their estimate too easily; this may result in increasing the bandwagon effect instead of reducing it as intended.

Almost all of the early Delphi studies used written questionnaires. One great drawback to this practice is that the exchange of information is time-consuming; another is that constructing unambiguous questions that do not require further explanation is extremely difficult. Personal interviewing is often used to handle this second difficulty. To alleviate both these difficulties, it is possible to have the respondents communicate with the steering group by typewriter or graphic consoles connected through an on-line time-sharing computer system. Inputs and outputs can be in natural language.

The Delphi written question-and-answer method also has some virtues we have not mentioned: It keeps the attention directly on the point at issue; it is democratic; and it produces documented records. It also has some disadvantages: the lack of personal communication between experts and analyst and the stimulation of face-to-face confrontation.

The Delphi procedures–anonymous response, iteration, controlled feedback, numerical estimates, statistical "group response"–promise to become a highly effective means for group information processing. The anonymous debate among experts as conducted by Delphi procedures, in the many instances where a valid comparison can be made, has proved to be superior to the same experts engaging in a face-to-face discussion in arriving at a group position on a given question. Whereas a true consensus is not always achieved, a convergence toward a consensus almost always takes place. Often, when a clustering around two distinct answers occurs, the reasons given by the participants reveal the causes of such dissension–in particular, whether it is factual or merely semantic.

An underlying premise of any Delphi procedure is that, by and large, a respondent is the better equipped to answer a question the more information he has about it, including how other individuals have answered the same question. This is provided through feedback. Feedback may be either verbal or numerical. Unless the number of participants is small, designing verbal feedback presents a problem. If one includes comments from all participants, the volume becomes prohibitive. Editing necessarily involves a certain degree of arbitrariness. When opinions are aggregated and condensed, certain participants will inevitably feel that their opinion has not been adequately represented in the edited version.

Also a basic issue that has not been dealt with adequately in experiments is the integration of existing "hard" information (e.g., in the literature) with the Delphi exercise. The tendency is to use the panel itself as a primary information source, and to add minimal externally supplied information during the exercise. This is clearly nonoptimal, but the procedures needed to improve the situation have not been worked out. A major part of the activity of most ad hoc panels, committees and commissions playing an advisory role to government agencies has been the information-collection stage. Information collection (and ingestion) is an onerous task. In most panel activities, it is spiced up with visits to interesting organizations or places, verbal reports by other interesting experts, and a certain amount of division of labor, in which each member of the panel is expected to be especially knowledgeable about one or another aspect of the problem. Some of this is window-dressing, but the kernel of truth is very important–namely if there exists available information of a relatively hard sort, the panel ought to know about it. N. Dalkey has done some experimental work on this topic.

Experiments in which the respondents seek answers to "factual" questions (with the answers known to the experimenter) pretty well indicate that (Dalkey, 1969a,b).

1. Face-to-face discussion is not as efficient as the more formalized Delphi procedures.
2. Improvement in accuracy of estimates may be expected with an increase in the number of respondents, with iteration, and with the requirement for estimates of range rather than simple point estimates.
3. Improvement can be obtained by using self-rating information to select more accurate subgroups (Dalkey *et al.*, 1970).

For discriminating among levels of knowledge or confidence, one approach is to accompany each question with a "confidence" self-rating–for example, with a scale running from 0 ("this is a sheer guess") to a high rating of, say, 3 ("I am prepared to defend this statement publicly."). Respondents in the lower groups might then not be considered in determining a consensus.

In view of the accelerating use of Delphi procedures by a wide spectrum of public and private institutions, two uses are of immediate practical concern; for forecasting technological and social events and for value judgments.

With regard to forecasting, experimentation (Dalkey, 1969) with short-range predictions suggests that the conclusions from factual estimation experiments apply to them as well; but this presumption needs confirmation. Not much can be done experimentally with long-range forecasts as far as checking on accuracy is concerned, but one may be able to investigate the reliability of such forecasts, in the technical sense of consistency of judgments over similar groups of "experts."

In the area of value judgments, the introduction of some objectivity can have extensive and important repercussions. As with long-range prediction, there is not too much that can be done with regard to "accuracy," but the reliability and stability of group evaluation can be investigated experimentally (Dalkey, 1972). There is evidence from applied exercises that iteration produces convergence with value judgments, but whether this convergence is stable or capricious is not visible from the uncontrolled exercises. Finally, a large amount of diffuse experience with Delphi suggests that the structural properties of the procedure lead to an enhanced acceptance on the part of the individual participants beyond what obtains with more conventional (e.g., face-to-face) procedures. If true, this is clearly a valuable characteristic, especially if the group is made up of decision-makers or others whose concurrence is required for the implementation of the policy being considered. This characteristic is amenable to study.

Delphi is not without criticism; for example, Pill (1971) and Sackman (1974). Much of this criticism is justified for the procedures have often been inappropriately used.

Delphi is not an opinion-polling technique. In general, a pollster is not interested in the correctness or incorrectness of the responses he gets from the sample. The responses are treated as data, not assertions, that allow the pollster to draw conclusions about the respondents, and thus, if his sampling procedures are correct, about the population from which they come.

This difference is basic; criteria that are crucial for pollsters are secondary for Delphi exercises. The purpose of a Delphi exercise is not to furnish the investigator with data about the respondents. It is, rather, to estimate the answer to an uncertain question for which there is no well-defined way to find a definitive answer at the time of the exercise. Delphi techniques should thus be contrasted with the customary, informal types of individual and group utilization of experts that are prevalent in the advisory community today.

Imperfect as it is, the Delphi process or some further modification appears to promise a way to investigate many problems with a high social and political content. Because it can substitute for a conventional model (by estimating the consequences of alternative actions), Delphi offers a hope of introducing a systems approach into a range of problems where such models cannot be formulated.

References

Adelson, Marvin (Ed.), Planning education for the future. *American behavioral scientist,* March 1967, **10**, (7).

Baran, P., *Potential market demand for two-way information service to the home.* Institute for the Future, R-26, Menlo Park, California, 1971.

Brewer, G. D., *Dealing with complex social problems: The potential of the decision seminar,* The Rand Corporation, Santa Monica, California, P-4894, August 1972.

Brown, Seyom, H., 'Scenarios in systems analysis.' Chapter 16 in Quade, E. S., and W. I. Boucher (Eds.), *Systems analysis and policy planning: Applications in defense.* American Elsevier, New York, 1968, p. 305.

Campbell, R., *A methodological study of the utilization of experts in business forecasting,* Doctoral dissertation, UCLA, 1966.

Dalkey, N. C., An experimental study of group opinion: The Delphi method. *Futures,* **1**,(5) September 1969b, 408–426.

Dalkey, N. C., *Delphi, some basic considerations,* UCLA, Tech. Memorandum CCBS-37, Los Angeles, California, Feb. 1973.

Dalkey, N. C., *Studies in the quality of life: Delphi and decision-making.* Lexington Books, (Heath) Lexington, Massachusetts, 1972.

Dalkey, N. C., *The Delphi method,* The Rand Corporation, Santa Monica, California, RM-5888-PR.

Dalkey, N., Brown, B., and Cochran, S., Use of self-ratings to improve group estimates, *Technological forecasting,* 1970 **1**(3) 293–299.

DeWeerd, H. A., *A conceptual approach to scenario construction.* The Rand Corporation, Santa Monica, California, P-5084, 1973.

DeWeerd, H. A., *Political-military scenarios.* The Rand Corporation, Santa Monica, California, P-3535, 1967.

Enthoven, Alain C., Operations research and the design of the defense program, *Proceedings of the 3rd International Conference on Operations Research.* Dunod, Paris, 1964, pp. 530–534.

Goeller, B., *System impact assessment: A more comprehensive approach to public policy decisions.* The Rand Corporation, Santa Monica, California, RM-1446-RC, 1974.

Gordon, T., and Helmer, O., *Report on a long-range forecasting study.* The Rand Corporation, Santa Monica, California, P-2982, September 1964.

Helmer, Olaf, *Social technology.* Basic Books, New York, 1966.

Helmer, Olaf and Rescher, Nicholas, On the epistemology of the inexact sciences. *Management science.* October 1959, **6**, (1), 40.

Jones, William M., *Fractional debates and national comments: The multidimensional scenario.* The Rand Corporation, Santa Monica, California, RM-5259-ISA, March 1967.

Kahn, Herman, *On escalation, metaphors, and scenarios.* Praeger, New York, 1965.

Kennedy, J. B., *A display technique for planning.* The Rand Corporation, Santa Monica, California, P-965, October 1956.

Lasswell, H. D., Technique of decision seminars. *Midwest journal of political science.* August 1960, **4** (2), 213–236.

McKean, R. N., Cost and benefits from different viewpoints. In F. J. Lyden and Ernest G. Miller (Eds.) *Planning, programming, budgeting.* Marcum, Chicago, 1967, pp. 199–220.

North, H. Q., Technology, the chicken–Corporate goals, the egg, In *Technological forecasting for industry and government,* J. R. Bright (Ed.), Prentice-Hall, Englewood Cliffs, New Jersey, 1968.

Pill, J., The Delphi method: Substance, context, a critique and an annotated bibliography. *Socio-Economic Planning Sciences,* 1971, **5**, 57–71.

Sackman, H. *Delphi assessment: Expert opinion, forecasting, and group process.* The Rand Corporation, Santa Monica, California, R-1283-PR, 1974.

Turoff, M., *The design of a policy Delphi,* National Resource Analysis Center, Systems Evaluation Division, Executive Office of the President, Office of Emergency Preparedness, T.M.123, 1970. (See also *Technological Forecasting and Social Change,* 1,(2), 1970, 149–171.)

Williamson, J., Defining universe of health professionals to provide prognostic epidemicologic estimates. Department of Medical Care and Hospitals, The Johns Hopkins University, 1970.

Chapter 13

OPERATIONAL GAMING

The Concept of Gaming

An operational game is a simulation involving humans as simulators. It is an exercise in which an attempt is made to learn something about a problem by having the participants interact by simulating the actions of individuals, or factions in a society, or even such things as sectors in an economy. Operational gaming is an outgrowth of military war-gaming, a procedure that has had a long history of usefulness for training and for testing war plans (Young, 1959) and, more recently, has become a research tool to study future weapons and potential conflict. The extension of gaming and gamelike activities to the investigation of public policy problems is, however, in its infancy (Shubik and Brewer, 1972).

Games are played for many purposes. Few dissent from the valuable role of games in educating and training participants, in improving communication among players with diverse backgrounds, and in generating hypotheses. The extent to which the results of games can be used to describe outcomes and make policy recommendations, however, is still the subject of controversy.

The term "gaming" is often applied, moreover, to the use of any analytic model or computer simulation to investigate a problem involving conflict, whether or not humans take part other than as analysts. Nevertheless, our usage of operational gaming is restricted to exercises that involve humans as simulators. Because their activities in such an exercise usually bear some resemblance to playing a game, to term it "gaming" may be reasonable.

Operational gaming, in this sense, may be one-sided and the play may be against a computer program or analytic model in an attempt to obtain the best outcome possible. Because the critical feature that qualifies a simulation as a game, in my view,[1] should be the presence of two (or more) separate decision centers–neither with complete access to the other's intelligence or plans–that compete or cooperate in separate efforts to obtain their own objectives, we are not discussing gaming in a strict sense.

Gaming should be distinguished from game theory as well as from simulation in general. Game theory is a mathematical discipline used to study situations of conflict or cooperation. It involves analytic models and its solutions are found

[1] My view follows that of Schelling; see later quotation from Schelling (1964).

by analytic methods. Although gaming may share with game theory concern as to how best to play a game and with computer simulation an enthusiasm for detailed representation of reality, gaming is unique in its active involvement of human players in decision-making.

There are two broad classes of operational games: man–machine games and the so-called "manual" games.

Man–machine exercises involve a computer and people playing roles in a system being studied. For example, the operation of some parts of a military logistics system have been studied in a laboratory by having analysts act out the role of decision-maker in the sections of interest, supplying them with data similar to those generated in the real system by means of a digital computer, and observing how they handle the problems that arise (Geisler and Ginsburg, 1965). An operational game of this type is essentially an experiment in which we attempt to learn something about the behavior of decision-makers and how to improve that behavior, by observing the actions of their simulators under controlled conditions. People are used in such exercises because human factors (particularly judgment) are important in the situation being analyzed, but sometimes also because they are cheaper than software.

Manual games involve one or more teams and a referee group operating within the framework of a scenario. (One form is described in the next section.) Typical examples of two-sided manual minimal rule games would be the politicomilitary exercises[2] conducted at and for the U.S. Joint Chiefs of Staff's Studies, Analysis, and Gaming Agency (SAGA). Among the military games, games of this sort are possibly both the least expensive and the most useful (Shubik and Brewer, 1972). Little use of games involving more than one side, or decision center, has yet been made in any civilian setting except for teaching and training. According to Shubik and Brewer (1972) there are about 500 business games, the costs of which vary widely from a few thousand to hundreds of thousands of dollars.

There are four levels of manual gaming that might be distinguished.

THE INFORMAL GAME

This is a conscious attempt to try to take into account an opponent's reaction. Two or a few more people take sides and try to use the spirit of gaming without the formal paraphernalia. It can even be played inside one man's head. One simply asks himself: What would the opposition do if I were to resist, or what does he think I will do if he does such and such? In an informal game one tries to look at the situation from a symmetric point of view, taking account, in a reasonable way, of one's knowledge of the opponent's knowledge about oneself or what he is likely to know.

THE MINIMAL-RULE OR "FREE" GAME

The classical "three-room" game described in the next section and a great preponderance of the business games mentioned above are of this type. The idea

---

[2] For a description of politicomilitary exercises, see Goldhamer and Speier (1959) or Bloomfield (1965).

is to have rules and planning factors to handle the estimation and evaluation of routine events but to leave great freedom of action to the referees and players. Computers may be used to speed up play by helping the players with the planning and control through evaluation.

THE FORMAL RULE GAME

In this class of games, the game directors try to abstract from the real world a definite set of rules that all parties must observe. If it is too difficult to write down all of these rules explicitly, then some may have to be left for a referee to invent at the time a ruling is needed. The idea is to design a game that could be played as well by two opponents who read the rules as by players who had substantial knowledge and experience in the issues.

THE "REALISTIC" GAME

For the military this would be a rehearsal, maneuver, or training operation. The military have found exercises of this type to have a very important role in evaluating plans; they have a tendency to generate substantive information that may have a very sobering influence upon people who put their faith in official planning factors.

Classical Gaming

Before considering an application, let me outline the classical two-sided or three-room gaming process.

The typical two-sided operational "free" game involves three sets of players: a Control team that directs the game, acts as referee and is the ultimate authority for all decisions; a Blue team, representing the friendly side in military games; and a Red team, representing the unfriendly forces. The player teams, Blue and Red, have resources, objectives, and courses of action open to them. They operate in a game setting or context based on a scenario that specifies the environment (political–military in a war game) along with the series of events and conditions that lead up to the situation to be investigated (usually reaching a confrontation of some sort in a military game).

The game begins with the presentation by Control of a description of last-minute events (an "intelligence" briefing in a military or political game) that calls for some sort of action by one of the teams. The two teams then prepare plans for action. These plans are usually asked for by Control in the form of a detailed estimate of the situation, including an analysis by each side of the "interesting" courses of action available to them and to the opposing side and a discussion of why the plan selected was chosen. Of course, only one plan from each side is really necessary to get the game started, but the listing of all the alternative possibilities given serious consideration by the teams gives Control a chance to see that the game develops in a way that will contribute to the research objective, possibly by rejecting the plan selected and asking the team to choose an alternative.

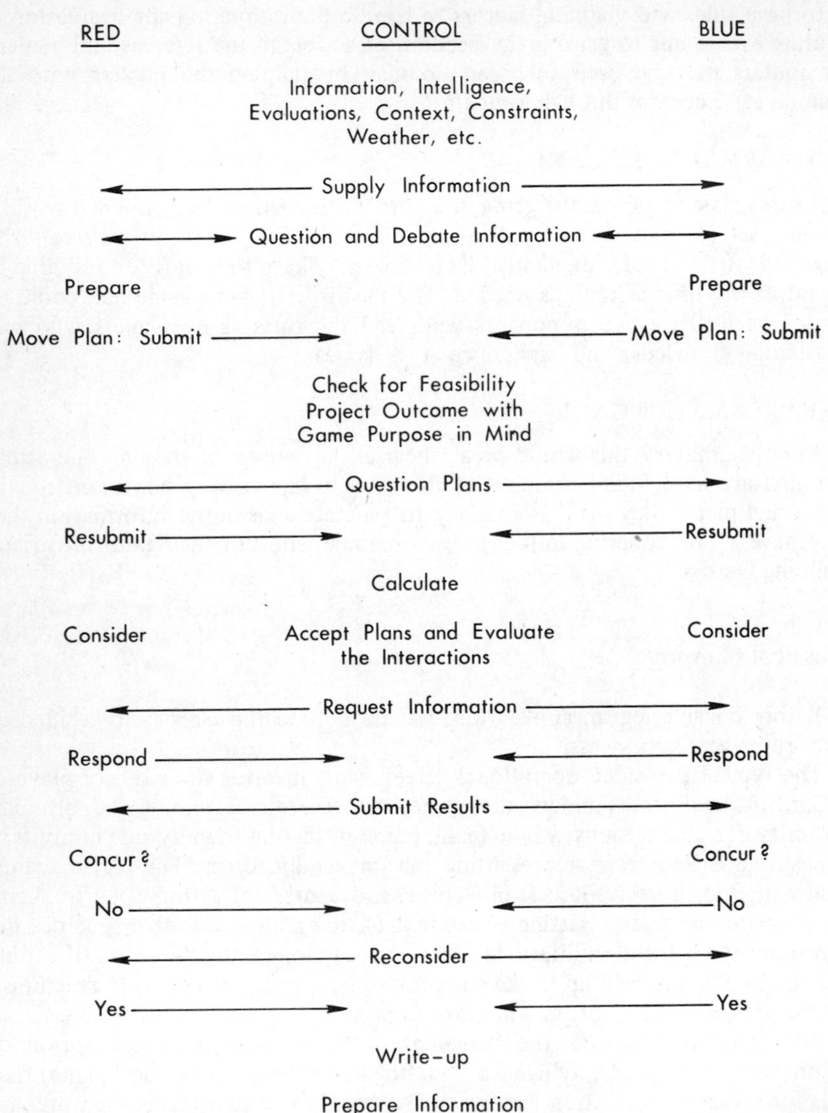

FIG. 13-1. Move cycle.

The functions of Control are (1) to provide political control of each team and thus of the game; (2) to introduce effects such as weather, actions by third parties, and so on; (3) to act as the referee and adjudicate the credibility, relevance, and materiality (to the purpose of the game) of the plans submitted by the teams and request changes if these seem necessary; (4) to evaluate the operational and logistics feasibility of the proposed moves; (5) to supply information and intelligence to the teams at the appropriate times; (6) to evaluate move outcomes, calculating the degree of success, etc.; (7) to maintain realistic time, space, and decision "pacing"; and (8) to maintain the "integrity" of the game (e.g., so that destroyed bridges are not crossed before they are repaired).

The general pattern for the exchange of information during the move cycle is given by Fig. 13-1.

While plans are being evaluated and assessed by Control and after the outcome has been made known to the teams, there is constant communication back and forth between Control and the two sides[3] in order to clear up confusion, replay events that have to be changed, and so on. When the sources of confusion, error, and argument have been reconciled, the outcome of the move is documented. This completes the move cycle. The next cycle then begins as before with a new intelligence briefing or message from Control.

Not much more can be said with respect to the technique of operating a game. There is a great deal of variation in practice, for the procedures are subject to considerable modification depending on the problem and context. For example, there may be other arrangements—in a political game, a White team may represent a third party, for instance, the United Nations—or Control may merge with Red or Blue. For more detail and other arrangements see Weiner (1959).

The realism and speed-of-play of free games of this type can be increased immensely by making extensive use of on-line time-shared computing. The most sophisticated and best executed game at this time know to me is XRAY (Paxson, 1972, pp. 78–81), a politicomilitary exercise. To my knowledge games of this caliber have not yet been applied to public problems other than those associated with national security.

The XRAY games are played with four teams: Blue (United States, Red (Soviet Union), Yellow, (China), and Green (Control). The first three times simulate the core position at the highest decision-making level in their respective governments. Green (1) serves as staff for the other three, preparing special studies on request, (2) plays the bureaucratic and public surrounds of the decision cores, and (3) represents all nations allied to Red, Blue, and Yellow, as well as the uncommitted and neutral nations.

A typical exercise has two major phases. The teams plan a total defense posture for the next 12 years, staying under a year-by-year budget ceiling that rises in proportion to gross national product, and choosing weapon systems from a rich array of existing and proposed systems. After individual intelligence information and a common scenario are presented, the teams are allowed to update their postures. The next phase then starts with the teams exchanging political and military blows of any intensity, constrained only by Green in its roles of staff, domestic sector, and international actor. Coercive bargaining, offers to negotiate, and delivery of ultimata can share the play with military events.

---

[3] In rare cases the sides are allowed to communicate directly.

The employment of on-line time-shared computing is vital to the exercise. The philosophy is to have previously constructed and stored independent modules (programs) for posture-planning, missile-defense management, aircraft deployments, bomber stream penetrations, fallout from surface bursts, and the like. The modules are called up when required, the desired parameter values inserted, and the results obtained almost instantaneously.

The posture-planning model (Fisher, 1971, page 286ff) is important and possibly the most dramatic of these modules. The program has, in memory, cost descriptions of programmed and feasible future weapon systems. These include not only the costs for research and development, production, and annual operation but also the time schedules associated with bringing these weapons into operation. The posture planner can select dates to phase in new systems or phase out old ones. Phase-in rates and final force levels can be prescribed. The program allows for production learning–lower average cost per unit the more purchased, and readjusts prices if a major component like an airplane is common to two or more systems. A total posture may involve 30 or more weapon systems.

Observations on Methods

The class of problems that might be considered as ideal candidates for study using gaming techniques have the following characteristics: complexity; contain elements of competition; involve the interests of large numbers of people; and be neither a completely unstructured plethora of data, beliefs, and issues, nor a situation the internal interactions of which are generally well understood.

> It seems to me that games have their major utilities in the following two areas. Given a problem that is so complex and with which there is so much data involved that large numbers of people will be required to handle various parts of it in a study program, a game becomes a useful device for organizing the group and the partitioning of specific efforts. I like to say that a game is a way of organizing a large group study effort that is better than the formation of an *ad hoc* committee. The reader with experience in *ad hoc* committee operations will recognize this as rather faint praise.
>
> A closely associated situation is a study effort in which large groups of people are to be involved not because the subject is too complex to be handled by small groups but rather that political or inter-organizational politics requires that the study be conducted with representatives from many elements of many organizations. Here again we have the problem of organizing a large group in a study effort and again a game fits this bill very handily. (Jones, unpublished, 1964)

If the situation is too fuzzy, some approach such as scenario writing or Delphi may be more economical of time; if the structure is well enough understood, then a mathematical model or a computer simulation can be used.

Often a device is needed that puts more pressure on the participants to work together than an ordinary committee, a decision seminar, or a Delphi exercise. For this purpose vicarious experimentation in the form of simulation by operational gaming is a powerful technique. Olaf Helmer (1966, pp. 18–19) explains

> Past experience with simulation models suggests that they can be highly instrumental in motivating the participating research personnel to communicate effectively with one another, to learn more about the subject matter by viewing it through the eyes of persons with backgrounds and skills different from their own, and thereby, above all, to acquire an integrated overview of the problem area. This catalytic effect of a simulation model is associated, not only with the employment of the completed model, but equally with the process for constructing it. (In fact, the two activities usually go hand in hand. The application of the model almost invariably suggests amendments, so that it is not uncommon to have an alternation of construction and simulation.)
>
> The heuristic effect of collaborating on the construction and use of a simulation model is particularly powerful when the simulation takes the form of an operational game where the participants act out the roles of decision- and policymaking entities (individuals or corporate institutions). By being exposed within a simulated environment to a conflict situation involving an intelligent opposition, the "player" is compelled, no matter how narrow his specialty, to consider many aspects of the scene that might not normally weigh heavily in his mind when he works in isolation. . . .
>
> We note, in passing, that a player's assignment in an operational game may be either optimization or simulation. In the first case, he is to attempt, within the constraints of the game rules, to maximize a personal score (his "payoff" in game-theoretical terminology). This tends to put the verisimilitude of the game model, which after all is intended to be only an abstraction of the real world, to a severe test and to suggest amendments in the underlying assumptions. The second mode in which a player may function, namely as a simulant, is more likely to utilize his expertise properly; for in this role, he is required to contribute constructively to the developing scenario by feeding in such simulated decisions which, in his estimate, would most faithfully reflect the decisions that his actual counterpart would make in the corresponding real situation.

Ideally, of course, the inputs into a game should be established facts, and, if they were, one might hope that the output would be equally reliable. In fact, this is not so, nor is it to be expected. Judgment enters crucially into every model at two levels: firstly, in selecting the factors to be included in the model, i.e., into the structure of the model; and secondly in the estimates and guesses involved in establishing those inputs into the model where precise information is not available. An operational game, moreover, introduces judgment at still another level—namely, in the selection of strategy.

Now a scientific model is not without value just because it involves intuitive judgment. What is necessary is to recognize the presence of such judgment. An idea coming out of the application of such a model must indeed be justifiable in the sense that it should stand up under a critical review of the judgmental assumptions underlying the model. Similarly, an idea produced by a game should stand up under critical review, except that the latter must now include the strategic choices made by the players.

A game, by the very presence of such strategic choices, is essentially different from a mathematical model and in this respect cannot be replaced by one—at least not until we have succeeded in constructing a theory that defines and permits us to compute rational behavior in a much wider class of situations than we are now prepared to cope wlth. Therefore, unless we are dealing with situations in which game theory may provide a model, there are no existing or imminent available means of constructing a complete scientific model in the accepted sense. Therefore, if we wish to plan, predict, and operate in an environment that involves political and tactical choices, we are forced, whether we are willing or not, to resort to other procedures. Among these procedures gaming at least provides a quasi-model, differing from the real thing mainly by having certain inputs manufactured on the spot by expert judgment rather than prefabricated.

The following points about the methodology of gaming should be noted.

1. Careful preparation is required. A significant aspect of preparation is the requirement for a definite and precise statement of the purposes of a game. There is a tendency to hurry to get the play started. But the time devoted to clarifying the purpose or the objective of the game has always turned out to be well spent. Gaming can be a great consumer of man-hours and not all problems are ideally suited to the gaming approach. For this reason, thought should be given before a game is actually started. Ideally the game directors should have some definite hypothesis in mind for which the game would provide a test. A large amount of background data is essential before a game can be conducted at a level of detail where specific decisions and their consequences can be examined. This requires data of many different types depending on the context and purpose. Because it is unlikely that even with careful preparation everything can be anticipated, consulting services should be available and staff to assemble such data on hand, otherwise there may be expensive delay.

2. Staff requirements are flexible. The size of the staff depends on the objectives of the game, the level of detail required, and the amount of time available. In many cases doubling up of staff members is possible. This seems to work out much more satisfactorily in competitive situations than devices like combining functions of control with those of one of the teams.

3. Strong game direction is necessary. This is probably obvious from the very nature of the exercise. In their effort to best the other side, players can easily lose sight of the purpose of the game, which is to learn, not to "win." Interest in the latter can lead them to study the Control team and attempt to manipulate its rulings to their advantage.

4. A complete record should be kept. A review of the game after it has ended with analysis of various moves and situations is likely to be valuable. This is impossible without full records.

Conventional gaming suffers from rather severe constraints: (1) Unless the game is a very simple one, each play-through is likely to be very time-consuming. (2) In the effort to maintain realism, so many parameters are involved that isolation of the key factors is not easy. Repeated plays, based on the same scenario, could help here. (3) The teams, to assess a potential move, and Control, to evaluate it, must process a tremendous amount of information. Time constraints and lack of analytic support sometimes force this to be done too superficially. (4) After a series of games is over, it is hard to reproduce the data and carry out a thorough analysis.

These constraints can be countered in various ways. The use of modern on-line, time-shared computing with multiple consoles, graphic displays, and natural languages can do much to speed up play and thus permit repetitions. It can also take over much of the recordkeeping and supply analytic substance both during the game for planning and evaluation, and afterward in assisting in the analyses.

In addition, gaming can be carried out in conjunction with other analysis. For example, Delphi procedures can be used after a game to uncover and summarize what the players thought they learned from the game and to explore what might have happened had other plans been implemented.

For an application of gaming to economic development, see Helmer and Quade (1964).

The Value of Games

No matter how carefully they are designed, games fail to achieve realism in a great many respects. For instance, a game, unless specifically structured to do otherwise, by the nature of the way in which a team of a few people interact, represents the decision-making bureaucracy as an organization with a well-defined, consistent set of objectives, reflecting a clear interpretation of intelligence, coherent policy, and the ability to eliminate ineffective alternatives rapidly. Therefore, unless the game designer specifically guards against it, the usual assumptions are likely to lead to plans or postures that are far more efficient in their use of resources than are found in real life. In other words, the decision-making process as modeled in the ordinary game turns out to be far too rational to reflect the many limitations on the decision-making process found in real-world bureaucracies.

Operational gaming, like other policy analysis techniques, undoubtedly would be most fruitful when applied with a clear objective in mind to well structured problems about which there are abundant data. But other less time-consuming techniques are usually available when these conditions hold. Its major utility lies elsewhere, particularly for complex situations where there are two or more adversaries, possibly with some common interests. It also excels as an educational device, providing both ideas and insights to the participants, useful for generation and preliminary comparison of alternative policies.

In the analysis of major questions of public policy, it may be well worth the sacrifice of precision in handling some of the elements that can be readily quantified to gain other benefits. Among these would be the possibility of getting some indication—although perhaps with inadequate emphasis—of the effect of many relevant political, economic, social, and psychological factors that one might otherwise be forced to overlook or to treat very inadequately outside of the model proper. Another benefit would be to provide the analysts with a greater opportunity to take into account "feedback" of the type that might lead one to modify the model in accordance with changes in opinions that occur as the analysis progresses. Controversial parts of the model, which are likely to be buried and forgotten in a computer program, remain visible. One virtue of such an unsophisticated simulation as a game is that it may give some clue as to how to model the situation it simulates more adequately later.

The formal structure of an operational game automatically subjects any notion or theory that is introduced to detailed critical review: a game requires the players to take active roles, to take specific and concrete actions in particular situations, and to examine the consequences whereas a man sitting in his office or participating in a discussion might fail to consider the full range of possibilities or to carry the argument beyond the opening steps. It is easy to be vague in talking about theory or doctrine, but a game shares with the analytically formulated computer model the quality of concreteness—there can be no vague moves in a well-formulated and well-run game.

In short, the technique of manual gaming, as a way of bringing experts together, can do much to facilitate a policy study. Admittedly, the predictive quality of such an exercise is very clearly a function of the quality of intuitive insight provided by the experts involved. In contrast, by allowing for the introduction of judgment at every step, the game provides an opportunity to take into account intangible factors often considered completely outside the scope of analysis. This is true both of the player, who can let his decisions be influenced by his appraisal of the human effects of the simulated environment, and of the expert on the control team. For example, the success or failure of a plan may depend upon assumptions about cooperation from the population or flexibility in the command structure. For an analytic formulation or a computer simulation, as we said earlier, decisions about these things must be made in advance; in a game they can be made seriatim, as the need arises.

A great disadvantage of a simulation using human participants is the time required to work through each particular case. A computerized simulation, once it has been programmed, can run through hundreds of thousands of cases in far less time. As we noted earlier, the gaming process can be speeded up somewhat by introducing a computer for routine and well-understood phases; whether this is economical depends, of course, on the scale and nature of the exercise.[4] Unfortunately, when the costs of purely manual gaming are reported, the personnel and opportunity costs are frequently ignored.

One additional point: Even when more conventional and quantitative techniques, such as computer simulation, can provide correct guidance, they may still be unpersuasive. Any solution to a problem that appears to have been formulated exclusively by "outsiders," using what is essentially a "black box," may not be accepted readily as a solution. In contrast, an important aspect of an unsophisticated simulation by gaming—and one that has not been much exploited—is that the decisionmaker or his representatives can actually participate.

---

[4] For an idea of the dollar costs associated with models, simulations and games, see Shubik and Brewer (1972).

Inasmuch as a game falls so far short of being a scientific model, it is sometimes held that its role can be no more than pedagogical and stimulative. Thus, Kahn and Mann (1956) in remarking about game results, state

> . . . the fact that they came out of a game is almost completely irrelevant to their reliability. One has to justify that in the same way one justifies any idea.

But the point to emphasize here is that this need for justification is not limited to games but applies equally to the output of any model including the conclusions of our most elaborate computer simulation.

The important questions regarding the use of gaming as a tool to investigate policy are: When does gaming lead to error? and When is it inferior to other methods?

Essentially what gaming can do to obtain data of a sort about ideas you may have about the organization of that portion of the world you are interested in. You may thus get some clue as to whether or not your ideas are correct. In a sense, a game can be made analogous to a scientific experiment, for it can be conducted in isolation and under controlled conditions similar to a laboratory experiment. As in the case of an experiment, the objective of the game must be clearly in mind.

There is no way for a game to simulate reality in all respects. After all, even the sciences, instead of dealing directly with the elements of the real world, are forced to examine a model that is an approximation, the operations of which are close enough to those of the real world to provide a basis for reasoning about real-world events; it is hoped that the game does the same.

One of the advantages of free-play gaming is that although it uses a model, that model is very flexible. A rigid model makes it hard to practice ingenuity, except in looking for ways to beat the model. Ingenuity consists in thinking of the unexpected. When the rules of a game are written down in rigid form it usually involves tightly closing off all the little doors that might lead a person into a place where he could practice ingenuity in private.

A game provides an excellent framework in which people may interact. The individual players study the problem associated with their side with interest and sometimes even passion. The referees can let the players do a great deal of arguing about the interpretation of what is reasonable or unreasonable. Moreover, when discussion tends to drag out, the referee has the privilege of making a flat decision that cuts off controversy and allows the game to continue. He is in a much better social situation than the chairman of a conference for everybody has to accept his decisions with good grace. The only retaliation open to the player is to write a paper. In fact, the game is a very good environment for instigating all kinds of detailed papers on policy, hardware, or context questions, which no one would be moved to write under ordinary circumstances. One is more likely to get feasible suggestions from a gaming context than from a conference or from a person working in isolation because game suggestions are made in a very concrete situation.

The value of the gaming principle does not necessarily lie in some elaborate technique but merely in the sense of bringing in the devil's advocate, the man or group who takes the opponent's point of view and capabilities and forces them on our conscience.

Gaming can do a number of things: it can uncover errors or omissions in concept; it can explore assumptions and uncover the implicit ones; it can check coordination measures; it can develop the contingencies on which a plan depends; it can draw out divided opinion; it can examine the feasibility of an operational concept. It automatically, or at least easily, draws attention to areas that are particularly sensitive or in which information is lacking. But, of course, it can give very limited indication of failure or success of particular strategies.

Leaving aside the results for which the game is organized, nearly every game may yield by-products worth the cost of people's time. Let me mention some of these.

First, games can be intensely stimulating: people involved are very active; ideas and conjectures get tossed around and analyzed by a highly motivated group of people. Because a great deal of expertise may be collected in a single room, people discover facts, ideas, possibilities, capabilities, and arguments that do not in any strict sense depend on the game but nevertheless may emerge from it. In an urban simulation someone may discover that in a certain city garbage is collected by a one-man crew at a great savings in cost—or that acute jurisdictional problems exist among contiguous cities in a certain county—or that certain operational plans have embarrassing by-products that could not be known to the people who drew up the plans. There is absolutely no reason why such little "lessons" depend on a game; as a matter of fact, they do not. But a game as a social and intellectual occasion tends to be highly productive of things of this sort, little by comparison with the game objective, but possibly not so in relation to the investment of time.

Second, people probably learn more about the geography, the distribution of population, the telephone system, the recent history, the political personalities, the economy, the weather, the street layout, the political and ethnic groups and all the other "tourist" information about an area or situation by going through a game than by any course that could be devised for a comparable period of time.

If a game involves policy-makers and their assistants from different agencies, it can provide an intense common experience. In circumstances in which a bit of humor is allowable and mistakes do not lead to real disaster, the arguments in the game typically lead to remarkably good relationships afterward. People find out about the other fellow's problems and learn to appreciate the limitations of his activities and how they impinge on the others. Speaking from experience, Schelling (1964) said:

> Let me turn now to what I think it is that "game-organized' research or planning can do that nothing else can. To begin, let me say what I take to be the critical feature of a game, the thing that makes it a "game." This is that at least two separate decision centers are involved, neither of which is privy to the other's planning and arguing, neither of which has complete access to the other's intelligence or background information, neither of which has any direct way of knowing everything that the other is deciding on. I am willing to call it a game if it is done by questionnaire, by

> people sitting at consoles, or by teams sitting in separate rooms, whether it lasts five minutes or five months whether it involves continuity over time or not. The point is that a "game" as I understand it involves two separate decision centers (and is technically non-zero-sum). What this mode of organization can do that cannot otherwise be done is to generate the phenomena of understanding and misunderstanding, perception and misperception, bargaining, demonstrations, dares and challenger's accommodation, coercion and intimidation, conveyance of intent, and uncertainty about what each other has already done or decided on. There are some things that just cannot be done by a single person or by a team that works together.

One of the most useful features of the gaming operation is that it permits a very detailed and descriptive demonstration of a proposed policy or plan in a more-or-less rich context. The view and understanding of a policy that one can develop from a written description can be much poorer than the view and understanding developed when one has seen it demonstrated in a dynamic, although admittedly highly synthetic, on-going situation. One thing that games are not alone in doing, but awfully good at is demanding careful sequential analysis of plans, decisions, events, and intelligence.

Two of the most salient features of any sort of team game are (1) the rapidity with which the bits of information on the problem area, known at the outset to individual participants, enter the common information base of the entire group; (2) the rapidity and accuracy with which individual members of a game team achieve an understanding of (although not necessarily an agreement with) the feelings of their teammates about the situation/problem being simulated. These two features are, I believe, the main basis for the noticeable post-game improvement in the ability of the participants to communicate with each other in a meaningful fashion. One comparatively short game can do more to educate the participants of a large team in the thought processes, the beliefs and desires of their teammates (and this can be a fairly large number of people) than innumerable meetings, volumes of written material, etc.

References

Bloomfield, L. P., The political-military exercise: A progress report, *ORBIS,* Winter 1965, 8, 854–870.

Fisher, G. H., *Cost considerations in systems analysis.* American Elsevier, New York, 1971.

Geisler, M. A. & Ginsberg, A. S., *Man-machine simulation experience.* The Rand Corporation, Santa Monica, California, P-3214, August 1965.

Goldhamer, H. & Speier, D., Some observations on political gaming. *World politics,* 1959, 12, 71–83.

Helmer, O. & Quade, E. S., An approach to the study of a developing economy by operational gaming. In *Recherche opèrationnelle et problèmes du tiers-monde,* Dunod, Paris, 1964. Also The Rand Corporation, P-2718, 1963.

Helmer, Olaf, *Social technology.* Basic Books, New York, 1966.

Jones, Colonel William M. (USAF, Ret.), *One view of games, simulations and analogs* (unpublished) 1964.

Kahn, Herman & Mann, Irwin, *War gaming.* The Rand Corporation, Santa Monica, California, P-1167, July 1956.

Paxson, E. W., Computers and national security. In *Computers and the problems of society,* H. Sackman and H. Borko (eds.) American Federation of Information Processing Societies, AFIPS Press, Montvale, New Jersey, 1972, 65–92.

Schelling, Thomas C., *An Uninhibited Sales Pitch for Crisis Games,* September 1964, (unpublished).

Shubik, M. & Brewer, G. D., *Models, simulations, and games–A survey.* The Rand Corporation, Santa Monica, California, R-1060-ARPA/RC, May 1972.

Shubik, M., Brewer, G. D. & Savage, E., *The literature of gaming, simulation, and model-building: Index and critical abstracts.* The Rand Corporation, Santa Monica, California, R-620-ARPA, June 1972.

Weiner, M. G., *War gaming methodology.* The Rand Corporation, Santa Monica, California, RM-2413, July 1959.

Young, John P., A survey of historical developments in war games. Operations Research Office staff paper 98, August 1959, 116 pp. (later Research Analysis Corporation).

Chapter 14

UNCERTAINTY

No aspect of policy analysis is more pervasive or more difficult to handle than uncertainty. Uncertainties (the unpredictabilities in factors that affect the success of a course of action) are a *sine qua non* of any decision-making process; they are, in fact, the issues.

Uncertainty about economic, technical, and operational parameters that can be listed, measured, or at least estimated and treated statistically complicate analysis, but we can usually find ways to take them into account. Uncertainties about future environments and contingencies and any activities that depend on the actions of people,[1] now as well as in the future, are more intractable; to find ways to guard adequately against the unexpected here is not always within the capability of analysis.

Let me illustrate how uncertainty can shape decision-making by tracing an historical example[2] that shows how changing expectations of the future and the subsequent hedges against uncertainty led to modifications in the development of a military aircraft.

The United States bomber, the B-36, was conceived in April 1941, after the fall of France. It was thought that if the United States were forced to enter the war, a long-range bomber would provide a hedge against the loss of England; that is, in the event of this loss the United States could still administer damage to Germany. Major Vandenberg and a small study group in the Army Air Force laid down requirements for a bomber of 10,000-mile range, capable of delivering 10,000 lb of high explosives. A design competition was held and in the fall of 1941, two prototypes were ordered.

By the summer of 1943 it was clear that the United Kingdom would not be lost. The United States, however, was now at war with Japan as well as with Germany. An island-hopping campaign was under way in the Pacific, but the outcome was uncertain, and there was trouble with the B-29. To shorten the development–procurement cycle, 100 B-36's were ordered in advance of delivery of the prototypes. The B-36 was again a hedge–this time against failure of the stepping-stone campaign in the Pacific and against failure of the B-29.

---

[1] Although I am unable to recall the exact words, a remark by one of Damon Runyon's characters is relevant here: "Nothing what depends on humans is worth odds of more than eight to five."

[2] Adapted from Wohlstetter (1966).

Later, the Pacific campaign succeeded, the B-29 became a great success, and it became clear that the B-36 would play no role in World War II. Because there was now an aluminum shortage, further development was stretched out. With the postwar period the Soviet Union began to look like a potential enemy. The United States had some overseas bases, but negotiations regarding others that were needed were long drawn out and the outcome uncertain. Vandenberg, now General Vandenberg, recommended buying the B-36 as a hedge against the failure to obtain these bases.

None of the original analyses (and few of the later ones) which were carried out to determine such characteristics of the B-36 as range, payload, altitude, speed, and armament considered the possibilities that

- Nuclear weapons might be developed.
- The enemy might be the Soviet Union.
- The jet engine would become operational.
- The defense would include jet fighters and ground-to-air missiles.
- There would be overseas bases.
- Air refueling would be possible.

Four jet engines were eventually added to the six-propeller engines, but the plane was on the way out before almost anyone took seriously the argument (Wohlstetter *et al.*, 1954) that its prelaunch vulnerability had rendered it obsolete.

What lessons can be drawn from this story? One, certainly, is that the decision-maker's lot is not an easy one. Nor is the analyst's; designing alternatives for an uncertain future is a chancy business. The second is the value of hedging, of taking insurance against the unexpected by designing flexibility into the system.

Types of Uncertainty

Uncertainty varies in nature and degree and hence must be taken into account in analysis in various ways. Some uncertainty is remediable—meaning that by continuing information gathering one could be rid of it. We are concerned largely with the more genuine uncertainty associated with future events that cannot be eliminated except through the passage of time.

One distinction we can make is between those uncertainties resulting from random events and those resulting from ignorance, incomplete knowledge, and the deliberate actions of other people. Uncertainties due to random events are called *stochastic uncertanties;* they are random in the sense that the associated probability can be given a relative frequency interpretation as can the flip of a coin or the roll of a die. If we were designing a health insurance program, an example of this type of uncertainty might be the number of people expected to be hospitalized with influenza in a certain community at any given time. Stochastic uncertainties whose probability distributions are known are called *risks.* Uncertainties that are not stochastic, for which probabilities cannot be given a relative frequency interpretation, we call *real uncertainties.* An example might be the date on which the so-called Women's Rights Amendments to the United States constitution is ratified.

Among the real uncertainties, I would like to make a further distinction between *strategic uncertainties* that are due to ignorance about factors under the control of another decision-maker, usually considered an adversary, and *environmental uncertainties* under the control of nature, so to speak.

We thus have the following classifications.

○ Stochastic Uncertainties	○ Real Uncertainties
Risks	Environmental
Nonrisks	Strategic

A situation of risk, or a decision under risk, is one in which all possible outcomes are known together with the odds that each will occur. What is not known is which particular outcome will occur. To illustrate, suppose in an analysis to determine pension requirements, the ages and other significant characteristics of those who are eligible are known. We then have a situation of risk. Insurance company and government actuaries have made observations over a sufficiently long time period and with so many differing groups that the probability with which any fraction of the number of people in the group will survive to a given age is known. Although we cannot predict exactly how many of those eligible will be alive to receive pensions at any specified time, we can use our knowledge of the probabilities to estimate the requirement for funds with considerable confidence of being reasonably close to what will be required.

There are other situations involving random events, however, in which the underlying probability distribution is unknown. For example, the successful operation of a new type of communication satellite, not yet fabricated, is obviously governed by a probability, the reliability of the communications equipment. Yet with no data and little theory, two people with equally good judgment are likely to assess the reliability of this yet-to-be-built equipment in different ways. In a systems analysis involving the satellite, one or the other or some average of the two estimates might be used. To make estimates about the behavior of the satellite, one might impute to the reliability in this case a subjective probability distribution, that is to say, a probability distribution that is not empirically or theoretically derivable, but that is imputed by the analyst to govern the contingencies. The situation can then be treated as one of risk.

There are other stochastic uncertainties[3] for which not only are the probability distributions unknown but also even the full set of possible outcomes.

---

[3] In statistics and economic decision theory, the stochastic uncertainties that I am calling nonrisks (namely those in which the probabilities associated with the various outcomes, and perhaps even all the possible outcomes themselves, are not known) are said to belong to a class called simply "uncertainties." I am calling these uncertainties "nonrisks" because the word uncertainty has a broader and more common meaning in ordinary discourse.

Risks and those nonrisks for which we can impute the underlying probability distributions (and thus reduce to risks) can be handled in analysis by actual calculation of the probabilities or by Monte Carlo methods or less precisely by using means or expected values. Such uncertainties, like most of those that occur in relation to cost or reliability, involve possible outcomes whose chance of occurrence can be estimated and the dangers accepted, provided steps are taken to guard against disaster. We can, for instance, build duplicate circuits to increase reliability or make allowances for additional costs. Stochastic uncertainties are therefore among the least of our worries; their effects are swamped by uncertainties about the state of the world and human factors for which we know absolutely nothing about probability distributions and little more about the possible outcomes.

The second type of real uncertainty, which we classify as "strategic" uncertainty, is usually uncertainty that is injected by an intelligent opponent. It can sometimes be handled by calculation also. It arises whenever two or more independent decisionmakers having different goals can influence the outcome. This type of uncertainty occurs most obviously in competitive situations such as military conflict or labor bargaining, where we have to deal with the difficulties generated by an enemy or by an intelligent competitor whose planned actions and objectives we can know only in part. Particularly perplexing are the problems associated with a substantial but incomplete community of interests on the part of the opposition. The type of quantitative analysis that is most appropriate for dealing with competitors or an opponent's strategy is the theory of games. Unfortunately, except in relatively simple situations, the difficulty in formulating a game and calculating its outcome is so great that game theory has afforded little more than conceptual guidance for handling competitive situations. As a consequence, we frequently resort to operational gaming in attempting to analyze situations in which the policy outcome to be selected is influenced by the interests of two or more independent decision-makers.

Uncertainties that can be calculated in one way or another tend to absorb the attention of the analyst all out of proportion to their importance. One reason is that to take them into account properly can be such a considerable challenge to his ingenuity. This sometimes leads to the neglect of the more serious uncertainties that are not stochastic in nature; the "real" uncertainties.

Uncertainties about human behavior, which do not necessarily have any logical basis, or about the future state of the world, are beyond the practical ability of analysts to predict and belong to the class of uncertainties that cannot, within the present state of knowledge, be reduced to risks. Under such uncertainties, we consider events—a major earthquake some time in the 1980s, for instance—to which individuals may attach subjective probabilities in which we have little confidence. With regard to such an earthquake, for example, real uncertainty involves such questions as: Will science find a means to warn of such quakes by 1980?; Will we get warning of this quake?; If we get it, will we believe it? For such uncertainties, there is frequently widespread disagreement about what probabilities, if any, are pertinent, and even confusion and vagueness with any one individual. For the stochastic uncertainties, such as risk and uncertainties reducible to risk, anticipation is possible, and appropriate calculations, even if subjective, can be made. In contrast, how real uncertainty will be resolved is impossible to foresee. The assignment of probabilities to actions or to perceived possible states of the world will be misleading inevitably, because the chances are so very great that the action or state that does materialze will almost certainly be one that was not perceived in advance.

That does not mean that uncertainties that can be handled by statistical techniques are easy to work with or that the results of analyses in which they occur can necessarily be accepted with confidence. For instance, consider the estimation of total system reliability for an electronic system. Often this is represented by the mean time between failures (MTBF), calculated by taking the reciprocal of the sum of the reciprocals of the subsystem MTBFs. The exponential distribution is then used to obtain the probability that no system failure will occur in a certain period. This simple scheme involves at least four tacit assumptions.

- The time between failures is exponentially distributed.
- Failures of subsystems are independent.
- A subsystem failure implies a system failure.
- Subsystems are utilized equally in time.

Ideally, the equations that express reliability should account for subsystem failure rates, redundancies, dependencies, and utilization. Although complicated, the calculations are not beyond the capabilities of a computer. But the estimates of the subsystem failure rates themselves depend on partial measurements and intuitive judgments influenced by temperature, humidity, dust, shock, stress, vibration, operating cycle, and the environment. The end result may be that predictions from the reliability model are highly uncertain.

Uncertainty might also be discussed in terms of the aspect of analysis with which it is associated. Therefore, uncertainty might be (a) *conceptual:* What precisely is the problem? (b) *factual:* What are the relevant facts associated with the alternatives and the current situation? (c) *predictive:* What changes in the situation are likely to occur before any decision can take effect? And what are the likely consequences and reactions to the alternatives between which a choice must be made? (d) *strategic:* What counteractions may be expected to be taken by opposing interests? (e) *ethical:* What should the goals be and which of the potential outcomes would be preferable in the light of those goals?

In order to deal with these uncertainties (the subject of the next section) we would need various approaches–fact-finding for (b), intuitive expertise, possibly Delphi or simulation of the opposition's decision-making apparatus for (d), ethical analysis ("soul searching") for (e).

Dealing with Uncertainty

What can the analyst do to take account of the proliferation of uncertainties? The most important advice is: don't ignore them. Ignoring uncertainties is a chronic disease of planners, but to base a projection on a single set of best guesses is absurd, and can be disastrous. For example, suppose there is uncertainty about 10 factors and we make a best guess for all 10. If the probability that each best guess is right is 0.6 (a very high batting average for most best guesses), the probability that all 10 are right is about six-tenths of one percent. If we confined the analysis to this one case, we would be ignoring a set of possibilities that had something like a 99.4% probability of occurring.

This matter of uncertainty raises the question of how to prepare for future action. The analyst (or the rational decision-maker) has two ways to cope with uncertainties. One way is to resolve them; the other is to try to formulate a scheme or contingency plan that renders them irrelevant.

For dealing with uncertainties, Schlesinger (1968) suggests that there are essentially two disparate points of view.

> . . . A first group, whom we might call the contingency *planners,*[1] has felt some confidence in our ability to chart in advance successful policies for the unknown future. Their method has been to designate a system which can deal adequately with each of them. A second group, whom we might describe as *contingency* planners,[2] has tended to emphasize the uncertainties and our limited ability to predict the future. Those who hold this view have consequently stressed the need for sequential decisionmaking, for improvisation, for hedging, and for adaptability. Heightening awareness of inevitable change should tend to make us more sympathetic to the latter approach.
>
> ---
>
> [1] That is, those who believe that an array and character of future contingencies can be specified in advance, and that *detailed advance* planning can be done to deal with whichever one does occur.
>
> [2] That is, those who believe that future developments will have a large element of the unforseen, that contingencies cannot be specified precisely in advance, and *that whatever planning one does must be done so that it may be adapted to the contingent and the unforseen.*

In the attempt to resolve uncertainties one can

1. *Buy time.* One can simply, or not so simply, defer his decision until he has better information. We say buy time because delay in achieving an objective is usually a cost.
2. *Buy information.* One can alleviate uncertainty by data collection and further research, but this requires time and money.
3. *Buy flexibility as a hedge.* If you are not certain one system will work, buy a second of different type as insurance in case the first should fail. It is also frequently reasonable to deliberately pay the cost of building flexibility into a design.

Consider highway construction as an example. In planning a highway, we are uncertain regarding numbers of vehicles and relative usage for we cannot know enough about population growth and movement and possible new means of transportation. We do know that a major cost in expanding existing highways is the tearing down and rebuilding of overpasses, bridges, and tunnels that constrain traffic flow. For example, a simple, if underutilized, hedge against expanded traffic flow is to make an original underpass sufficiently wide so that additional lanes of traffic can be provided without demolition. Such a hedge adds measurably to initial construction costs, but provides a hedge, insurance against the much higher cost of rebuilding the overpass much sooner than contemplated in the original plans.

4. Make use of *a fortiori* and "breakeven" analysis. To make an *a fortiori* comparison we bend over backwards to make assumptions about the uncertainties in ways that will not help one or a set of systems or alternatives but that may help the other options. The alternative or set of alternatives for which we made such assumptions, and thus handicap in the comparison, is usually the alternative or set of alternatives that previous analysis has indicated as preferable. For example, if reliability is an important parameter in a comparison and we are informed that it will lie somewhere between 0.8 and 0.9 for all systems, then we might assume it is 0.8 for the system we think best and 0.9 for the others. If, after we have done this, the handicapped system still comes out superior on whatever criterion we are using, we are in a stronger position to prefer it than before. Unfortunately, when we try to make an *a fortiori* comparison, it frequently happens that the system we think best no longer comes out ahead. We might then try a "breakeven" analysis. That is, we find what assumptions about important values have to be made in order for the performance of, say, the two systems that look best to come out essentially the same. We then ask those who must act to judge how the assumptions favor one or the other and to what extent.

5. Take a "conservative" approach. In its most extreme form this is the "maximin" approach. Here one attempts to choose the alternative that gives the best result if the environment is maximally mean to him. In other words, one resolves uncertainties by making the blanket assumption that the worst will happen.

In military applications a frequent practice has been to treat "strategic" uncertainty in terms of several sets of assumptions, most often three–a "worst" case (the most pessimistic set of assumptions for us that seems reasonable), the expected case, and an optimistic case. The alternative that was most cost-effective under the "worst" case almost always did well on the other two and became the "preferred" one. In the design of an analysis, it is important, however, to avoid structuring it in such a way that only very conservative or worst case assumptions are dominant in dealing with uncertainty. This deprives decision-makers of vital information and in effect precludes them from being able to distinguish between descriptions or estimates of uncertainty and attitudes or preferences about uncertainty.[4]

6. Use decision-making techniques that employ judgmental probability; that is, employ decision theory and what is often called a Bayesian approach.

Decision-theory techniques are well-described elsewhere (Raiffa, 1968, or Schlaifer, 1968), for example. The key steps are to investigate the available decision options and the related uncertainties. For those uncertainties that do not have probability distributions, use judgment, e.g., call on experts through

---

[4] For a discussion of this last point and of the next approach see Hammond (1971).

Delphi or some other procedure, to assign probability distributions. Thus, if we were studying transportation, for example, we might be required to assign a probability distribution to the demand for air travel between Los Angeles and New York in 1980 or to the earliest date that Congress might bar the internal combustion engine in interstate travel–in other words, to whatever uncertainties are of interest. The various aspects of the problem are then combined into a model to permit determination of the risks and other consequences inherent in any alternative. In any except the simplest cases this will require Monte Carlo simulation. The approach has much merit but there are so many difficulties in execution that in the end many uncertainties may still have to be handled as certainties.

7. Use sensitivity and contingency testing to examine the results for the range of possible values of the incompletely known parameters. A "sensitivity analysis" or test is an attempt to determine how sensitive the results are to variations in the parameters. A "contingency analysis" basically does the same but is a term sometimes used when we change the assumptions about the context or environment rather than just those related to the numerical parameters.

Sensitivity testing is a powerful and illuminating method. The major drawback of this approach is that it may require the computation of an extremely large number of cases or parameter changes to give an adequate coverage. With an extensive simulation, the computation time required to make these many runs may be prohibitively expensive.

In traditional sensitivity testing, just one parameter is varied at a time. Therefore, we might compare urban bus and subway transportation with respect to variations in cost per mile, frequency of service, capacity, trip time, and so on, one parameter at a time. We might discover, for example, that bus use depends critically on trip time and that a 10 percent increase in speed would change the order of preference–provided that everything else stayed the same. This is not good enough. In a better approach to sensitivity testing we use a Monte Carlo sampling process to examine sensitivity to the simultaneous variation in a number of parameters.[5] Suppose we can obtain an estimate of system performance by calculation with each set of specified values of the parameters, say, the means or best guesses. Suppose further that the range and distribution of the possible values of each uncertain parameter are known or can be estimated. To test for sensitivity, we select values of the uncertain parameters for each one of a series of trials by sampling at random from the frequency distributions of the parameters. If this is done enough times and the system performance is calculated for each set of selections, the results obtained give an indication of the likelihood that the alternative indicated as superior when best guesses were used, will continue to be so indicated when uncertainty is taken into account.

---

[5] For an example, see Emerson (1969).

Drawing Conclusions When Uncertainties Are Great

When the uncertainties are great, it is well to look for gross differences among the alternatives and, specifically, for differences that have a chance of surviving many resolutions of the various uncertainties and intangibles. The important question is which system has a clear advantage rather than precisely how much better one system is than another (Fig. 14-1).

Let me add emphasis to this with a discussion from Fisher (1966).

> The structure of the analysis was an equal cost comparison of several alternative future courses of action; that is, for a specified budget level to be devoted to a particular military mission area, the alternatives were compared on the basis of their estimated effectiveness in accomplishing the stipulated task. The final quantitative results took the following form. [Fig. 14-1.]

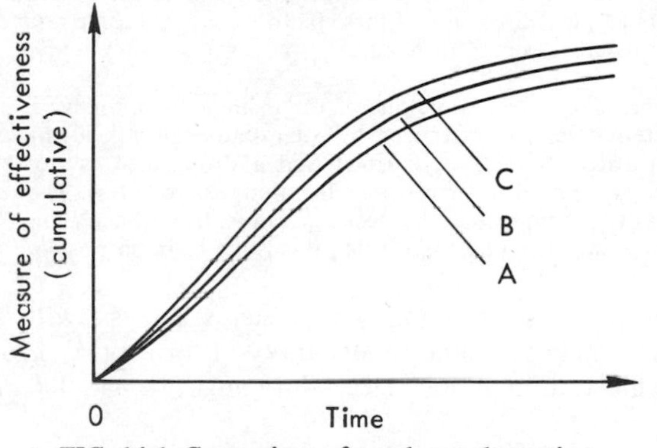

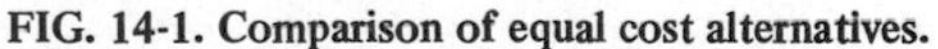
FIG. 14-1. Comparison of equal cost alternatives.

> The stated conclusion of the study, based almost exclusively on these quantitative results, was that alternative C is preferred over A and B for a wide range of circumstances and contingencies. (The context of the study, I should point out, involved a time period some 10 to 15 years into the future.) Yet the differences in estimated effectiveness of the alternatives (for a constant budget level) was *at most 15 percent*! Now my point is simply that the context of the problem was clouded by so many uncertainties and the model used in the analysis was so aggregative, that calculated differences among the alternatives averaging less than 15 percent *just cannot be regarded as significant.* Thus, the stated conclusions of the study, if taken literally, could in a real sense be misleading to the decisionmakers.[6] In decision problems of this type where uncertainties are very great, the analyst is generally looking for much larger differences among the alternatives being examined. How great? There is no general rule. However, I can say that in the past when experienced analysts have been dealing with problems of this type, differences in the neighborhood of a *factor* of 2 or 3 have been sought. I personally feel that in most long-range

---

[6] Needless to say, the Secretary of Defense and his analytical staff were not misled in this case. They are too experienced in interpreting the results of analytical studies to be overly impressed by small differences.

planning problems where major uncertainties are present, quantitative differences among alternatives must be *at least* a *factor of two* before we can even begin to have any confidence that the differences are significant. In any event, when they are smaller than that, the analyst must exercise extra caution in interpreting the results, and he must not make statements that are likely to mislead the decisionmakers.

There is another side to the coin, however. When quantitative differences among alternatives fall within a relatively narrow range, does this mean that the study is of no use to the decisionmaker? Not necessarily. If the quantitative work has been carried out in a reasonably competent manner and the differences among alternatives do tend to be relatively small, this fact in itself can be of considerable interest to the decisionmaker. This is especially true if sensitivity analyses have been made showing that as key parameters in the problem are varied over their relevant ranges, the final results are still within relatively narrow ranges. Given results of this kind, the decisionmaker can be less concerned about making a mistake regarding the quantitative aspects of the problems, and he may then feel somewhat more comfortable about focusing more of his attention on the *qualitative* factors—political, psychological, sociological considerations. In fact, if the analyst has done a reasonably thorough job, he might include a discussion of these factors in a qualitative supplementation to the purely quantitative part of the study.

The main point here is that while one of the main goals of analysis is to search for "preferred alternatives" characterized by quantitative results *significantly* different (better) from other alternatives, the fact that a strong case cannot be made for a preferred alternative does not mean that the study is worthless. The results, and the sensitivity analysis supporting the results, can still be very enlightening to the decisionmaker. And again I emphasize that this is the main purpose of analysis.

A good discussion of uncertainty in systems analysis with a more mathematical treatment may be found in Madansky (1968). Fort (1966) gives a good (hypothetical) illustration of how uncertainty might be handled in an agriculture context.

References

Emerson, D. E., *UNCLE–A new force exchange model for analyzing strategic uncertainty levels,* The Rand Corporation, Santa Monica, California, R-480, November, 1969.

Fisher, G. H., *The analytical basis of systems analysis.* The Rand Corporation, Santa Monica, California, P-3363, May 1966, pp. 12–14.

Fort, Donald M., *Systems analysis as an aid in air transportation planning.* The Rand Corporation, Santa Monica, California, P-3293-1, March 1966, pp. 12–15.

Hammond, J. S., III, *Defense decisionmaking: Prudent versus excessive conservatism.* The Rand Corporation, Santa Monica, California, R-715-PR, June 1971.

Madansky, Albert, Uncertainty. In E. S. Quade and W. Boucher (Eds.). *Systems analysis and policy planning.* American Elsevier, New York, 1968, pp. 81–96.

Raiffa, Howard, *Decision analysis: Introductory lectures on making choices under uncertainty*. Addison-Wesley, Reading, Massachusetts, 1968.

Schlaifer, Robert, *Analysis of decisions under uncertainty*. McGraw-Hill, New York, 1968.

Schlesinger, James R., The changing environment for systems analysis. In E. S. Quade and W. Boucher (Eds.), *Systems analysis and policy planning*. American Elsevier, New York, 1968, p. 359.

Wohlstetter, A., The analysis and design of conflict systems. In E. S. Quade (Ed.), *Analysis for military decisions*. Rand-McNally & Co., Chicago, 1964, pp. 106–111.

Wohlstetter, A. J., Hoffman, F. S., Lutz, R. J., and Rowen, H. S. *The selection and use of strategic air bases*. The Rand Corporation, Santa Monica, California, R-266, April 1954.

Chapter 15

EVALUATION AND EXPERIMENT

I. General Remarks

The effective delivery of public services requires more than the discovery and installation of what the policy-makers have decided is the best program they can devise. In recent years analysts have come to realize that the policy-maker needs also to determine how these programs actually behave—to measure the outcomes under operating conditions and to find out whether the program is accomplishing what was intended and if not, to put himself in a position to discontinue or to suggest improvements. The required investigation has come to be called an evaluation. Evaluation in this sense has just come into general prominence during the past few years, although it has a fairly long history in the fields of education and medicine. With the spread of new social programs there has been wide acknowledgement by public officials of the need for evaluation. Federal legislation now requires it; money has been provided, evaluation staffs have been created and strengthened, fairly major evaluation studies have been undertaken, some using rigorous experimental methods, and attention has been paid to the results.

Evaluation, in its most general usage, is a process carried out to determine worth. It is thus a part of almost every public policy analysis. Despite this broader usage, in this chapter the term is applied as above in a more restricted and technical sense as an investigation to measure how well the actual acomplishment of an on-going program (or, less frequently, a completed program) matches the anticipated accomplishment. It is not completely retrospective, however, for the purpose may be to suggest a change in resource allocation, to improve operations, or to plan future activities. The key element that leads us to distinguish evaluation from, say, the cost-effectiveness or cost-benefit analysis of a proposed program is that the latter do not have to cope with an activity that is on-going and thus with the people participating in or being affected by the activity. People interact with and affect an evaluation in ways that must be taken into account by the analysts.

Evaluations cover a wide range as the following titles suggest:

HEAD START: Final report of the Evaluation of the Second Year Program, L. M. Vogt et al., Urban Institute, Washington, 1973.

Evaluation Results for the Los Angeles Woman's Job Corps Center, S. Carrol et al., R-745-OEO, The Rand Corporation, Santa Monica, 1971.

The Indianapolis Police Fleet Plan: An Example of Program Evaluation for Local Government, Donald M. Fisk, The Urban Institute, Washington, October 1970.

Evaluation of the JOBS Program in Nine Cities (Final Report) TM-WD-(L)-313/001/000, Systems Development Corporation, Falls Church, Virginia, September 1969.

Final Report: *Evaluation of Office of Economic Opportunity Legal Services Program,* Volumes 1 and 2, The John D. Kettelle Corporation, Paoli, Pennsylvania, 1971.

The essence of evaluation is the assessment of the outcome of a program or other activity—what happened that would not have happened in its absence—and of relative effectiveness within sets of programs or activities—what individual projects or types of projects work best? The reason for doing an evaluation is to provide information either to policy-makers on the cost and effects of their programs and projects and to aid in the efficient allocation of resources, or to program managers to assist them in the effective management of their programs or both. In other words, evaluation of government programs or projects is a process of assessment designed to provide information about present operations and their effectiveness in order to assist in making decisions about the future.[1] Unfortunately much activity classed as evaluation has been irrelevant because it did not produce the kind of information needed by the decision-makers.

The following example, suggested by Kathleen Archibald, may help to illustrate the difficulties that arise.

Assume you are a public official, a decisionmaker, and that you have gotten worried about the high unemployment rate of Latin Americans in San Jose (or blacks in Albuquerque or aircraft workers in Seattle). You manage to get some

[1] It may be worthwhile to make a distinction between evaluation and the traditional post audit examinations, even though auditors are changing their point of view (Staats, 1973). They were carried out for different purposes—in some ways almost opposite. The conventional post audit sought to check on legality and propriety of financial transactions and to ensure that budget ceilings were not exceeded. It was not really evaluation because it did not tell the policy-maker or program manager what the activities in the project being investigated had accomplished. It dealt with inputs, not outputs. Furthermore, the conventional post audit tended to be backward-looking; it attempted to place blame. It contributed to improvement only in the sense that it served a deterrent function, probably preventing some abuses from happening. An evaluation, on the other hand, should be primarily forward-looking and it should help management to decide what to do next.

The key to the distinction is the decision that it is hoped the investigation will affect. If the investigation is designed to have impact on a decision concerning program size, or strategy, or operation, it belongs in evaluation proper; if it is to check adherence to regulations, guidelines, workloads or administrative practices, then it is compliance control. Investigation of these latter things do not usually shed any light on the worth of the project since compliance with legislative intent and administrative regulations may have little to do with performance.

federal money to start a manpower training project for them—the only one in the community. Being genuinely concerned about the effectiveness of the program, you contract for an evaluation or hire an evaluator. You decide, and the evaluator agrees, that you want to know whether the program is having an impact. A competent, thorough study is done, which incidentally costs you a lot of money. Suppose it finds out for you one of the following:

1. The unemployment rate among the participants in the program is the same as that in the control group, i.e., there is no evidence that your program is having any impact.

2. The unemployment rate of the participants is x percent less than that in the control group.

With this information, what are you supposed to do? If the finding has no impact, it does not tell you how to change the program. If the rate for participants is lower, does this show you are doing a good job? a mediocre job? a bad job? Even if the unemployment rate for participants turns out to be zero, it would be nice to know if you could decrease services, add more participants, and still come up with as good a result. The point is that information on possible changes to improve performance or even how the program is doing in some relative sense is likely to be more helpful than information in an absolute sense.

Because evaluation is a form of policy analysis, policy analysis procedures are applicable. The objectives of the program to be evaluated must be identified, measures of effectiveness found, and criteria for comparison determined. The alternatives to be compared are then investigated, data gathered, and the costs and other consequences estimated. There are always at least two alternatives—to continue the program as is or to discontinue it—and usually more, as interest should also lie in improving the operation.

In the comparisons, we would like to compare what is actually happening, or what will happen, as a consequence of the program with what would happen were the program to be discontinued or changed. Various schemes can be used to make the comparisons. For example, Hatry[2] *et al.* (1973) remark

> Ideally, we would like to compare what "actually happened" to what "would have happened if the world had been exactly the same as it was except that the program had not been implemented." Since it is impossible to determine exactly what "would have happened if . . . ," the problem is to use procedures that approximate this.

With this goal in view they suggest five approaches for identifying and quantifying effects due to the program

1. *Before vs. after program comparison*
   Compares program results from the same jurisdiction measured at two points in time: immediately before the program was implemented and at some appropriate time after implementation.

---

[2] Their definition of program evaluation excludes the process of developing alternatives to existing programs and analyzing the likely future effects and costs of these alternatives, focusing instead on actual past performance of existing or completed programs.

2. *Time trend projection of pre-program data vs. actual post-program data*
   Compares actual post-program data to estimated data projected from a number of time periods prior to the program.
3. *Comparisons with jurisdictions or population segments not served by the program*
   Compares data from the jurisdiction where the program is operating with data from other jurisdictions where the program is not operating.
4. *Controlled experimentation*
   Compares pre-selected, similar groups, some of whom are served and some of whom are not (or are served in different ways). The critical aspect is that the comparison groups are pre-assigned before program implementation so that the groups are as similar as possible except for the program treatment.
5. *Comparisons of planned vs. actual performance*
   Compares actual, post-program data to targets set in prior years—either before program implementation or at any period since implementation.

Controlled experimentation (4) is the most powerful, but it is also the most expensive and the most difficult to carry out. A major problem with all these schemes is to identify and take account of the influence on the comparison of external factors unrelated to the program—for example, a change in the economy or other programs that might be going on simultaneously.

There are differences in the way one must go about gathering the data to estimate the performance of alternative programs while such programs are still in the proposal stage and the evaluation of a program after it is in operation. Most people do not like to have their work evaluated even when they may agree in principle that evaluation is a good idea. And the more they suspect that they may be doing a bad job, the less they like to be evaluated. At times, the fear of being evaluated amounts almost to "paranoia." People associated with a program, both managers and participants, are concerned that the analyst may not be on their side (particularly when he is from outside the organization and especially if he works for a budget agency). Even when his announced purpose is to improve operations, they are not sure whether to take his stated purpose for being there at face value. In fact, the word evaluation alone is enough to trigger a negative reaction and this can cut down access and make data difficult to obtain.

Before a proposal or program is funded and in operation, the supporters believe in it—they have not seen its faults to the extent that someone who has actually implemented or participated in a program sees them. They know their proposal is good and will work to convince the analyst.

In terms of self-interest, an executive has a stake in seeing that a good preprogram analysis is done. He want to look carefully at the alternatives before posing a new program; he has a better chance of getting the program approved in that way. When it comes to carefully evaluating existing programs, then the balance of feelings in the executive may be against the evaluation, especially if the results might become a matter of public knowledge. This may be a reason that evaluation is still one of the weaker activities in the public sector and in the federal system. Also, it is not surprising that Congress may have played a stronger role or play a stronger role than the executive branch in pushing for evaluation activities.

We first consider evaluation to assist an outside decision-maker with resource allocation. We follow with consideration of evaluation, largely for internal consumption, to bring about improvement. Following that, we consider some problems of experimentation[3] and then conclude with a few observations about evaluation in general.

Evaluation to Affect Resource Allocation

Evaluation to affect resource allocation is designed to assess the worth or effectiveness of an on-going program or project in order to help determine the funds (or possibly other resources) it should be assigned. It sometimes involves a choice between using funds to continue or to end a program, but more often the decision is resource allocation at the margin—adding a little to the programs that seem to be doing well and cutting back, or not increasing, the others.

The criterion problem is a major stumbling block here just as it is in all analysis. To assess relative worth, the impact or effectiveness of a program may be compared with unmet needs, with specified performance objectives, with past program performance, with the performance of programs competing for the same assets, and with its own costs.

To indicate improvement, governments frequently use a comparison with past performance. This can be misleading. A health program that cures 70% of its cases this year but only 20% last year may look good on this basis but not if new techniques that made the increase possible should cure 98%.

Evaluation to compare alternative programs like every policy analysis needs to be very clear about the objectives of the programs considered and the ways to measure attainment. If measures of effectiveness can be chosen that are clear-cut and closely correlated with program goals, evaluation can be straightforward. For example, the objective of the Vera Foundation's Bail Bond project was to demonstrate that arrested persons released without having posted bail would appear in court. A count to establish the fraction of persons who had been released on their own recognizance and later appeared in court when compared with the fraction who appeared after paying bail, demonstrated the feasibility of the procedure. As a consequence it was widely adopted for its obvious cost-effectiveness.

---

[3] One way for a decision-maker to choose between alternative social program is on the basis of experimental programs. Experiments are treated with evaluation because experiments, to be useful, must be evaluated. On the other hand, to give valid results, evaluation, or at least outcome evaluation, may have to use experimental or quasi-experimental methods.

The goals for on-going programs are just as likely to be as ambiguous, multiple, and conflicting as the goals a policy analyst faces when he undertakes almost any study. It is therefore necessary for the evaluator to press for a clear definition of goals and the priorities among them before he can accomplish very much.

For example, a program for working with teenage gangs may state its function is to improve "social behavior" on the part of its participants. Success should be measured in terms of what the participants do differently after they have participated. But what social behavior should be measured: number of arrests? number of convictions? number of fights? times at home after 10 P.M.? school attendance? attendance at program sessions? Part of the answer lies in the hard realities of time, money, and access. To decide what aspects to use is not likely to be an easy job. To pick one aspect, or a combination of several without knowledge of how they are related, is likely to be misleading. Another consideration is the use to which the evaluation is to be put; how much success is required? Measurement is desirable, as well as elegance and precision, but indicators of success that are not strictly quantitative may have to be used, including opinions obtained from teachers, truant officers, police, neighborhood storekeepers, and so on. A profile or "scorecard" made up of a host of indicators may have to be presented to the decision-makers and judgments made.

Unfortunately, because the criterion problem is so difficult, evaluation projects tend to measure work load rather than measure true performance. Workload measures may not tell you whether the program is attaining its goals or how well. They tell you that some job is getting done, but it may not be clear that it is the one the program wants to do. Archibald (1970) illustrates

> For instance, let's say one manpower training project is providing training for 200 of the hardcore poor per year at a cost of $1,000 per trainee—just to keep it in round numbers. Another very similar project is training 400 a year and the total cost is the same, so this project is managing on only $500 per trainee. Marvelous—they're handling double the volume at the same cost. But that isn't an evaluation because the objective isn't to put people through a training program—presumably it's to increase their chances of getting and holding a job.
>
> To decide which project is giving you more for your money, you're going to have to know how many of their graduates are getting jobs. Even when you have that information, it still may not be easy to decide between the two projects. Say 60% get jobs in the one where 200 trainees are handled at a cost of $1,000 per trainee, and 40% in the one that handles 400 at $500 per trainee. This would mean that 120 graduates from the first project get jobs and 160 from the second project. The total cost of the two projects is the same, so if cost per job obtained is your criterion you'd take the one that trains 400 at a cost of $500 each. This project gives you more in jobs, but it also gives you more "failures"—240 don't get jobs compared to only 80 in the other project. And you may want to worry about the effects of failure, the consequences of raising expectations and not meeting them. If you're willing to invest some to avoid this, you might prefer the $1,000 per trainee project.
>
> This is a very simple example, there are several other things that should be taken into account. For instance, perhaps the $1,000 per trainee project is dealing with a different clientele—who are more disadvantaged, say. Or if the projects are running in different localities, there may be important difference in the job market. Also you probably want to look at how long jobs are held.

If the decision to be made is of the go/no-go variety, that is, continuing or discontinuing the program, then the evaluation should be designed to compare the costs and benefits of the program with the costs and benefits of no program. It is not safe to assume that both the costs and benefits of having no program are zero.

As one might expect, cost-benefit and cost-effectiveness analysis have application here. If the program results can be expressed in terms of dollars (for a vocational rehabilitation program this might be the lifetime gain in earnings or the taxable returns therefrom), then the resulting difference between the benefits and the costs of the program can be used to help make allocation decisions between the program being evaluated and all other uses for the dollars (including tax reduction). Similarly, cost-effectiveness comparisons can be made between programs for which the same effectiveness measures can be used (e.g., the average percentage increase in reading scores on standard tests to compare alternative programs that use teaching aids of various sorts). In general, the methods to be used here for comparisons are the same as those used elsewhere but there are additional complications peculiar to the "on-going" nature of subjects for the analysis, which are likely to be heavily "people" intensive.

Evaluation to Improve Operations

Evaluation to improve operations is frequently done internally since its purpose is to investigate possible changes in the program with a view to improving performance, not to see how the program is doing in comparison with similar programs or in any absolute sense. A particular evaluation project may, of course, provide information on both overall impact and for program improvement. For example, from Wholey (1972)

> The Westinghouse evaluation of the Head Start program therefore produced generally negative findings. The negative findings, however, did not significantly reduce the budget level of Head Start. Powerful constituencies would have fought any reduction in funding for Head Start. Results that seem to have come from the Westinghouse Head Start evaluation are (1) the "hold" placed on the program–increased funding levels would not be sought; (2) the diversion of some Head Start program funds into experimental child development program, "planned variation," designed to test whether there are better approaches than those that were being used in the national Head Start program; and (3) the reduction of the proportion of Head Start funds now going into *summer* Head Start (the Head Start component with the *least* apparent value).

The decision which the evaluation is intended to assist must be considered in the design of the analysis. Levine and Williams (1971) make this clear.

> . . . . For instance, an evaluation project might be designed to help project managers determine what makes the best projects so good; if so, it would be an improvement evaluation. Or it might be designed to help program managers determine which projects should be cancelled; if so, it would be an allocation evaluation. While in a

> brief description the two evaluations might sound very similar, they would in fact be quite different if appropriately designed to meet the differing decision needs. The one designed for improvement would concentrate on the mix of techniques within the projects and on environmental conditions confronting good and bad projects. The object would be to find out what techniques work best under what conditions, with a view of spreading the use of the best techniques under the appropriate conditions. The one designed to determine relative worth would concentrate primarily on the worst projects, and would attempt to compare the costs of cancellation with the costs of continued operation. It will frequently be economical, however, to look carefully at both sets of factors in order to provide inputs for both types of decisions.

Evaluation to improve operations imposes certain requirements on data collection: (1) the data collected must be relevant for administrative purposes—thus the administrators and program operators probably have to have a considerable voice in deciding what data to collect; (2) the data must be processed, interpreted, and fed back to those responsible for operations *quickly* enough to be useful. The quality of an evaluation, in fact, depends to a large extent on the design of the program being evaluated. A good design for evaluative purposes provides for the collection of adequate and timely information.

The type of data relevant for the improvement of operations may be low-level, routine, and short-range, such things as: number of persons applying for admission to the program; number and kind accepted; number actually participating; number dropping out and number completing successfully. These items may serve as indicators or proxies for more abstract concepts—the numbers associated with applications, admissions, and attendance may indicate the attractiveness of the program, the effectiveness of recruiting, and the level of interest. Evaluation may have its most useful role here. Again quoting Wholey (1972).

> . . . . Over the past two years, members of the evaluation group at the Urban Institute have become more and more convinced that the primary evaluation payoff (in terms of decisions actually influenced) may be in evaluation that is done in enough detail to get at the effects of operational changes within operating programs. Many program managers really want to know what works best, under what conditions. There is a market, a use, for this type of detailed evaluation information.

There are sometimes advantages in having an evaluation of this sort done externally provided it can be made clear that the objective is to improve the program and not to cancel it. Outsiders frequently carry more weight with the program administration than do the program employees. Also participants and employees may feel freer to discuss the problems and shortcomings of the program with an outside consultant than with someone closely associated with the program director. Although the outside evaluator may not know any more about tests and measurement and research design than the directors of research in a program or agency, the extra program position may give him credibility with the program director and his superiors that an inside man does not have (Haggart and Rapp, 1973).

If changes in the program are to be suggested, a model of the way the program works is needed—a means to trace the course of the program from input to output from which the effect of variations can be estimated. Using an example from Weiss (1966) to help make this clear, if a project giving group counseling to girls identified as potential problems as a means of reducing delinquent behavior is to be improved, one needs to know the causal chain by which the counseling input is expected to achieve its goal. Questions such as: Is it by changing the girl's self-image? by providing other opportunities for self-expression? by motivating them to greater in school work? by providing models for alternative behavior? or by what? A model of what goes on in a program is needed to make clear what to look for. If we know what is supposed to happen at various intermediate stages in a program, we can find out where things go awry and point to the stage where adjustments are needed.

Experimentation

Experiments offer an opportunity to do a much more valid form of evaluation but at the same time present the analysts with a whole new set of problems. These problems are of two sorts: those associated with the techniques of the experimental design and those associated with the public context in which the experiment must be conducted. As part of their training, most analysts, or at least analysts trained in statistics, have learned something about how to cope with the experimental design problems. The design problems for large-scale social experiments, such as the current (1973) HEW-funded effort to examine alternative health insurance plans, however, are likely to be very much more complex than those found in college texts. In any event, they are too specialized for treatment in the book. Therefore, we shall not discuss that aspect further.[4] But problems of the second type are also likely to be troublesome, maybe even more so.

Program officials and policy-makers often do not appreciate what is required if an experiment is to be carried out properly. For example, the random assignment of participants (or communities) to treatment is fundamental to experimental design. A program administrator, however, may insist on responding to the ethical imperative to serve those who most need service. What is so wrong, he may argue about choosing the people most in need of housing to participate in a housing allowance experiment? Legal problems may also be raised by unequal distribution of publicly provided services; the courts have not yet given clear guidance on what Rivlin (1971, page 110) speaks of as the "equity of inequalities deliberately created for experimental purposes."

Control groups are needed, but this aspect may be hard for the analyst to sell. Lack of extra subjects may interfere with obtaining any control group at all. There are difficulties in getting cooperation from people to whom you give nothing. And, of course, money and staff are always in short supply.

---

[4] General references here are Cochran and Cox (1957) and Campbell and Stanley (1966). For the health insurance experiment, see Newhouse (1972).

Wholey (1972) points out another difficulty.

> *Time* also presents an enormous problem for the evaluator of experimental programs. As soon as there is sufficient legislative support to fund a series of experiments, there may be enough support to enact such a program nationwide. The concern that legislation will be enacted before the experiments have had time to produce reliable results may lead to pressures for the release of early, less reliable findings. The New Jersey Income Maintenance experiments experienced this pressure. Some early tentative results from the study were released with reluctance and heavy qualifications. If experimentation is to become a major vehicle in policy research, then ways must be found to anticipate and deal with these types of pressure.

As we said, random assignment of subjects to experimental and control groups is essential, but the program operators or political pressures may prevent this. In the anxiety of demonstrate that the program being examined by experiment will be a success, participants may be chosen from those most likely to profit from the program. Also, even though the analyst may sometimes have control over who enters the program, he is likely to have almost none over those who leave.

In spite of these problems, the addition of some sort of comparison group, even if it is only a partial equivalent to the subject group, can help with finding explanations for what happens. Without any comparison group, the analyst must search for all other explanations for whatever happens, other than the program, that are not excluded by the experimental design.

In addition to the possibility for being unfair, experimentation is, by its very nature, risky. A good many experiments should therefore fail. Failure costs time and money and, if people are the subjects, there may be other costs. Such other costs may extend to administrators. Rivlin (1971, page 112) remarks

> Another reservation about the desirability of social experimentation concerns the honesty with which experimental results will be reported. No one likes to fail. Rightly or wrongly, the administrator of a successful experimental project will receive more acclaim and greater opportunities for advancement than the administrator of an unsuccessful project. Under these circumstances will there not be a temptation to cheat a bit—to choose the most favorable measuring instruments, to "lose" the records of children who fail or patients who die, to coach participants on what to say to the evaluator or how to beat the test?

A so-called demonstration project is by its nature an experiment of sorts, but unless the results can be evaluated properly, all that may be demonstrated is that it is possible to spend public funds in a particular way. A demonstration is not a true experiment unless control is exercised over inputs, participants, and parameters, and the results are measured to determine the extent to which the program does what it was designed to do.

Despite the differences, evaluation of experiments generally has a great deal in common with the evaluation of on-going programs. In both cases, the analyst must remember that evaluation is an applied discipline, carried out to help with a current problem. It is not scientific research in search of truth (although sometimes it might do that also), but is intended to provide a basis for action. As a consequence, the evaluator must resist the temptation to search for answers to questions that interest *him,* but which may not be high on the list of questions the decision-maker wants answered.

Observations

Until recently, political considerations ruled out large-scale social experiments. This has changed; we now have OEO's negative income tax experiment under way in New Jersey; HEW's income maintenance experiments in Vermont, Gary, Seattle, and Denver; and OEO's experiments with performance contracting in elementary education and numerous others.

Some of the problems with evaluation are pointed out in Wholey (1972).

> Let's turn now to some of the real problems in getting useful evaluation. From the point of view of decision-makers, evaluation is a dangerous weapon. They don't want evaluation if it will yield the 'wrong' answers about programs in which they are interested.[5] On the other hand, decisionmakers are more advanced in their ability to ask pertinent questions than evaluators are in their ability to provide timely answers at reasonable cost. Valid, reliable evaluation is very hard to perform and can cost a lot of money.[6] Evaluators have real problems in detecting causal connections between inputs and outputs–and in doing so in timely enough fashion to be useful to decisionmakers. The structure of a program can have an important influence on the technical feasibility of separating the effects of the program from the effects of other, often more powerful, forces not under control of the program. To the extent that a program is run as a controlled experiment, for example, the evaluator's chances of separating out casual connections may be greater.
>
> Our reviews have found typical federal program evaluation studies marked by certain design characteristics which severely restrict their reliability and usefulness:
>
> (a) They have been one-shot, one-time efforts, when we need *continuous* evaluation of programs.
> (b) They have been carried out in terms of national programs and are very weak on process data.
> (c) They have been small sample studies working with gross averages, when we need studies large enough to allow analysis of the wide variations we know exist in costs and performance among projects within programs.

[5] In some cases, political pressures will simply override the empirical evidence without the formality of a methodological argument. Here, the only recourse open to the evaluator is to publish the results and hope that some other more enlightened or less pressured decision-maker with similar problems will make use of the results.

[6] The Stanford Research Institute evaluation of the Office of Education Follow Through program, for example, has already cost approximately $7 million.

As in other analysis done for public bodies, the evaluator must be aware of the decision-making process and the operative constraints of a particular evaluation. The organizational structure of the agency may be a major determinant of the services that are delivered. The insider–outsider issue can be particularly troublesome, especially when evaluation results are negative. The project personnel, the sponsoring organization, and groups with a "self-interest" in the appearance of success can react with hostility that can easily turn into personal attack on the outside analyst's research abilities and personal interests.

Evaluation can be full of pitfalls. Often a new public activity or program that sets up a scheme for supplying a service is offered by means of a new organizational structure or by a change in the old. In evaluating one must decide when it is the service and when it is the organization one wants to evaluate. A program can be evaluated in terms of its effectiveness and costs. An organizational structure, however, cannot be evaluated solely by its success in running a particular program. Its ability to detect program failures and maintain success over time must also be taken into account. Hence, one should not evaluate an organization by an initial program failure, but by its capability to "learn" from that failure and change the program policies to improve operations.

Consider performance contracting in education. A school district enters a contractual arrangement with an independent firm to provide all or some part of the educational service, for a school, or set of schools. The contractor is paid a specified sum for improving student performance by certain increments, or for bringing achievement up to a certain level. If he is not successful, he is not paid, with success or failure being measured in terms of something like the reading and mathematics scores on standardized achievement tests.

This practice clearly represents a different organizational approach to public school education. It has not always been so evaluated, however. Because most contractors have emphasized innovation, making heavy use of such techniques as programmed learning, computerized instruction, and audiovisual materials, the performance contracting approach is sometimes evaluated as an innovative approach to teaching, based on technology. In other cases, it is viewed as a device for introducing change to be replaced as soon as the teachers learn enough to run such a technology intensive program themselves. In this latter case, one tends to judge a program a success if the school system adopts the methods used by the contractor.

For example, the OEO evaluated performance contracting by selecting six private firms. These firms apparently were chosen because they used a variety of different instructional approaches, including various teaching "machines," incentives, and teacher-training methods. The formulas for payment were essentially identical and performance for a one-year period was considered. The conclusions were that the students who participated in the experiments did not do significantly better than those who did not. As a consequence, it is doubtful that this evaluation told us anything about organizational arrangement and adaptability, since the firms were essentially constrained to use the same methods throughout the year.

Another pitfall can be to attribute an outcome to the wrong cause. The incentive can make a difference. Consider one case of performance contracting in which the contractor was to be paid on the basis of the number of students in the program who exceeded national norms on two standardized tests. To maximize his profits, he divided the students into three groups: a high group that needed no further instruction to exceed the norm; a low group that was hopeless; and a middle group for which the issue was in doubt. He concentrated on the middle group, leaving the other two essentially uninstructed. The middle group did well; the other two went down enough so that the average of the whole did not rise. Educators judged this case a failure for performance contracting.

Another possibility for error lies in determining the real gainer from a program. Archibald (1968) points this out as follows.

> Some attention in evaluation should be paid to who gains what? In some cases training may be little more than an indirect subsidy to the employer. I am thinking of those cases where the employer hires from minority groups anyway and has customarily provided on-the-job training. Jewelry making provides a good example . . . Jewelry makers in New York City have been hiring Puerto Ricans for a long time; in fact a number of the shops are Spanish speaking. The kids get a number of weeks or months of training at . . . then are hired at a minimum legal wage since the employers argue they still have to provide considerable training on the specific tasks the kids will be doing. The employer gets subsidized from NYC OJT money, so he is not even having to pay the full minimum wage for the first period of employment. The end result is that the kids get some social support and employment facilitation, but the employer gets a nice healthy subsidy with savings on finding employees, training employees, and a payroll subsidy. If one evaluates only placement percentage in a case like this, the effectiveness of the program would end up being grossly overrated.

Another aspect of training programs warrants attention in the evaluation process (Archibald, 1968).

> I haven't seen anyone worrying, at either the operating level or the planning level, about the consequences on clients of failure of programs. If 'new careers' turn out to be dead-end jobs, if training proves to be essentially irrelevant and useless, and the client has taken the promises of the agency seriously, the main consequence will be an increase in alienation. The costs of such failures may be particularly high if the view is correct that the problem for the poor is already one of not relating means to ends in the sense of deferring gratification and working towards a goal. The failure of a training problem will again confirm for them the lack of any predictable connection between institutionalized means and ends. An evaluation should take the risks and costs of failure into account. I think this is a serious worry in a program like New Careers.

In evaluation, more than in most policy analysis, the analyst may have to attend to acceptance problems. When the results are negative, suggesting that the program be terminated or reduced, because this means some or all of the staff would lose their jobs, resistance is to be expected. Weiss (1966) points out that if the analyst "can get across the idea that 'we' are appraising the project" and that negative results can have constructive implications for improving the program, he may avoid trouble later on. An analyst must recognize that if his evaluation indicates that the project is deficient and not doing what it is supposed to do, he has created a problem for the director and his staff. In their view the most likely cause of that problem, and the one that should be investigated first, is whether there is something wrong with the evaluation itself.

Concluding Remarks

In an informal sense, evaluating government programs is an activity in which almost all of us indulge. We are coming to realize, however, that more systematic attempts to determine the positive and negative effects of such programs on the part of the public are needed. Also, that the objective should not be merely to grade the programs with a view to eliminating the worst, but, more importantly, to improve those worth continuing.

Evaluation has a rapidly growing literature; in addition to the references cited, other references might be Guttentag (1971), Rossi and Williams (1972), Wholey *et al.* (1970), and Weiss (1972). The report by Hatry *et al.* (1973) is an excellent attempt to provide practical guidance to state and local officials who are interested in either developing a program evaluation capability or in improving their existing capability.

References

Archibald, K. A., *PPB systems and program evaluation.* The Rand Corporation, Santa Monica, California, P-4534, December 1970.

_______, from an unpublished Rand trip report commenting on the problems of evaluating manpower training programs, February 1968.

Campbell, Donald T. and Stanley, Julian C., *Experimental and quasi-experimental designs for research.* Rand McNally, Chicago, Illinois, 1966.

Citizens Conference on State Legislatures, *State legislatures: An evaluation of their effectiveness.* Praeger, New York, 1971.

Cochran, W. G. and Cox, Gertrude M., *Experimental design.* Wiley., New York, 1957.

Guttentag, Marcia, Models and methods in evaluation research. *Journal for theory of social behavior*. April 1971, **10**, (1), 75–95.

Haggart, Sue A. and Rapp, Marjorie L., *The appraisal of educational alternatives–Making evaluation work in choosing future courses of action.* P-4990, The Rand Corporation, Santa Monica, California, April 1973.

Hatry, Harry P., Winnie, Richard E., and Fisk, Donald M., *Practical program evaluation for state and local governments,* The Urban Institute, Washington, D.C., 1973.

Levine, R. A and Williams, A. P., Jr., *Making evaluation effective: A guide.* R-788-HEW/CMU, The Rand Corporation, Santa Monica, California, May 1971.

Newhouse, J. P., *A design for a health insurance experiment,* The Rand Corporation, R-965-OEO, Santa Monica, California, November 1972.

Rivlin, Alice M., *Systematic thinking for social action.* The Brookings Institution, Washington, D.C., 1971.

Rossi, Peter H. and Williams Walter, (Eds.), *Evaluating social programs.* Seminar Press, New York and London, 1972.

Staats, Elmer B., Postauditing, an aid to the legislative overview function. *State Government Administration,* March/April 1973.

Weiss Carol H., *Evaluation of demonstration projects.* Presented at Workshop on the Role of Demonstration Projects in Delinquency Prevention and Control, April 18–20, 1966, Washington, D.C.

_______, *Evaluation research.* Prentice-Hall, Englewood Cliffs, New Jersey, 1972.

Wholey, Joseph S., What can we actually get from program evaluation. *Policy Sciences,* 1972, **3**, (3). 361–369.

_______, Scanlon, John W., Duffy, Hugh C., Fukumoto, James S., and Vogt, Leona M., *Federal evaluation policy: Analyzing the effects of public programs.* The Urban Institute, Washington, D.C., 1970.

Chapter 16

FORECASTING AND PLANNING

Forecasting and planning are everyday activities, not confined to people called "Forecasters or Planners." Many people would deny that there is any particular methodology involved in these activities and few would question their value. But there are questions about both value and method.

For the individual, forecasting and planning are almost automatic. He inspects the sky on a cloudy day and, as a consequence, he may take his raincoat to work. Man looks ahead for many reasons; because he is inquisitive, because he may want to get his mind off an unpleasant present, but more importantly because he believes it useful to do so. If he forecasts correctly or reasonably so, and plans and then acts accordingly, things seem to work out much better both for individuals and for organizations. Not surprisingly, forecasting sometimes has its highest payoff when events do not turn out as forecast; a forecast of disaster may lead people to change their actions and thus avoid the disaster, or at least mitigate its consequences. Without forecasting, without anticipation, actions taken today may lead to consequences that will materialize so quickly that it will be impossible for man to do anything to protect himself or to profit from them.

Forecasting, planning, and programming are all aspects of decision-making. A decision is not made without some idea of what to expect; if that idea is made explicit we clearly have a forecast. Plans and then programs (which are simply more detailed plans) are developed on the basis of the forecast to increase the chance of getting what we want. Once a broad plan is developed then a more detailed plan or program of operations can be devised, that is to say, a statement that specifies the resources to be committed and the sequence of actions that have to be taken in order to carry out the plan.

For decision-making, the value of a forecast does not necessarily lie in whether or not it comes true, but in its utility in helping the decision-makers to choose a satisfactory course of action and to do it in time. That is to say, one cannot always judge the value of a forecast by how close it comes to what actually happens. One reason is that the act of making it may be self-defeating and may enable the decision-makers or others to counter what is forecast whenever they have sufficient control over the situation that develops.

Forecasting is a prerequisite to planning, and planning should be a prerequisite to action. Like policy analysis, planning is directed toward improving decisions. When an organization plans, it determines its aims or objectives and then seeks the best means of achieving them. In an active organization, planning must be flexible. The organization, its policy, and its environment are in a constant state of flux. This means that planning must be continuous and changing. It must consider political constraints and pressure and the appropriate ways of operating in their presence. To this end one must consider operating procedures, supervisory techniques, personnel development, coordination, communication, problems of structure. In other words, planning must be closely tied in with management, with resource allocation, with budgeting, and therefore with decision-making.

In this chapter we first discuss the role of forecasting in decision-making and how it is done, next planning and how that is and might be done, and finally programming, particularly with reference to Planning–Programming–Budgeting. We then conclude with a brief discussion of the question mentioned earlier: Is planning worthwhile?

Forecasting

THE NEED FOR FORECASTING

To forecast is a verb with a number of interpretations. My intention is to use it in its dictionary sense: to anticipate or predict some future event or condition, usually as the result of rational study and analysis. An explicit probability statement, often a conditional one, may be associated with a forecast.

Decisions simply are not made without any regard whatsoever to future consequences or with no attempt to determine what is likely to follow. Those who say they do not forecast are sometimes said to mean that they assume the environment will not change very much or that the situation is so uncertain and uncontrollable that anything can happen and that nothing can be done to anticipate or influence what may take place. But if this latter is what they mean, why make a decision at all? If a decision-maker has several alternatives open to him, he will try to choose the one that he finds most satisfactory. To make that decision rationally he needs to consider by calculation or otherwise what its impacts will be. Thus his decision is based on a forecast; he does not have a choice of not forecasting; his choice lies in how explicit he makes the forecast and how it is obtained. The alternatives are really whether to make the forecast with rational and explicit means or not.

When a decision-maker has no control over the outcome of a forecast, that is to say, when none of the actions open to him can alter the predicted outcome, he tailors his actions to whatever outcome he believes will occur. Knowing that the element of uncertainty is large, he hedges against it. At the other extreme, if the decision-maker had complete control over the outcome, he would have no need to hedge, for then he could make the outcome what he wants it to be. The intermediate cases are most interesting, as Martino (1972, page 12) explains

> In the in-between cases, where the decision-maker has partial control over the outcome, he does need a forecast, but its usefulness, surprisingly enough, may have little to do with its eventual correctness. The reason for this lies in the paradox of self-fulfilling and self-defeating forecasts. A self-fulfilling forecast is one which comes true only because it was made; a self-defeating forecast is one which fails to come true only because it was made. If the self-fulfilling forecast had not been made (or at least not been made public), the event forecast would not have occurred. If the self-defeating forecast had not been made, the event which was forecast to occur but didn't would have occurred. How do these paradoxes come about?
>
> The essence of this paradox is that it arises from a rational response on the part of people who hear and believe the forecast. Suppose a noted economist makes a forecast that the U.S. will have a depression the following year. Suppose this forecast is widely circulated and accepted. What is the rational response of an individual who believes a depression is coming? If he is rational, he will pay off his debts, not contract any new ones, sell any stock he may own, etc. If enough people act this way, a depression will come inevitably. Here we have a clearcut example of a self-fulfilling forecast. This example is somewhat unrealistic, insofar as people are unlikely to act this way on the forecast of a single person. However, there are many situations where a forecast from a reputable source will be accepted and, as a result of people's rational responses, will either come out true or be forestalled. . .

One possible instance of the latter, are the forecasts, after Sputnik, that the United States would, in 10 years or so, face a drastic shortage of engineers and scientists. Whether as a consequence or not, for a shortage in a field usually leads to greater demand and higher salaries, graduate school enrollments in the fields where shortages were predicted went up, and the shortages never materialized.

Forecasting plays an important role in the planning process. It helps to identify limits beyond which it is not sensible to plan, it can suggest rates of progress, and it can indicate possibilities that might be achieved.

Enzer's comparison (1971, page 5) of a forecast to a "predictive display," which is used as a navigational aid on many modern ships, is an apt one. Such a predictive display shows the surrounding terrain and the present position of the vessel and projects or forecasts its future position, assuming that the controls, the weather, the wind, and the currents remain the same. A forecast also attempts to identify what the surroundings are, what the environment is, where we are, where we appear to be going, what the limits of our control system are, and which of these controls should be used.

Forecasting usually requires analytic help because the methods that are used in forecasting associated with policy analysis are mainly analytic in nature. Techniques such as trend projection, curve-fitting, cross-impact analysis, and computer simulation are used.

In fact, planning and forecasting may be viewed as special types of analysis and as tasks for analysts. Although a forecast is not ordinarily something that a decision-maker makes a decision about, a plan is. The analyst makes the decisions about the forecast. He must decide what forecast to use, for example, when he is preparing a plan. He uses the forecast to alert others to dangers that lie ahead in order that they may be avoided and to discover future opportunities of which advantage can be taken.

This view of how forecasting should be used is very likely not new, but its actual use in this way may be. It reflects a change in attitude toward the future among "planners." The old view of the future was more fatalistic, seeing the future as unique, unforeseeable but inevitable. Current view has it that we can influence the future we desire through planning. Admittedly this influence is very weak and marginal. We have learned from economists, however, that small marginal adjustments in planning affairs can make a great difference in outcomes.

The techniques of forecasting, particularly technological forecasting and the use of cross-impact analysis, are well treated in Martino (1972), Gordon (1971), and Dalkey (1972).

STEPS IN FORECASTING[1]

The steps in a forecast should

1. Survey the alternatives, that is, anticipate what occurrences are possible and estimate their probabilities.
2. Investigate the interactions among the possible alternative occurrences. Identify the situations over which some control seems possible and estimate the extent of that control.
3. Analyze our preferences among the possible occurrences and estimate how varying degrees of intervention can change the associated probabilities.
4. Convert the results of the analysis into a display from which the impact of possible actions can be assessed.

Planning

In contrast to forecasting, planning is somewhat difficult to define acceptably; there are far too many interpretations of what is meant by the word planning.[2] It involves forecasting because it suggests ways to deal with the future on the basis of decisions made today and is thus concerned with the future impact of those decisions. Planning cannot eliminate risks; at best it can give some assurance that the risks taken are the right risks and suggest "hedges" against them.

Planning (perhaps plan-making would be a more appropriate word) involves two stages—an analytic or problem-solving stage and a scheduling stage. The first, to *solve the planning problem,* requires two steps.

1. A specification of objectives or goals
2. Finding satisfactory or acceptable programs for achieving these objectives or goals

[1] The idea is from Olaf Helmer.

[2] Wildavsky (1973) discusses a number of these.

The second is to *prepare a plan* of operations based on the solution.

The first stage is therefore a problem in analysis: the problem needs to be clarified, objectives determined, alternative courses of action generated, costed, and compared, and a preferred one chosen. The second stage is more a problem in scheduling: the chosen course of action has to be fine-tuned into a time-phased program compatible with available resources, organizational constraints, and so on. A plan is therefore a set of decisions for actions to be taken sequentially and intended to produce a desirable set of circumstances at some future time.

With reference to administration and management, Dror (1963) defines planning as follows: "Planning is the process of preparing a set of decisions for action in the future, directed at achieving goals by optimal means." Among the implications Dror draws from this definition are the following.

First, planning is a continuous process taking place within an organizational unit and requiring resources to do so. As a process, it is to be distinguished from its product, a "plan." Second, a "plan," defined as "a set of decisions for action in the future," can be arrived at by any means whatsoever. It need not be the product of planning. Third, "achieving goals by optimal means" implies that the planning process must not only take into account the decisions which result from the planning but also the resources that went into that planning. (This latter, we know, is seldom possible to do adequately.)

Planning involves preparing alternative sets of decisions, not actually making those decisions. Hence, the planning activity can be carried out by persons other than those responsible for the ultimate decision or for operational implementation of the plan. Also, because the set of decisions in a plan are directed toward achieving a set of goals, somebody has to determine as part of the planning process what these goals are to be. Actually setting goals is a policy-level activity and the responsibility of the decision-makers (although planners may assist); finding the means to meet those goals, however, is an analytic activity.

Planning also implies preparing a set of documents. Abert (1974) suggests at least three are desirable.

1. Options documents which describe possible broad policy approaches, not yet constrained by feasibility and budgetary considerations containing imaginative and far-out ideas not yet screened by those who must implement whatever plans are accepted.
2. Documents which restrict the range of alternatives to those that the operating and staff offices consider feasible and desirable.
3. A final set of plans with the top decisionmaker's endorsement.

In recent years planning has become fairly analytic. A few years ago transportation planning, say, was carried out by someone drawing numerous routes on a map with a ruler and pencil. Today, all the resources of simulation, computers, and systems analysis are applied. The results are somewhat better, but sometimes we still run into difficulties, find roads inadequate, or have to abandon freeways before construction is complete. We may do better, but because the future is not predictable, there will always be similar difficulties.

The future is uncertain; actions specified today to be taken at a time in the future may no longer appear appropriate when that time arrives. This implies that even if a plan is accepted, it is understood to be flexible and subject to revision and that, in fact, it will be revised. A plan that is not used, however, is not a waste of resources for ordinarily it provides the basis from which someone can prepare a better plan.

To many people, nevertheless, a plan is something that must be approved at a high level and stuck to, no matter what happens. Schlesinger (1966) has a name for this sort of planning; he calls it Cook's Tour planning and advocates, instead, a second approach that he calls Lewis and Clark planning. In the first the assumption is made, usually implicitly, that one can chart a straightforward course possibly years in advance. The second acknowledges that obstacles and forks in the road will appear the character and timing of which cannot be clearly anticipated. Only limited confidence is placed in predictions as to which options will appear most promising or when the choices must be made. The idea is not to chart a precise course of action but rather to keep oneself in condition to cope with what lies ahead, determine indicators that will tell when a change is needed, in order to respond in timely and appropriate fashion. The concept is that of facing, and hedging against, uncertainty, not of pushing it aside.

This idea is summarized very nicely by Rittel and Weber (1973), although they have reservations about even this sort of planning:

> Many now have an image of *how* an *idealized* planning system would function. It is being seen as an on-going, cybernetic process of governance, incorporating systematic procedures for continuously searching out goals; identifying problems; forecasting incontrollable contextual changes; inventing alternative strategies, tactics, and time-sequenced actions; simulating alternative and plausible action sets and their consequences; evaluating alternatively forecasted outcomes; statistically monitoring those conditions of the publics and of systems that are judged to be germane; feeding back information to the simulation and decision channels so that errors can be corrected—all in a simultaneously functioning governing process. That set of steps is familiar to all of us, for it comprises what is by now the modern-classical model of planning. And yet we all know that such a planning system is unattainable, even as we seek more closely to approximate it. It is even questionable whether such a planning system is desirable.

Judging by its performance in controlling the consequences of our actions, planning, particularly national planning, has been a failure. The knowledge required is too great. We need to know, among other things, the interactions among elements of the plan itself, the incentives for those involved to insure compliance, and the resources that are needed and when.[3]

Planning should not be confused with the existence of a formal written plan, for planning may take place in a very informal manner. National planning or urban planning, nevertheless, usually is done by people called planners in institutions with the word planning its official title, something like a planning

---

[3] For discussion of the reasons for lack of success, see Wildavsky (1973).

commission, who produce something called a plan to direct the economy, or zoning, or some other activity. If government planning is to be more than an academic exercise, it must actually guide the making of government decisions. To plan, therefore, is to govern.

The distinction between planning and programming is essentially a matter of degree. One ordinarily thinks of a program as being more specific and more detailed than a plan. In other words, a program is a specific blueprint for a definite course of action, a plan to implement a plan. For example, a government planning commission might plan to increase farm production by consolidating small marginal farms into larger, more efficient units. Before the Department of Agriculture could implement such a plan it would need a detailed program to specify what steps to take.

Relation of Planning to Analysis

Planning is not analysis, although a plan may be based on rational thoughts and actions and may be the result of a policy analysis. Also forecasting not identical with analysis although forecasting, of course, can be based on analysis. An essential element in planning, however, is the forecast of future events.

Planning, like analysis, is related to decision. Planning can facilitate decisions by making choices more clear cut, the alternatives more concrete. Not all decisions are planned nor can they be planned. Kaplan (1973) lists four constraints on decisions and planning that deserve special attention.

> (1) *The unplanned.* Not all decisions are planned decisions, nor can they be. The urgency of a decision may be so great that it allows no time for planning. The problem may be such as to yield, in a given state of knowledge and scientific techniques, only to the unanalyzed judgment based on long experience. A decision may not be isolable as such, and so provide no occasion for the intervention of planning. The execution of a decision inevitably shapes it in some degree; there is no fixed policy content for which administration is only a neutral form: In short, from the nature of the case, planning cannot apply everywhere in the policy process.
>
> (2) *The unanticipated.* However careful a forecast we make of the consequences of our actions, some are always unanticipated, and sometimes these are of more importance than those that entered our calculations. Moreover, much of what is to be anticipated depends not on our own actions, but on those of others, which are obviously more difficult to assess. And there are always the acts of God, Whose ways are unfathomable.
>
> (3) *The uncontrolled.* Planning presupposes control, i.e., it assumes that what is planned comes within the range of possible decisions. But the area of what can be decided on is always only some part—usually a small part—of the entire range of possible outcomes. Thus Puerto Rican economic planning, however great the controls to which it subjects its own economy, must recognize that it is subject in turn to trends and patterns in the American economy, which lie beyond the range of Puerto Rican planning.
>
> (4) *The inappropriate.* Planning itself may lead to the imposition of limits to its own activities. The ends aimed at by the planner may be of a kind which preclude planning for their attainment. A society which repudiates totalitarianism will recognize a zone of privacy, for instance, which the government planner is not to invade.

More generally, there are goods which cease to be such when they are the outcome of planned effort: the florist who undertakes to send flowers in your name every year on a given date, in planning to preclude forgetfulness, destroys the meaning of remembering. But though a planner must recognize such limits, he is not under obligation to yield passively to them.

Forecasting is an important activity in analysis. What is dangerous for the analyst is not so much that the occurrence of an event he considered and decided upon was unlikely, but the occurrence of one he did not even conceive of and thus whose possibility he did not take into account.

Decision-makers plan but analysts also plan for them. In this sense, planning is a device for allowing many people of varied skills to contribute to decision-making, rather than leaving it only to the great skill of a small group of leaders. By using a formal procedure many people can contribute.

Because a plan is no better than the objective toward which it is directed, the problem of determining these objectives is just as much a part of the planning process as the final preparation of the plan. The planning problem is thus a problem in analysis. The problem needs to be formulated, the available resources determined, alternative course of action delineated and compared so that, finally, a preferred course or pattern of action may be designated by the decision-makers. Because the uncertainty is so great, however, and the alternatives so many, one cannot isolate this preferred course of action by the standards that one would ordinarily like to use in a policy analysis of this sort; it must be done by methods that are largely judgmental and only partly analytic, picking on a few key criteria for emphasis.

Another reason that analysts must be involved in planning is that the amount of detail needed to spell out a plan or program can be elaborate and voluminous, although it may involve fairly straightforward and relatively unsophisticated computation. There are, however, so many things to be considered and the procedure is so time-consuming that it is likely to tax the decision-maker's capacity to consider everything; hence, he needs assistance. Also, because a detailed program requires detailed estimates of costs and resources used, analysts are required.

Methods of Planning

If it were not for uncertainty, planning would be a routine matter. A large amount of data might have to be processed, but computers could probably handle that.

Consider educational planning, for example. The planner works with, or has under his control, at least partially, (or the man he works for has[4]) teachers, teaching methods, buildings, administration, curricula, etc. Somehow the planner has to devise a scheme that will control these elements and achieve

[4] This is true to the extent that he has authorized funds.

certain desired goals and values in face of uncertainties about such things as the general economy, people's leisure activities, their feelings about integration, property taxes, bussing, the labor market, and other factors, many of which can change much more rapidly than the items that are under control.

One additional reason for the difficulty in planning is that goals and values are involved and it is difficult to get an agreement on goals. Fortunately, as we remarked earlier, there may be more agreement among people and nations on action and planning for such action than on the philosophical basis that may be put forward by each as the justification or taking that action.

Kaplan (1973) has proposed a set of planning principles that may help with some of these problems.

> As criteria, principles are formal or procedural. They do not determine the ends of governmental action in terms of which a planner can formulate goals and objectives. Rather, they set limits to such ends, and to the programs by which ends may be realized. They function in planning as necessary conditions which are to be satisfied by other criteria, and so allow us to reject plans without providing a sufficient basis for accepting others. . .
>
> The following tentative set of principles in planning is far from exhaustive, but may provide a point of departure.
>
> 1. *The Principle of Impartiality*
>
> There is no prior specification of persons or groups who are to benefit or suffer from governmental policy. There is no place for love and hate in policy, i.e. discriminatory treatment of some sectors of the population only because they are the people that they are, without regard to acts or traits that others might conceivably share. Examples of violations of this principle are the Nazi policies against the Jews, or colonial exploitation of a native population on behalf of the master-class.
>
> 2. *The Principle of Individuality*
>
> Values are finally to be assessed as having their locus in the individual. No groups or abstract aggregates, like the State, provide proper substantives for value adjectives–save indirectly, and with ultimate reference to individuals. A "strong State" or "wealthy State" is not a value save insofar as it implies, sooner or later, individuals who enjoy security or a high standard of living. The State is made for man, not man for the State.
>
> 3. *The Maximin Principle*
>
> Improvements in a value distribution consist in cutting off the bottom of the distribution, not extending the top. The achievement of a policy or program is appraised by its minima, not its peaks. We assess a technology, from the standpoint of social planning, by the price of shoes rather than the achievement of a Sputnik. Equivalently, the principle dictates that those with least of a particular value should have the first priority for more of it.
>
> 4. *The Distributive Principle*
>
> The more people that have a good thing, the better. This is the principle that declares against elite formation, part of that aspect of democratic thought that gives weight to sheer number. It implies the use of the method of summation in assessing the values for a set of individuals. However attractive a particular configuration to a philosopher of aristocracy of Plato's stamp, we can never have too many enjoying a particular good.
>
> 5. *The Principle of Continuity*
>
> Changes in patterns and practices are of no value for their own sake, and are subject to established procedures of change. No merit attaches to a break with tradition merely because it is a break. However revolutionary the changes made, their value lies in the substance of the changes, not in the fact of their having been made.

It is this principle, in effect, that distinguishes immature rebelliousness from the achievement of mature independence.

6. *The Principle of Autonomy*

Government is to do for people only what people cannot do for themselves. This is another basic component of democratic theory, repudiating paternalism, dictatorships of whatever benevolence, and the like. It is not necessarily a distrust of government but rather a faith in the governed.

7. *The Principle of Urgency*

"If not now, when?" The rate of progress toward social goals is to be maximized. The presumption is in favor of dealing with present needs; postponement on behalf of future goods requires explicit justification.

The formulations just given do not pretend to exactness, of course. Moreover, just as other criteria, Kaplan's principles are contextual. In some situations they will have less force than others, and circumstances may even rise when they should be set aside.

Planning should be adapted to the degree of confidence that can be assigned to whatever forecast of the future we have in mind. The higher the confidence level, the more explicit the plans can be.

Planning and Budgeting

Planning has many important implications for budgetary decisions and for the use of existing resources and thus for day-to-day operations. Therefore, it must be tied to budgeting and management. A great many schemes, formal and informal, have been designed to do this. Program budgeting or PPBS (Planning–Programming–Budgeting System) as adopted by the Federal Government in 1965, is one of these.

Program budgeting, ideally executed (something that may never be achieved) was supposed to require each agency to provide

(1) A set of program options, presented in a format that emphasized the goals these programs were designed to achieve

(2) An analytic process to discover and design alternative programs, estimate their costs and effectiveness, rank them on various criteria, and supply arguments pro and con

(3) A data information system to tell the policy-makers how their programs are getting along and to provide material for analysis

The new format for the presentation of the agency's budget request was supposed to indicate, for each alternative over an extended time period, what the money would be spent to accomplish . . . say, recreation, job-training, fire-prevention, and so on—as opposed to how it would be spent—for things like rent, transportation, printing, and paint. The need for choice would thus force the decision-makers to consider explicitly whether or not they were selecting the best policy and directing their resources to best use.

Thus, a program budgeting system, particularly PPBS, was intended to be a framework in which agency-wide policy analysis was to be performed, forcing the organization to define its objectives clearly and to project the full costs of obtaining those objectives. It was also meant to integrate the results of the analysis, which became its programs, with conventual budgets, serving as a means of communicating to individuals inside and outside the organization what the organization was trying to do and how it was going about it.

The important characteristics–emphasis on objectives, on a number of options, and on an extended time horizon–forced, and in the view of some were actually designed to force, a dependence on analysis. This system was in marked contrast to then conventional United States government budgeting which, until the introduction of the new system, tended to present a single plan, no analysis, a short time horizon, and emphasis on what was to be bought rather than what was to be done.

Deciding first on the program and then on the budget rather than first on the budget and then on what can be done is a useful idea. But merely to know, say, that one is planning to spend $100 million to improve junior high school reading skills does not in itself provide information for the Office of Education as to whether the results of such activity are going to be worth the expenditure or how the money should be allocated for salary, wages, travel, transportation, and equipment for its employees. Here is where policy analysis comes in. Without it, the planner can, at most, have good intentions. The decision-maker in the Office of Education who seeks to spend $100 million say for retarded children, is using $100 million that, through reduced taxes, could be used for private expenditure, or for other educational programs. Analysis can help him compare what the $100 million would buy in each alternative. Still higher decision-makers can then use further analysis to consider what that sum would buy if spent on other programs, say defense, mental health, or scientific research.

In 1969 President Johnson was still enthusiastic about PPB, crediting the system with substantially improving the basis for decision-making within the executive branch. This was reflected in his January budget message.

> The introduction to PPB has provided an impetus toward increased use of formal analysis in the decisionmaking process. The development and consideration of alternatives has been stepped up, both in the programming stage and at the budget decision stage. The emphasis on cost effectiveness analysis as part of the analytical effort has drawn attention to ways of achieving given objectives at least cost, or attaining maximum analysis, which had been previously practiced chiefly in the military agencies and the water resources field, is now underway on various programs in most major agencies of Government.
>
> As experience has been gained, the various elements of the PPB approach and the annual budget process gradually are being more effectively interrelated, so that the analytical results of PPB are playing a greater role in decisionmaking for the annual budget.

This was far too rosy a picture; by this time the requirement for specification of goals, probably the most important aspect of the whole idea, had been essentially abandoned and by mid-1971 the whole idea was dead in the United States as a government-wide activity at the federal level.

There are numerous factors that contributed to "the demise of Federal PPB" (Schick, 1973). According to Schick, among other causes, "PPB failed because it did not penetrate the vital routines of putting together and justifying a budget." The practices and traditions of budgeting were firmly established and not enough leadership, support, and resources were invested on the behalf of PPB. Good analysts and data were also in short supply and those available did not always give sufficient consideration to budgetary traditions and institutional loyalties.

For an agency to really want to cooperate with a new budgetary system, it must anticipate some gains. For some of the older established agencies, it must have seemed clear that PPB would retard it in achieving its "true" objectives–to fulfill the public interest, as that interest was expressed by Congress when it established their program many years ago. Goals that were originally desirable have not always remained so, and the agencies set up to achieve them have become subject to criticism.

The U.S. Bureau of Reclamation is an example. It was established to encourage the settlement of the West through irrigation. For the last several decades, however, nongovernment economists have evaluated the economic benefits of the Bureau's investments at far less than their economic costs. According to the economists, clearing and draining marginal land in the Southeast would have been far more economically desirable than the investment in western high-desert lands. When pressed to justify their investment in what appeared to be uneconomic projects, the Bureau has replied that its assigned mission is not economic farming but the settlement of the West. One charge is that Bureau policies have handicapped southern agriculture, especially cotton growing, moving it to California and Texas, causing unemployment among Negro workers, sharecroppers, and farmers, and leading to their movement into urban ghettos.

Program budgeting, as implemented in the Federal PPB systems, may have had some defects too. PPB seems to be designed for a world of separable goals rather than one of interdependent goals[5]–a characteristic of today's world. It works best, or only, when any given program or means affects one rather than several goals. Also, PPB was seen primarily as a tool of the executives and was administered in that way; the Legislatures, who control the allocation of funds, consequently took little interest in it.

Is Planning Worthwhile?

Planning is certainly desirable for the individual. I think all of us would agree that it is helpful and can be done successfully in our daily lives. As the entity for which we plan increases in size, however, our confidence in the value of planning

[5] This was pointed out by R. E. Bickner.

falls off rapidly. One possible reason is that, for the individual, his plans have almost no effect on the environment of other people except for his family and business; for a nation, in contrast, planning can have great effect on the lives of many people, and consciously or unconsciously they may do things that thwart it. Aaron Wildavsky (1971, 1973) makes a good case that planning at the national level is an attempt to control the future consequences of present actions is a failure. He remarks (1971)

> . . . the record of planning has hardly been brilliant. For all we know, the few apparent successes (if there are any) constitute no more than random occurrences. Despite the absence of evidence on behalf of its positive accomplishments, planning has retained its status as a universal nostrum. Hardly a day goes by in some part of the world without a call for more planning as a solution to whatever problems ail the society in question. Doubts as to the efficacy of national economic planning are occasionally voiced, casually discussed, and rarely answered. Advocates of plans and planning, naturally enough, do not spend their time demonstrating that it has been successful. Rather they explain why planning is wonderful despite the fact, as it happens, things have not worked out that way. Planning is defended not in terms of results but as a valuable process. It is not so much where you go that counts but how you did not get there. Thus planners talk about how much they learned while going through the exercise, how others benefited from the discipline of considering goals and resources, and how much more rational everyone feels at the end. . .

If planning on a large scale is a failure, must it be a failure or can we by better methods of analysis and forecasting attain much better results?

References

Abert, J. G., Defining the policy-making function in government: An organizational and management approach. *Policy Sciences,* 1974, **5**, (3), 245–255.

Ament, R. H., Comparison of Delphi forecasting studies in 1964 and 1969. *Futures,* March 1970, **II**, (1), 35–44.

Campbell, R. M., Methodological study of the utilization of experts in business forecasting. Doctoral dissertation, University of California, Los Angeles, September 1966.

Dalkey, N., An elementary cross-impact model. *Technological forecasting and social change,* 1972, **3**, 341–351.

Dror, Y., The planning process: A facet design. *International review of administrative sciences.* 1963, **29**, (1), 44–58.

Gordon, T. J., The current methods of futures research. Menlo Park, The Institute for the Future, P-11, August 1961.

Kaplan, Abraham, On the strategy of social planning. *Policy sciences.* March 1973, **4**, (1), 41–62.

Martino, J. P., Technological forecasting for decision-making. American Elsevier, New York, 1972.

Rittel, H. W. J. & Webber, M. H., Dilemmas in a general theory of planning. *Policy sciences,* June 1973, **4**, (2), 155–169.

Schlesinger, J. R., *Organization structures and planning,* The Rand Corporation, Santa Monica, California, P-3316, February 1966.

Schick, Allen, A death in the bureaucracy: The demise of Federal PPB. *Public administration review.* March/April 1973, pp. 146–156.

Wildavsky, Aaron, If planning is everything, maybe it's nothing. *Policy sciences,* June 1973, **4**, (2), 127–153.

Wildavsky, Aaron, Does planning work. *The public interest,* Summer 1971, **24**, pp. 95–104.

Chapter 17

ACCEPTANCE AND IMPLEMENTATION

When a decision-maker (or his agent) who has a decision to make commissions the study of a public issue, he hopes for more than an academic exercise. He would like to discover, with (or without) help from the analysis, a course of action that he can endorse and for which, if he does not have full authority, he can win acceptance with the confidence that the decision, once implemented will not be vitiated. That is, he seeks a policy or action that will not be so modified by his fellow decision-makers or by the implementing bureaucracy, or constrained by the courts, or repudiated by the public, or otherwise so changed that it can no longer accomplish the objectives he had in mind. This implies, therefore, that the ultimate goal of policy analysis is not just to help a policy-maker discover what might best be done in some ideal or abstract environment but to help in a practical sense, by taking into account the problems of acceptance and implementation associated with the real context.

Widespread attainment of that goal is a long way off. To keep it in mind and to strive for it in principle is one thing; to do very much in practice is quite another matter. The present state of the theory of behavior in bureaucratic organizations and in the political milieu leaves much to be desired. We simply do not know how to examine alternative acceptance and implementation strategies systematically or exhaustively. Political and organizational factors are difficult to bring into analysis except through judgment and intuition. Moreover, sophisticated political insight is not only rare but what there is is even more rarely available to those doing the studies. Also, excessive concern for what is politically feasible tends to stifle anything that does not fit the status quo.

In addition, in many instances, it would be folly for policy analysts to spend much of their time on matters of acceptance and implementation. These are cases for which it is clear from the start that any of the policies that could bring about significant improvements are "politically" infeasible in the foreseeable future. But it is vitally important that these currently impractical alternatives be examined; otherwise the policy-makers, the politicians, and the public will not know the real cost implications, that is to say, the opportunities foregone, of their present politically dominated policy. This information is one of the important gains that can come from policy analysis; and for this purpose, consideration of how to win acceptance and secure implementation may not be very relevant.

Nevertheless, often a lot can be done. At a very minimum, for instance, the analyst can conduct his investigation in a way that will not so alienate the client's staff that anything he proposes will be viewed as unacceptable. Some objections to a proposed alternative can always be discovered and rebuttals prepared. Checks on political feasibility can often be made. In some circumstances it may even be possible to do such things as devise incentive schemes for the implementing bureaucracies.

There are thus three stages associated with policy analysis. First, discovery, attempting to find an alternative that is satisfactory and best among those that are feasible; second, acceptance, getting the findings accepted and incorporated into a policy or decision; third, implementation, seeing that the policy or decision adopted is implemented without being changed so much that it is no longer satisfactory. There are problems with the first stage. How to help in overcoming these problems is the primary responsibility of the analyst and has been the subject of most of what we have said up to this point. But there are also difficulties with the last two and an analysis can help with these too. What can be done and how to do it is the subject of this chapter.

Here I am reminded of the first item in that remarkable selection of readings, compiled as part of their investigation of Planning–Programming–Budgeting by the Senate Subcommittee on National Security and International Operations (1968) under the title *Specialists and Generalists.* "The Mice in Council" from Aesop's *Fables* illustrates how crucially important it is that analysts consider more than the design of a solution.

> A certain Cat that lived in a large countryhouse was so vigilant and active, that the Mice, finding their numbers grievously thinned, held a council, with closed doors, to consider what they had best do. Many plans had been started and dismissed, when a young Mouse, rising and catching the eye of the president, said that he had a proposal to make, that he was sure must meet with the approval of all. "If," said he, "the Cat wore around her neck a bell, every step she took would make it tinkle; then, ever forewarned of her approach, we should have time to reach our holes. By this simple means we should live in safety, and defy her power." The speaker resumed his seat with a complacent air, and a murmur of applause arose from the audience. An old grey Mouse, with a merry twinkle in his eye, now got up, and said that the plan of the last speaker was an admirable one; but he feared it had one drawback. He had not told them who should put the bell around the Cat's neck.

Acceptance

The client, the decision-maker, or the policy-maker, and their staffs to whom the analyst brings his study have backgrounds that are likely to be far different from that of the usual analyst. It includes the following.

1. *Experience.* They "know" from past work the characteristics of decisions and policies that have succeeded or failed and many of the reasons why.
2. *Responsibility.* Even the lowest bureaucrat believes that through his deci-

sions and actions, he has some influence on the future welfare of his organization, of the government of which he is a part, and possibly of the world.

3. *Different perspectives.* Individuals may be open-minded or biased by their organization, have their views warped by a hard-driving superior, or just be hide-bound conservative. Moreover, the man who sits on the other side of the desk may often (although it is becoming far less frequent) lack the academic and scientific background common to most analysts that leads an analyst to present the results of his study in the precise terminology (and the imprecise jargon!) common to his particular discipline, modified somewhat, of course, by his effort to make his meaning intelligible to a lay audience.

Presentation

When, and if, an analyst presents[1] his findings, they should not come as a surprise to the client and the staffs. For if he has been sensible, he will have consulted with them during the course of the study and given them a good idea of what to expect. Nevertheless, the audience accepts or rejects his ideas, at least initially, depending on how conveniently the information he is presenting fits their personal framework. The operating staff members, for example, tend immediately to foresee almost insurmountable difficulties in getting the indicated actions started (and there is a good chance they are right unless changes are made, particularly when they have not been consulted frequently during the study). When the organization that accepts and uses the analysis to determine policy also has the responsibility for implementing it, winning acceptance helps with implementation.

Presentation is important. Everyone knows this but not all analysts seem to realize its significance. This goes for both oral and written efforts. Each member of an audience has a certain tolerance for things he does not understand, but when that tolerance is exceeded he is likely to quite listening or reading. A study presented by a professional—a mathematician, economist, behavioral scientist, or what have you—with other professionals in mind can be exceedingly difficult for a layman to follow. A determined reader is willing to do a certain amount of digging, but most readers are not very determined and they are likely to give up before they understand. The worst that can be done, however, is to have the results of a study presented by someone who does not understand what is behind them and does not know the subject.

Few decision-makers can absorb all the information prepared for their attention nor can they remember all that they may have once absorbed. But what is considered and retained is the key to whatever help is provided. This, in turn, depends on the way the analysis is presented and on the decision-makers' confidence in the man and organization from which it comes; it also depends on the kind of response given to their questions. There should, therefore, be a

[1] Unfortunately, he may not get the chance for a face-to-face presentation. Reports are often made by mail—and, when received, are merely filed.

general admonition to the analyst to keep his client relations good. If they are not, he may never win acceptance for his analysis and may not even be listened to attentively. In his presentation he must answer questions carefully and avoid expressions of condescension or a style of communication too didactic for his client's taste.

The analyst must carry the open-mindedness that he presumably exploited during his study into his presentation. He should concede (and give the appearance of conceding even if he is convinced he is right) that he may be wrong. The audience will realize it even if the analyst does not. A study is very seldom the last word on a subject—only the latest word. He should expect to have some of his ideas rejected and others picked up and used out of context. (He has to remember it was largely he who defined the context.) Even in the rare circumstances in which he is recognized as being obviously right, it will take time to get everything accepted, if ever, and the action taken, if any, may differ widely from what he would have chosen.[2]

As noted by Kahn (1960), it is always easier for an organization to "sell" a policy or decision (say to the budgeting authority above it in the hierarchy that must allocate the resources to carry it out) if it can be presented as necessary to complete an existing program, or as being in support of such a program, or (a last resort) as a necessary modification. Unless there is a crisis or the idea is exceptionally glamorous, it is risky to present the recommendation as the expansion of an existing program and even more risky to let it look like the initiation of a new program.

Working with the Client

Work with a civil client, say, the executive branch of a city government, can be extremely frustrating for an analyst accustomed to the military or to the industrial world. There seem to be far too many interested parties and too much divided authority. There is the client, of course, but the others range from the career civil service and the municipal unions who see analysis as seeking ways to reduce the work force, to upper level officials who see it as a threat, not so much to jobs, but to tasks that had better be left to them. The state and the federal government have their interests there too, as well as neighborhoods, chambers of commerce, and action groups of every sort.

Based on his experience while head of the New York City Rand Institute, Peter Szanton (1972) has made a number of cogent observations that seem to be of wider applicability.

> *Agency Support.* First, there must be support for the analytic effort from potential decision-makers within the government. The support need not be whole-

---

[2] He has to fight if the proposed action is to, say, implement recommendation A when to do so will be counter productive without also implementing B and C.

> hearted or entirely free of skepticism, but it must exist. Where it is absent, the analyst will find his access to data restricted, and the participation of city staffs in the study foreclosed. Perhaps most important, he will be unable alone to identify the kinds of analyses which might affect policy, or the schedule on which they must be produced in order to affect it.

It may not seem important, but personal acceptance usually comes before the acceptance of ideas. For the latter, you have to establish credibility with the client's people by demonstrating that you have something to offer. One way to do this is to select problems initially that the operators see as relevant and that offer good prospects for a quick payoff. This may mean setting the most important problems aside temporarily.

The road to successful acceptance and later implementation seems to be to start, when possible, with the client's mundane problems–those that bother the staff every day. In other words, in working with a police department for the first time, one should not attempt to investigate recidivicism, organized crime, corruption in the department, or riot control but reduction in booking time, quicker ways to get a search warrant, or maintenance on police cars.

Again quoting Szanton (1972)

> *A Period of Grace.* Second, if the analyst is attacking problems of any scope or complexity, he will need a period of grace. The tougher the issue, the longer the period required. Grace may be provided by personal relationships with the client, by the reputation of analysts, by fiat from a Mayor's Office, or by extraordinarily good luck in solving quickly some other problem of concern to the agency. However derived, it will normally be necessary in order to sustain the effort during those opening months in which the presence of the analysts represents demands on the governmental agency–demands on time, attention, money, assistance in the acquisition and interpretation of data, and patience–which normally will far outweigh the assistance the analysts can then provide.
>
> *A Manageable Problem.* Third, the problems attacked analytically should be within the power of the client to deal with. The better and more conscientious the analyst, the more likely he is to be concerned with larger and more fundamental problems–and the more likely he is then to identify difficulties or to propose conclusions which the client, at least over the short run, has no substantial power to effect. A good example is the Institute's work for New York's Economic Development Administration (EDA), research which moved rapidly to an examination of the demographic forces at work in New York and to the economics of the labor market in the metropolitan area. These were issues of clearly fundamental importance to the mission of EDA, but they posed a challenge far beyond the capability of that agency, and indeed beyond that of the City government, to cope with. They consequently generated little interest and have had, as yet, no impact on policy.

"Educating" the Client

Whereas analysts may produce analysis easily and effectively, public officials may not really know how to use it and, even when they do, may feel it too difficult to teach the staff to use it.

C. Jackson Grayson, Jr., in speaking (1973) of his experience as Chairman of the Price Commission[3] pointed out a number of reasons why he, well trained in analysis himself, could identify no incident of a conscious, explicit use of a single management science tool in his activities.

> The third reason that I did not use management science tools explicitly is that educating others in the organization who are not familiar with the tools, and who would resist using them if they were, is just too difficult a task. Management scientists typically regard this problem as outside the scope of their jobs—at most, they remark on the need to educate more people and to change organizations so they become more scientific. Or, if they do recognize the problem, they grossly underestimate the time and organizational effort needed to "educate and change." I suggest that management scientists do two things:
>
> 1. They should build into their models some explicit recognition of the financial and emotional cost of change of this kind and make explicit allowance for the drag of change resistance. I am quite aware that some change techniques are being used: sensitivity training, Esalen-type devices, management by objectives, quantitative analysis courses for managers, and so on. I have used them myself, and I know that they help. But the magnitude of time and energy required to install them is not generally appreciated—certainly not by management scientists—and their impact is highly overrated.
> 2. They should get themselves some education and direct experience in the power, politics, and change-resistance factors in the real world of management so they can better incorporate the imperfect human variables in their work.

One of the purposes in first working with a client's lesser problems is to provide some of this education.

Effective transfer of information between an analytic group and a public agency occurs best when both groups work together. The agency thus acquires a vested interest in the outcome. Participation by the client is important throughout the analysis from problem definition to implementation, but particularly at both ends. Szanton (1972) emphasizes this as follows.

> And finally, where analytic work is intended to affect policy, officials and their staffs should be directly involved in the research itself. The Institute's work has been most successful where agency people themselves participated fully in the work—not merely providing data, but checking assumptions, challenging hypotheses, proposing alternative lines of inquiry, noting barriers to implementation. It is not always possible to find officials or staff members willing to involve themselves so deeply. But where this kind of joint effort does evolve, its benefits are enormous. The quality and relevance of the research are improved, officials develop stakes of their own in the success and utility of the studies; and, in the internal processes of the agencies, more analytic approaches to other issues develop.

The analyst should consider the social context in which a problem is to be solved as part of the problem, not something that can always be taken care of later by his client. The analytic process must embrace or at least take cognizance of the bureaucratic and organizational aspects of the context in which the problem is embedded. The relationships between the researcher, his client, his

---

[3] In Phase II of President Richard M. Nixon's Economic Stabilization Program.

client's organization, and the problem can be exceedingly complex. Conflicts are almost always present and some, possibly all, of the parties involved may be constrained to accept alternatives not wholly what they would like. The old cliché applies to what should be sought: Not the best possible result but the best result possible.

In brief, the argument for paying considerable attention to procedures for winning acceptance from the client and the staff as opposed to sole dependence on the logic of the analysis for that purpose is that, if the findings of a policy analysis fail to influence the relevant decision-makers, then that analysis, as a piece of policy-oriented research, did not accomplish its purpose no matter how good it might seem in the abstract or to other analysts.

Immediate acceptance of all aspects of an analysis, however, is rarely to be expected; acceptance of ideas takes time. To be listened to and carefully considered is a practical goal even though not a completely satisfactory one.

The Implementation Process

Any analysis designed to assist a decision-maker in initiating or changing policy, thereby affecting individuals and institutions, should consider whether anything can be done through analysis to help with implementation. Implementation is not a process that follows automatically once a policy has been formulated, but is often a difficult and frustrating affair, full of pitfalls that analysis can sometime help to uncover.

Many policy analysts seem to assume that once a decision has been made or a policy formulated by a public body, the results that will follow once the decision or policy has been announced and ordered implemented will be those desired by the policy-makers. Citizens generally make this same assumption. For example, when Congress passes a policy bill that appropriates sufficient money, and the executive branch approves and organizes a program that hires people to carry out that policy, we assume the effects of the policy will be what was originally intended. But unfortunately it does not always happen that way—prices go up when they are controlled and so on.

Problems with policy implementation, in fact, are widespread. In the United States there is difficulty in implementing certain kinds of programs, particularly those of a new, non-incremental nature. For a thorough illustration of these difficulties see Pressman and Wildavsky (1973). The War on Poverty programs offer other examples (Smith, 1973).

Difficulties with implementation led the Planning–Programming–Budgeting System, PPBS, introduced with great energy and presidential support, to be judged at least a partial failure five years later. The organizations that had to implement it were not prepared; the lower echelons resisted and lacked the talent to carry it out. Congress, although not directly involved, also might have lost influence relative to the executive were implementation to have worked as planned—and that did not help.

Interest groups, opposition parties, and affected individuals and organizations often attempt to force changes in policy during the process of implementation. The bureaucracy that has the responsibility for implementation may also cause difficulties.

The process of implementing a policy that forces complex organizational structures to change is not likely to be a smooth process. Archibald (1970) states it well.

> . . . . even if the policy alternative recommended by the analyst is accepted by top decision-makers, the program that comes out of the organization may have little resemblance to the alternative originally envisaged by the analyst and the top decision-makers. I am not merely saying that an alternative when implemented may not produce the consequences expected. Rather I am saying that the policy alternative actually executed is quite likely to have undergone radical revisions at the hands of (the) operating levels. And since a policy is no better than its implementation, this suggests that analysts need to pay attention to the feasibility of a policy alternative at operating levels as well as to its acceptability at the top decision-making level.

One reason bureaucracies may sometimes find it difficult to implement new policies resulting from policy studies is that such policies tend to be less incremental in nature than policies arrived at by the usual processes.[4] Policies that differ little from past policy in the area of concern are likely to run into less resistance.

There are costs associated with implementation, and with the failure to implement when an attempt fails, that must be explicitly considered in the analysis. One obvious one is the cost of the organizational changes, if any, that must be made if a new program *B* is to supplant on-going program *A*. More serious may be the costs that are generated if a decision-maker tries to implement program *B* and fails. We must recognize that some programs have a better chance than others of being successfully installed and implemented. Hence, the alternative suggested to the decision-makers should not necessarily be the one with the greatest potential for an excess of benefits over costs *unless* the probabilities of successful initiation and implementation have been estimated and the expected costs that would be incurred by failure have been taken into account. Peterson and Seo (1972) discuss this problem in the context of public administration planning in developing countries.

A model of the policy implementation process has been developed by Thomas B. Smith (1973). In his view

> Governmental policies have been defined as deliberate action by a government to establish new transaction patterns or institutions or to change established patterns within old institutions. *Policy formulated by a government, then, serves as a tension-generating force in society.* While policies are implemented, tensions, strains, and conflicts are experienced by those who are implementing the policy and by those affected by the policy. . . .

---

[4] Smith (1973) suggests that difficulties here are particularly significant in Third World nations where in the press for development they "have not been able to afford the luxury of incremental policymaking."

The tensions generated by the policy implementation process may cause protests, even physical action, and may require the establishment of new institutions in order to realize the policy goals. They may also cause changes in other institutions.

The policy implementation process is modeled by Smith (1973, pp. 202–205) in terms of four components.

1. The idealized policy, that is, the idealized patterns of interaction that those who have defined the policy are attempting to induce
2. The target group, defined as those who are required to adopt new patterns of interaction by the policy. They are the people most directly affected by the policy and who must change to meet its demands
3. The implementing organization, usually a unit of the government bureaucracy, responsible for implementation of the policy
4. The environmental factors, those elements in the environment that influence or are influenced by the policy implementation. The general public and the various special interest groups are here. Again quoting Smith (1973)

> . . . . Environmental factors can be thought of as sort of a constraining corridor through which the implementation of policy must be forced. For differing kinds of policy, differing cultural, social, political, and economic conditions may prevail. For example, in policies related to local self-government in Third World nations, the basic cultural and social life-styles at the village level may be an environmental constraint of great magnitude.

These components are diagrammed below (Smith; 1973, Fig. 3, p. 203). (The transactions are Smith's term for the responses to the tensions, stresses, and strains within and between the components of the policy implementation context.)

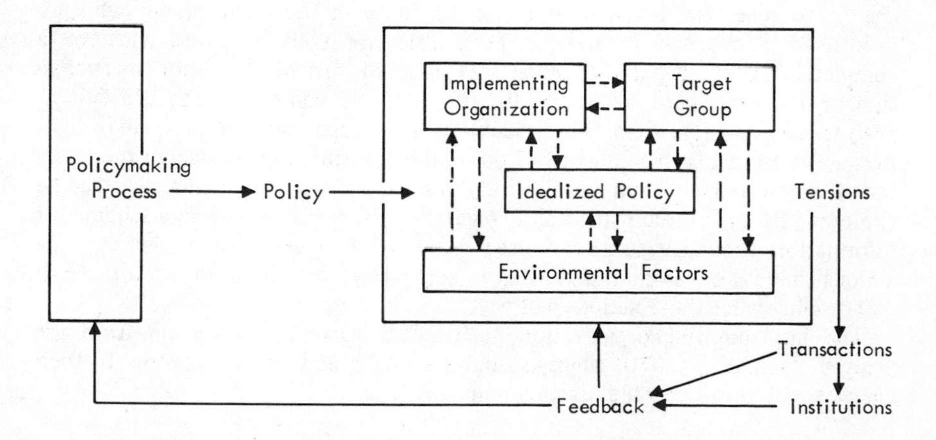

FIG. 17-1. A model of the policy implementation process.

In the past, analysts have concentrated largely on the first component, in an attempt to help determine the ideal policy, with minor attention to the second and almost zero to the last two. Some of the strategic bombing studies done at the time missiles were just coming into consideration as a strategic force were almost a caricature of this practice. Little consideration was given to the possibility that the enemy would act like an intelligent opponent and none was given to the changes that the replacement of aircraft by missiles would force on the implementing organization, the Air Force, or how the populace at large might react to consequences of a missile base in their neighborhood.

The practice has not been unique to military analyses. We have, for example (New York City Rand Institute, 1972), seen municipal outpatient clinics set up by a city for ministering preventive medicine in areas where the target population had no understanding of their purpose. In these low income areas, residents go to a medical clinic when they are sick and need help never when they are well. When they arrived at the new clinics they were typically turned away and referred to a hospital for treatment they needed. This resulted in confusion as to the value, even the relevance, of the services provided.

One change in the practice of policy analysis that would help to increase the chances that the implementation process takes place as the policy-makers would like it would be for the analyst to give some thought, sometimes considerable thought, to the problems that might accompany implementation. Another might be a deliberate effort to include more and more social scientists on policy analysis teams, particularly people knowledgeable in the processes of planned change, the diffusion of innovation, and organizational behavior. Other changes in practice might include the attempt to design alternative implementation strategies as well as the alternative policies themselves. Ways to provide incentives to the implementing bureaucracies might be an aspect of such strategies.

Consideration can also be given not only to the reaction of the target group—the people directly affected by a proposed policy and who may have to change to meet its demands—but also to those in the environment who are influenced in one way or another. Difficulties are likely to follow whenever a completely packaged external solution is imposed, arrived at without participation of those affected by its conditions. To avoid some of these difficulties, analysts can do more than merely identify the groups that stand to gain or lose. Because presumably the gains from the policy considerably outweigh the losses (or otherwise it would not have been adopted), ways to make this clear can be explored. In some circumstances, it might even be possible to plan a scheme for information exchange so that those who must live with, or implement, the actions that follow from analysis can in some way participate in the process of determining what those actions will be.

The implementing organization itself needs particular attention from the analyst. Archibald (1970) suggests analysts might add three concerns to their analysis with respect to the organizations involved.

1. A concern with the internal (to the organization) acceptability of decisions as well as with their potential effectiveness and political feasibility. This

would imply an investigation of the internal consequences of alternatives as well as of their external consequences

2. A concern with the internal structure and needs of the organizations involved, that is, a concern with their stability, viability, and cohesion, and with the creativity of the environment within them

3. A concern with the relationship between the personal needs and objectives of individuals in the organizations and organizational needs and objectives

There are other aspects of organizations that affect the proper implementation of a particular policy to which the analysis associated with that policy can contribute little. Archibald (1970).

> .... The nature of these problems can be illustrated by the story of a farmer who was one day approached by a young and enthusiastic extension agent, an agricultural whiz kid. The agent was pushing a new technique that was supposed to raise the farmer's output by ten percent. The farmer's response was: "I'm only farming half as well as I know how to right now." For whatever reason, the farmer wasn't interested in the performance gap.
>
> There are many areas within government agencies and other organizations where we all think they could do much better than they are now doing. Without any analysis most of us could point to examples of too little coordination between programs, redundant programs or procedures, use of clearly inefficient techniques, little use of available information, under-utilized personnel, etc. We all think we know about such things, especially in someone else's organization, but few of us have prescriptions for dealing with them effectively.

The above problems are people problems. System analysts have usually referred to them as "political" problems and possibly one should try to handle them as political problems. Systems analysts, until recently, made no attempt to deal with these bureaucratic and organizational problems, mainly grumbling about them, hoping that rationality would win out in the long run.

The fourth component that affects implementation, the environment, is replete with true political problems and observations on these are presented in the next chapter.

The client, particularly when he is the individual or group that makes the decision, may sometimes have to be considered as part of the problem.

Clients and Decision-makers

The client who commissions a study may not be the decision-maker, and the decision-maker who profits from the analysis may decide only indirectly about issues analyzed. That is, the analysis may investigate a choice of alternatives; as a partial consequence, the decision-maker may decide how to allocate his budget, or how much to ask the legislature to appropriate, or even whether or not to recommend new laws or tax changes or further analysis. The point is that decision-makers in taking action influenced by analysis must almost always consider things not considered in the analysis and that very likely could not have been considered there.

Moreover, the decision-makers to whom an analysis is presented almost always have information that may not be known to the analyst. In military studies, for example, the question occasionally arises whether under some very special desperate circumstances men in the ranks can be counted on to carry out a maneuver that means certain death. The officers who have to act on the analysis may be the best sources of information on this type of question. The analyst may not really know the value system of the men who are going to use his analysis for decision. Hence, he must be prepared to have any recommendations for action he has made modified by considerations that the policy-makers alone can apply or by differences in judgment.

There is, in turn, much that a decision-maker cannot do alone. He is dependent on others for much basic information and for help in the formulation of the specific detailed alternatives from which choices can be made. Also, others control many channels for policy implementation.

The question of how far an analyst should go beyond stating the conclusions of his analysis by making recommendations as to what should be done to a decisionmaker is a moot one. Whether or not some particular course of action should be followed usually depends on factors quite beyond those that can be quantified by the analyst. All such factors that the analyst considers significant may have been treated qualitatively and the analyst may have brought the best judgment available to bear on them. Nevertheless, there are these other considerations and almost invariably the decision-maker himself will have information or a last-minute perception of the situation of which the analyst and the experts he consults cannot be aware. Decision-makers may, in fact, resent being presented with a set of conclusions and recommendations for action. Some regard it as illogical. Major General Glenn A. Kent, USAF, speaking at an operations research meeting (1969) contends

> But of all our sins, the one that will finally hurt the profession the worst is the blurring of "analysis" on the one hand and "position-taking" on the other. By making no distinction between them we are compromising a very useful tool. Analysts should be recruited because they have the talent to expose problems—to collapse seemingly complicated problems to much simpler terms. They are to be graded on how elegantly and simply they were able to "model" some problem. One recruits such people from those who have been exposed to the theories of economics, logic and mathematics. You look for people who have exhibited an uncommon amount of ability to think and explain.
>
> It is probably permissible, although somewhat dangerous, for analysts to be allowed to take positions. But, I submit, these are two quite different functions and it is time we recognized they are different and acted accordingly. The position that is to be taken invariably hinges on far more factors than the analyst can include in his model. The analysis (the study itself) should not, repeat not, contain conclusions and recommendations. In the vernacular of "Completed Staff Work" the analysis is a subset of "Factors Bearing on the Problem." But the operating word is "Subset" as distinct from the "Whole Set."
>
> If the analyst feels compelled to announce his position to the world, then he should do so in a covering letter—not within the confines of the document that is allegedly an analysis. All of this is to get analysts into a frame of mind that promotes at least a modicum of objectiveness and relieves the reader of the unwanted burden of separating analysis from position-taking. If the analyst makes, as part of his

> analysis, the recommendation that we should buy A rather than B, then he is apt to go back through the analysis and turn every single input to the "buy A" position. He does this because he has been burned in the past by some reviewer who made the deathless charge that "The Conclusions and Recommendations were not supported in the body of the Analysis." If the unfortunate analyst had not fouled in the first place by including a "position," he would not have been open to the charge at all.

Because the analyst knows that his study will be subject to scrutiny, interpretation, and possibly further analysis, he should make his subjective judgments known.[5] Trust is essential because the client has to take the analysis and recommendation of any study team in large part on faith. The client usually cannot repeat the study, will very seldom have the time to review it in meticulous detail, and will be influenced by it depending on his belief about how the analyst reached his conclusions. He cannot hope to master the variety of specialized skills that frequently go into a complicated analysis. At best he can acquire enough background to identify really incompetent or patently biased work. But faith in the analyst's purely technical and scientific competence is not sufficient; what is required is a similar confidence in his subjective judgment. Trust requires disclosure; the client must know either how the analyst has disposed of the subjective elements in the study or whether he has merely accepted and used the client's judgments. If he does this latter uncritically, the analyst is not using the full potentialities of analysis.

In particular, the analyst must guard against the tendency to provide the client with answers that he would most like to hear. Kent in the talk quoted above (1969) notes

> Too many analyses seem constructed in the context that the purpose is to convince friendlies that the position that they already hold is a good position. The cons to the position are carefully avoided lest we shake the abiding faith in our own righteousness: "Don't put in the cons or the Chief may not buy our position;" "Don't bring up so-and-so, it will only open up Pandora's Box." But to be a persuasive advocate he (the Chief) needs to know all about the cons and the counters to these cons. Skeptics have a very nasty habit and a diabolical instinct to focus on the poorer aspects of any proposition you may make, as distinct from the better aspects. One point of the above is that even in the dirty business of advocacy it pays to be honest.

Kent's position is that a decision-maker should want the facts; he can supply his own distortions.

There is little reason to suspect that giving policy-making and management responsibility to analysts would improve the quality of decisions. Once he becomes an administrator an analyst becomes awfully busy with a lot of things besides analysis. He has to do a lot of listening to the opinions of other people without checking them through, so that many of his judgments become judgments as to the competence of other judges, and judgments of the extent to which he was able to understand what they might be trying to communicate to him. Computers, data banks, and sophisticated mathematical techniques are not

---

[5] But, of course, not for these reasons alone.

likely to have a profound effect on how decision-makers make their decisions. Their staff members may interact with computer terminals and work through policy studies and as a consequence more facts and better informed people may be available to them, so the quality of their decisions can improve. But the process still will involve compromises, hunches, and generalizations from incomplete evidence.

Let me draw an analogy between the decision-maker using a study team for advice and a medical doctor using a clinical laboratory. Suppose, for example, our doctor is trying to decide whether to send his patient to a surgeon to have his stomach resected or to treat him medically for a gastric ulcer. The doctor is influenced by

1. The technical findings of the laboratory crews: Like the decision-maker, he may or may not be able to carry out these investigations himself, but it would not be economic for him to do so. He depends, therefore, on laboratory reports, some of which will be on cold slips of paper without comment or nuance—numbers alone. Others from the laboratory might write paragraphs or talk to the doctor or bring x-ray plates to discuss with him.
2. Observations or analyses the doctor makes himself: Some of these are in the form of written notes; those he can't write out he holds in his head.
3. Impressions of the risks and possibilities of success with various treatments: Some of these impressions are from his experience, others from medical reports.

Finally, like the decision-maker, the doctor must make a judgment based on whatever facts or analyses he has. This judgment is the ultimate synthesis the doctor makes of the numerical tests, the written out but relatively diffuse notes, the unrecorded conversations with technicians, and his own introspection. It is not a mere calculation, but is made on intuitive grounds. Sometimes a factor is overriding, but on the whole he just does not know. He could do more analysis, sometimes even risk the patient's life in order to guard it—call for a liver puncture or other dangerous procedures—but this inquiry can never be complete. He also has the problem of deciding to what extent he should share the decision-making with his patient-target. His judgment, like that of every decision-maker, must be made with uncertainties in mind.

Many of these judgments are delicate, that is, judgments that are hard to make with certainty, or perhaps cannot be made with anything resembling near-certainty at the current state of the information available, and on which the life of the patient or the success of the enterprise depends. The decision-maker, having to delegate authority as he does, will think it particularly important to economize on the extent to which he delegates the authority to make delicate judgments. It is not good to have a lot of people making interacting delicate judgments. For instance, one may have situations in which five or six people report to the one above in chain, each allowing for the optimism or pessimism of the next so that a terribly distorted picture can arise. A single person can avoid this concatenation of error.

The danger is not that analysis will give the wrong advice; it may be wrong, just as a laboratory report may be incorrect, but without analysis or laboratory reports, the chances of doing the wrong thing are much greater. And for some questions analysis is essential; without calculation there is no way to estimate how a change in the law will affect the number on welfare, or how arms control may affect security. Analysis offers an alternative to "muddling through"; to waiting until one can see the problem clearly and then attempting to meet the situation. Delay can be hazardous; in the world today there could be a crisis or other situation that could not be handled in this way. This is not to say that every aspect of such problems can be quantified or that analysis is without limitations, but only that it is not sensible to formulate policy without careful consideration of whatever relevant numbers can be discovered.

It is easy, unfortunately, to exaggerate the degree of assistance that analysis can offer a policy-maker. In almost every case, it can help him to understand the relevant alternatives and the key interactions by providing an estimate of the costs, risks, and possible payoffs associated with each course of action. In so doing, the analysis may sharpen his intuition; it will certainly broaden his basis for judgment. This can almost always help the decision-maker make a better decision than he would otherwise make, but the inherent limitations of analysis mean that a study can almost never demonstrate, beyond all reasonable doubt, that a particular course of action is best.

To summarize the main implications of this chapter, because the purpose of policy analysis is to provide decision-makers with the relevant information needed to arrive at a well-informed decision, the analysis must seek to do more than merely discover an ideal course of action in some theoretical sense. If there are going to be troubles in winning acceptance for and implementing good policy, and these can be anticipated, they should be pointed out.

References

Archibald, K. A. Three views of the expert's role in policy making: Systems analysis, incrementalism, and the clinical approach. *Policy sciences,* 1970, **1**, 73–86.

Grayson, C. Jackson, Jr., Management science and business practice. *Harvard business review,* July–August 1973, 41–48.

Kahn, Herman, *On thermonuclear war.* Princeton Univ. Press, Princeton, New Jersey, 1960.

Kent, Major General Glenn A., USAF, *The role of analysis in decisionmaking.* Keystone Speech before the 24th Meeting of the Military Operations Research Society, 1969.

New York City–Rand Institute, The, *Operations research,* May–June 1972, **20**, (3), 509.

Peterson, R. E. & Seo, K. K., Public administration planning in developing countries: A Bayesian decision theory approach. *Policy sciences,* September 1972, **3**, 371–378.

Pressman, Jeffrey L. & Wildavsky, Aaron B., *Implementation.* Univ. of California Press, Berkeley, 1973.

Senate, U.S., 90th Congress, 2nd Session, *Specialists and generalists.* Subcommittee on National Security and International Operations, 1968.

Smith, Thomas B., The policy implementation process. *Policy sciences.* June 1973, **4**, (2), 197–209.

Szanton, Peter, Analysis and urban government: Experience of the New York City–Rand Institute. *Policy sciences,* June 1972, **3**, (2), 153–161.

Chapter 18

POLITICS, ETHICS, AND THE ANALYTIC ENVIRONMENT

Public policy is made in a political environment. It affects, to a greater or less degree, what problems are analyzed, who does it, how it is done, what decisions are made as a consequence, and how those decisions are implemented. Policy analysis must thus cope with politics.

This chapter considers three topics: political feasibility, some ethical problems for the analyst, and a few suggestions of what can be done, other than making analysis better, to make it more effective.

Analysis and Politics

Government did not develop in ways that make it easy for policy-makers to use analysis to help guide social and political decisions. Public officials face a dilemma; they may fully understand the advantages of using an analytic approach for allocating resources and otherwise guiding their actions, but at the same time they must face the realities of politics. Even if they are not politicians themselves, they depend on politicians. If a politician does not produce or appear to produce what his constituents demand, his time in office may be short. Decisions thus tend to depend more on politics than on analysis.

Where the aim of an analyst is to work out, as closely as one can hope to attain, the optimum use of resources for the well-being of the public, the politician may operate to minimize his loss of power or to maximize the votes he gets at the next election. In politics the appearance of effort, no matter how ineffective, may be far more remunerative for these purposes to an official than the implementation of a complete new program that is likely to be costly and may stir up an unpleasant reaction from those who pay the cost. Consequently, resources tend to be applied thinly over a wide array of programs. It is the symbol of concern that is important.

Thus, for example, a politician is likely to try to diffuse issues to get a broader and more flexible program so that diverse groups will find it possible to support him. An analyst, on the other hand, tries to make the issues sharper so that people will understand clearly what the choices are.

James R. Schlesinger (1970) points out two pitfalls that analysis seeks to avoid: (1) foot-in-the-door approaches that lead to large (and often unexpected) expenditures later, and (2) allocation decisions based on input rather than on output. In contrast, the exploitation of the foot-in-the-door is part of the art of politics and in that world the inputs may attract more attention than outputs. Furthermore, as (Schlesinger, 1970) put it:

> Political leaders are keenly aware that in formulating policy you must start from where you are. They also recognize the countless constraints imposed by and variables involved in working within the system. These considerations are not always evident to the professional observer who enters into the system—in a sense from the outside—with a coherent set of objectives, and who proceeds to develop a rational program for achieving those objectives, while ignoring all those political considerations that he regards as irrelevant or adventitious. These include local interests, personalities, habits, prejudgments, rivalries, and the like. He is likely to wonder why the society fails to get on as rapidly as he thinks it should toward achieving his very reasonable goals. But he has left out all of the elements which are involved in creating a political consensus and which so regularly constrain political decision.

Even the bureaucracy, which by its nature—continuity of office and function, accountability and so on—is far more receptive to analysis than most other political bodies—parties, community organizations, ethnic groups, and mobs, for instance—can be an impediment to the successful employment of analysis. An agency tends to become concerned primarily with its own survival, protecting its area of operations and maximizing its budgets. Instead of welcoming a penetrating analysis designed to uncover waste and inefficiency in its programs, such analysis is viewed as a threat to its existence and to be countered by all means possible.

It is almost never the case for a decision to be made purely on the basis of an analysis. For one thing, conflicting policy advice abounds, even that based on analysis. Consider some issue, say the Supersonic Transport. Some analyses say pollution from it will harm the environment irrepairably; others contend that, on the contrary, it will not measurably affect the environment and will be an economic blessing. With the same basic data, analysts here and elsewhere reach conclusions with different policy implications because they differ in their assumptions about values or risks, or because the data are inherently uncertain. Also, in making a public policy decision, the policymaker will want to consider the impact of his decision on the political environment: Will it restrict what he can do in the future and, if nothing else, how will it affect his chances for reelection? What will other decision-makers tend to do? In these situations, the policy-maker tends to work through a political model of the situation, a line of analysis that tells him what to do in terms of the bureaucratic and domestic political situations.

The political and policy analytical approaches to decision are basicly different. Dror (1971) attempts to make this clear with a simple metaphor.

> .... Policy sciences theory states that one should not leave the problem of crossing a river until the river is reached; rather, one should survey the territory in advance, identify rivers flowing through it, decide whether it is at all necessary to cross the river—and if so, where to cross it; and then prepare in advance the materials for

> building a bridge and design a logistic network so that the bridge is ready when the river is reached. But practical politics will often say that not only should the problem of crossing the river be left until the river is reached, but one should leave it until one is already up to the throat in water—when imminent danger will result in complete agreement that the river must be crossed at once and in the recruitment of energy to do whatever is necessary to get through the river alive. By leaving consideration of the problem of crossing the river up to the last moment, so a partly correct claim may go, one is at least sure of reaching the river rather than being bogged down through premature controversy on what type of bridge to build and on whether one shouldn't stop moving so as to avoid the problem of the river altogether.

As Dror points out, policy analysts must not ignore the essential features of politics. If they do, the policy-makers may more often feel that to get things done, it may be better not to attempt the use of analysis.

It has sometimes been said that one purpose in introducing analysis into the public policy area is to eliminate politics from decisionmaking. If so, the effort is not only doomed to failure but basically undesirable. To try to eliminate politics and bargaining from public decisionmaking in a nation that guards its freedom from authority is not only impractical but unappealing. Impractical because no policy decision can be based on analysis alone, divorced from considerations of political values; unappealing because it would imply a surrender of authority to analysts.

Analysts should not be upset because analysis becomes involved in political debate, for the analysis will (or at least it can) raise the quality of that debate. Compromises are the heart of political actions, but there are sensible compromises and others less so. Analysis can help determine what is a good compromise.

When an analyst remarks that the decision toward which his analysis was directed was eventually made on political grounds, it is seldom with approbation. But the political basis for a decision can be because the policymaker saw that what was needed was a decision that broadly served the public interest rather than one made on the basis of what might have been relatively narrow technical expertise. Whereas this might lead to the action that would bring him the most votes at the next election, the public interest may have been uppermost. People who favor a decision tend to think of it as objective; those who do not tend to think of it as controlled by politics. Usually it lies in between with some consideration paid to the petty as well as to the strong political pressures and considerations that are always party to public decision-making. Analysts sometimes look down on politics as a profession but this is without justification. The competition is tougher there than in analysis and it is harder to rise to high levels.

The public policy-making system, in which a large number of different kinds of elements interact in a great many ways is extremely complicated. There are a variety of paths leading to any action, some partially hidden and known only to a minority of the participants, and others open and surrounded by a great deal of noise. Proposals for change from any source must surpass many hurdles; those that result from analysis may have additional hurdles.

One such hurdle is communication. To be useful, a policy analysis must be brought to the attention of the people actually in the business of making policy. This is almost invariably a problem for analysts working outside the appropriate area of government. Even analysts within mission-oriented agencies sometimes find themselves communicating at the wrong level or with the wrong people.

Dror (1968) contends that the actions that may result from analysis should be examined in the analysis itself for political feasibility, that is, for the probability that the proposal will be acceptable to the various secondary decision-makers—the special interest groups, the public, and the bureaucrats who must translate it into action. If that probability is too low, compromises[1] must be made to increase acceptability. How to carry out that examination by anything other than judgmental means is not clear. Political feasibility depends on the power structure in which the parties affected are involved and on the ability of the policy-makers and the policy itself to recruit support. Dror argues that it is important that analysis cover all aspects of a policy, including political feasibility even if it requires the use of far less desirable means to compare alternatives and measure effectiveness. He would go so far as to measure output by means of input—for instance, by using the professional qualifications of the policy-makers to estimate the quality of their policy.[2]

Some Problems for the Analyst

It is rarely possible to carry out an analysis of any public issue in such a way that all those who hold various views of the issue involved will consider it fair and objective. Generally, this is unavoidable owing to the nature of the issues, uncertainty and differing views of values. In part, it is a defect on the part of the viewers and sometimes this may be chargeable to the analysts involved. Policy research is carried out in an environment ridden with conflict; a study, even one poorly done, can confer advantage in political argument. Loyalties are strong and analysts are not always as objective as they might be.

I do not mean that policy analysts consciously seek to slant their analysis (although this sometimes happens), but indisputably objective analysis is hard (probably impossible) to achieve in competitive situations. Openmindedness, willingness to follow evidence whereever it may lead, and readiness to reconsider conclusions when doubts arise are supposedly the marks of the scientist, including the policy scientist. But these are more ideals than marks. Policy analysts are people like anybody else. They do not like to have their ideas knocked over, or to appear stupid, or to be disloyal to their agency. Expecting to find these "marks" to the same degree in all analysts is too much. It is like talking of the marks of a Christian. There are certain noble sorts of behavior we characterize as

---

[1] Some compromises are likely to be preferable to others—those that increase acceptability without a proportionate loss in attainment of policy goals. Analysis can help find these.

[2] For an example where this may be justified see Citizens Conference on State Legislatures (1971).

Christian, but anyone who expects to find these in every person he sees on a pulpit is doomed to disappointment.

Moreover, willingness to accept another view is not categorically a virtue. A policy analyst has a duty to defend what he believes in against prejudice and even against counter evidence. Evidence is not really categorical.

Who Is the Real Client?

When the sponsor is a public official or agency, the analyst may sometime feel he should consider himself more than the agent of that particular official or agency alone. In other words, he may consider himself an agent of a wider society, of his country perhaps, or even of the world. Thus, he tends to view his role as one of influencing his sponsor to take the right path rather than merely viewing his role as that of helping the sponsor in his battles with the rest of public officialdom. This raises the interesting question of whose agent the analyst is and, in particular, whether or not he should be fully the sponsor's agent.

A committee of the New York City Council raised this question in protesting about the volume of various contracts for research and analysis let to the New York City-Rand Institute and several management consultant firms. As quoted in the *New York Times,* November 16, 1970, page 25:

> When a consultant works for government, who is the client? Is it the executive branch which engages him, or the legislative body which raises the funds to pay him, or the citizen whose stake in his work may be far more vital than a stockholder's interest in a private corporation?
>
> The consultants presently employed by the city and the agencies which retain them believe beyond any question that the client is the executive branch of government.
>
> We conclude, however, that the client is not solely the executive branch, but the legislature as well and ultimately the city's citizens. It cannot be otherwise in a free society.

One aspect the City Council objected to was the confidentiality of the reports prepared by these consultants. They wanted full disclosure to the Council—and therefore to the public at large—of all aspects of publicly financed research. Unfortunately, bureaucratic sensitivity may sometimes demand a certain degree of confidentiality, at least temporarily, if the analyst is to maintain access to real data and problems. The analyst may thus have to accept considerable restriction on publication to keep good relations with the sponsor. In the conflict environment of public policy, any criticism, even the most constructive, can be used as a weapon by the opposition and decisionmakers are much concerned about premature disclosure before their position is firm.

The sociologist Kathleen Archibald after taking a look at the literature of policy analysis, came to the conclusion that analysts talk about improving decision-making rather than about helping a client. They do indeed, in spite of pressures from their sponsors, tend to see themselves, not exclusively as agents of the organization, but of the broader polity; they are interested in the public welfare. Archibald (1970) puts it this way.

> . . . What is good for the nation or the world may not be good for the policy maker, yet for-the-greater-good arguments can scarcely be declared illegitimate by the policy maker. Thus even the policy maker must accept in principle the legitimacy of the expert not being fully his agent, although in practice, if he has commissioned a study, he will almost certainly object to 'end-running,' that is, the expert presenting his recommendations to other government agencies if they are perceived as disadvantageous to the commissioning agency.
>
> [*] Wesley W. Posvar has discussed the responsibility of experts from a somewhat different angle. His more philosophical discussion includes under responsibility the question of agentry, that is, in whose interests the expert is acting. He argues that 'the requirement to perceive and understand the interest of the state, and to act upon that interest, rests upon' expert and decisionmaker alike. This is a commitment 'to the state as an abstract repository of all the values of society' and Posvar feels that such a commitment is inescapable—perhaps morally inescapable—if one 'chooses to be a part' of the policy system. (*Strategy Expertise and National Security,* unpublished doctoral thesis Harvard University, Cambridge, Mass., 1964, pp. 263–266, 278–279.)

His responsibility to persons and organizations other than his client is one of the more difficult issues a policy analyst has to face. The executive branch of the government and the public are not equivalent and it is a nonequivalence that makes a difference.

It matters for at least two reasons. Information is an element of power. Analysts working for one branch of a government, where that branch can restrict to itself access to the work, disturb the balance of power. The second reason is that it also disturbs the balance of power between the government and the governed. Interest groups—ethnic and racial associations, labor organizations and neighborhood communities—are legitimate participants in decision-making. If these groups have no access to analytic support they will resent the advantages of the executive branch and take one of two courses, either to strike the instrument of analysis from the hand of the executive, or to provide one for themselves. The first course may appear easier.

In brief, then, an analyst must recognize he has a responsibility that may extend beyond the interests of his sponsor and act accordingly. He must press for the good, for what he believes in as the result of his analysis. But there is no hard-and-fast rule to tell him how much. One reason why policy analysis cannot be completely objective is that the analyst may have to compromise, to allow some bad to get the good accepted. Thus he is not completely objective, but an advocate.

Since policy analysts have professional responsibilities, a code of ethics is desirable. Such a code may help them handle difficult values, dilemmas, and ethical issues they are likely to have to face. The Operations Research Society of America has taken a step toward codifying the use of operations research as an applied science or engineering discipline (Caywood *et al.,* 1971). Dror (1971, page 119) suggests a code of ethics for policy scientists should include the following.

a. A policy scientist should not work for a client whose goals and values, in the opinion of the policy scientist, contradict basic values of democracy and human rights.
b. When the goals and values of a particular client contradict basic beliefs of the policy scientist, the policy scientist should resign rather than help in the realization of goals and values with which he intensely and fundamentally disagrees.
c. The purpose of policy sciences is to help in better policymaking, and not to displace legitimate policymakers and decisionmakers with policy scientists who become "gray eminences." Therefore, policy scientists shall try to preserve and increase the choice opportunities for their clientele, e.g. by always presenting a number of alternatives. In particular, a policy scientist should not hide an alternative because it contradicts his own personal values and preferences.
d. Policy scientists should explicate assumptions and should present clear value sensitivity analyses, so as further to increase the judgment opportunities for their clientele.
e. A policy scientist should refuse to prepare studies, the sole purpose of which is to provide a supporting brief to an alternative already finally decide upon for other reasons and considerations by his client.
f. Policy scientists should not work for clients who do not provide necessary access to information and opportunities for presentation of studies and their findings.
g. All forms of conflict of interest should be avoided, including utilization of information for private purposes and presentation of recommendations in respect to subject matters in which a policy scientist has a personal and private interest.

Ways to Increase the Value of Public Policy Analysis

Clearly, one way would be to improve the quality of the analysis that is done. There are also other actions that might help, although certainly not as much.

GIVE POLICY ANALYSIS A RECOGNIZED PLACE IN POLICY-MAKING

Scholars and other citizens have been analyzing government programs for centuries, usually without noticeable impact, even when their techniques were excellent and their recommendations sound. To lead to action, suggestions for improvement, whether they originate in a letter to an editor or from a well-constructed study by an established institution, must not only reach the right people but overcome their attention threshold. Even then, if the idea has not reached some critical mass, it can be neutralized by adjustments elsewhere in the bureaucracy before it can influence the system. To become a really effective part of the policymaking process, policy analysis must be established as part of that process so that issues are subjected to analysis before time becomes too short to collect the data and carry out the work with consideration of its recommendations, although, of course, not acceptance, almost automatic.

At the federal level, the introduction of the PPBS in 1965 was, if not designed to do just this, a large step in that direction. Its important characteristics—emphasis on objectives, on having a number of options, and on an extended time horizon seemed to force a dependency on analysis. Unfortunately, at the federal level it is no longer required and many agencies have abandoned it. However, many of its component parts are still very widely operational, mostly under different names, e.g., Program Analysis and Evaluation.

During its existence, PPB did much to spread policy analysis. Schick (1973, page 146) puts it this way.

> What died and what remains? The name is gone and so too are the burdensome routines that came to be the end products of budget innovation. But there survive cadres of committed and skilled analysts, a growing supply of analytic studies and data, and perhaps most significantly, a spreading consensus (among budgeters, program administrators, and congressmen) that the inherited budget practices are in need of improvement. Among those who believe that analysis always has been "the heart and soul of PPB"[2] it has become fashionable to interpret the abandonment of PPB as evidence of the new sophistication of reformers who after much sorrow have learned to distinguish the product from the package. Many analysts long have regarded the system and its bundle of techniques–in particular the program structure and the long-range plans–only as the necessary price for projecting analysis into federal decision making. When the price became too high and the technique got in the way of analysis the deadwood was cut away without impairing the analytic core. Thus they see the stripping away of PPB's mechanistic apparatus as evidence of the strength and durability of their analytic efforts.[3]
>
> [2] The term is Alice Rivlin's, though she may not agree to the interpretation to which it has been put here. See Alice Rivlin, "The Planning, Programming, and Budgeting System in the Department of Health, Education, and Welfare: Some Lessons from Experience," in U.S. Congress Joint Economic Committee, *The Analysis and Evaluation of Public Expenditures: The PPB System,* p. 915. In this article, "analysis" is meant to cover planning as well.
>
> [3] This was the view of a panel of PPB practitioners who gathered in Washington on March 2, 1972, under the auspices of the Association for Public Program Analysis to discuss "Is PPB Dead?"

Although PPB did create a demand for analysts and performed miserably when they were not available, budgeting, not analysis, was the central concern. Indeed, Wildavsky (1969) argued that there was a need to rescue policy analysis from PPB. Let me again quote Schick (1973, page 146).

> The Meaning of Failure
>
> With those who believe that the spirit–if not the essence–of PPB lives, I have no quarrel. But I believe that they err in regarding analysis as the central concern of PPB. What PPB tried to do, and the measure by which it must be judged, was to recast federal budgeting from a repetitive process for financing permanent bureaucracies into an instrument for deciding the purposes and programs of government. Analysis was to be a change agent; it would reorient budgeting by serving it. The linkup of analysis and budgeting was to close and direct, through plans, cost-effectiveness studies, and the other informational channels opened by PPB. Analysis was not valued for its own sake or structured, to operate independently of the budget process. Rather, PPB–as its initials attest–was designed with budget outcomes in mind. That is why it required an overlay of forms and routines.

PROVIDE TRAINING

Training both for analysts and for the people in the public sector who could use policy analysis would obviously help. If a department or an agency is to do analysis, a capable staff with varied background is needed, trained in how to provide information, to define criteria and objectives for programs, to design alternatives, and to develop new analytic tools. In addition, because effective analysis is not a one-sided proposition, but a dialogue that gets sharper as it goes on and on between the analysts and the policy-makers, the policy-makers themselves must have an appreciation of what analysis can do and what it cannot. This also takes some education.

The training of policy analysts is being carried out at various universities and the number of institutions where such courses are offered is increasing fairly rapidly so this aspect is being taken care of fairly well. Decision-makers, particularly at the highest and lowest levels, are still relatively uninformed, however.

SUPPORT AND MAINTAIN POLICY ANALYSIS CAPABILITIES

Some thought has to be given to where such capabilities are located and how they are to be maintained. It would not make sense, for example, to locate all within the government or to support all such capabilities by a single government agency. One would like to have the analysts free from institutional and organization biases, and able to think beyond, and maybe even advocate, what may not be politically acceptable or, if that is not possible, at least have the biases and political beliefs of one set offset by those of others. It probably does not make sense to try to locate such capabilities solely in the universities. Policy studies, as we have seen, must be problem-, not discipline-centered and that usually works best with an interdisciplinary approach with close cooperation between many skills, something very difficult to achieve in a university environment. But most importantly, it should not be a capability in a single institution, for institutional biases may then dominate. The answer is probably a mix. A number of university centers and independent research institutions variously located, but maintaining close relations with the government, industry, and with each other, plus special staffs attached to government bureaus, department, and legislative bodies at all levels, may be the answer.

It should be obvious, moreover, that analysts trained in methodology alone are not enough. The analyst, or his team, must know or be able to learn very quickly, a great deal about the subject matter of the analysis. Public policy problems are very broad in scope; they require contributions from many disciplines. Policy can be effective only if it is based on reliable knowledge about the actual state of the world. Experts and specialists of various kinds must be present or access to them must be readily available.

Various other actions that might, or might not help, to increase the value of analysis in public decisionmaking have been suggested.

A provision for competitive analyses is one. Analysts are not perfect. They err sometimes and they make different assumptions, reaching different conclusions. Except for the expense, competition is all to the good, with one minor caution. The client must be careful that the prize is awarded for the better analysis, not for the one that best upholds the client's preconceptions.

Independent review boards are another possibility. The theory is that independent reviewers with no vested interests would prevent falsification or stretching of the data to change the relation of costs to benefits. One difficulty is how to achieve true independence. Hovey (1968) feels that there might be another. For example, if the analysis followed all the guidelines and assumptions set up by officially constituted authorities, a review board might find it hard to reject the analyses, even if they believed these guidelines and assumptions should also be rejected.

The establishment of central guidelines for the conduct of analysis might be a third possibility. These guidelines could cover such matters as the appropriate interest rate to use in various circumstances and assumptions about such matters as the economy and future population growth. But such guideline would be a mistake. Conditions change rapidly and the guidelines would soon get out of date and therefore would be completely inappropriate. One virtue of analysis is that it offers a way to challenge current assumptions by showing their implications.

References

Archibald, K. A., Alternative orientations to social science utilization. *Social science information,* April 1970a.

Caywood, T. E., Berger, H. M., Engel, J. H., Magee, J. F., Miser, H. J., and Thrall, R. M. Guidelines for the practice of operations research. *Operations research,* 1971, **19**, 1123.

Citizens Conference on State Legislatures, *State legislatures: An evaluation of their effectivness.* Praeger, New York, 1971.

Dror, Y., *Designs of policy sciences.* American Elsevier, New York 1971.

———, *Public policymaking reexamined.* Chandler, San Francisco, California, 1968.

Hovey, Harold A., *The planning-programming-budgeting approach to government decision-making.* Praeger, New York, 1968.

Schick, Allen, A death in the bureaucracy: The demise of Federal PPB. In *Public administration review.* March/April 1973, pp. 146–156.

Schlesinger, James R., The uses and abuses of analysis, pp. 121–136 and Testimony, December 10, 1969, pp. 463–496 in the Senate report cited below.

Senate, U.S., 91st Congress, 2nd Session, Planning Programming Budgeting, Subcommittee on National Security and International Operations, Senator Henry M. Jackson, Chairman for the Committee on Government Operations, 1970.

Wildavsky, Aaron, Rescuing policy analysis from PPBS. In Joint Economic Committee, U.S. Congress, *The Analysis and Evaluation of Public Expenditures: The PPB System,* Government Printing Office, 1969, pp. 835–852.

Chapter 19

REDUCING FIRE ENGINE RESPONSE TIME: A CASE STUDY IN MODELING

A mathematical model is a relatively simple concept although, in a given situation, a usable one may be difficult or impossible to produce. Those without training, however, find the process mysterious. This chapter is a case study in mathematical modeling, a chronological narrative that relates how one analyst built and used a mathematical model to help the New York City Fire Department reduce the delays experienced between the time an alarm is received at a borough Communications Office and the time units are dispatched to that alarm.

Background

The problem of reducing fire engine response time is part of a larger problem. Some background information is in order.

The 1971 Annual Report of the New York City Rand Institute (where the work was done) describes the situation as follows.

> The New York City Fire Department faces complex and challenging planning problems. In one small neighborhood in the City there are more than 5000 alarms per square mile per year; in an adjoining neighborhood, there are less than 500. On an average day, between 6 a.m. and 7 a.m. there will be only 2 alarms in the Borough of The Bronx; between 9 pm. and 10 p.m. there will be 15, except on a hot summer evening when there may well be 30. In some areas of the City, the chance that an alarm reported by street alarm box will be false is as high as 50 percent; in other areas, the chance is less than 10 percent. A large fire can require 15 pieces of equipment at the scene, depleting fire apparatus from some regions and requiring another 6 pieces of equipment to move about the City to maintain balanced coverage. On a busy evening during the summer, 11 of the 27 ladder companies in the Bronx may be out of their houses responding to or working at alarms 50 percent of the time. If the traditional 3 engines and 2 ladders attempt to respond to every alarm during peak alarm periods, the supply of equipment will quickly become exhausted for the busy period.
>
> With such wide variation in the demand for service over location and time, the problem of matching the amount of men and equipment sent to an alarm to the size of the task represented by that alarm has become one of enormous complexity and great urgency. Solving the problem has required new tools and new management approaches.

After a fire of any significance occurs, it is eventually detected and reported by alarm box or telephone. The report is processed; after some delay, usually short, fire-fighting units respond to the alarm; some time later these units arrive; finally, the fire is extinguished and the units return to the station.

The primary objective of a fire department is to minimize loss of life and property. To do so, some attention is paid to fire prevention, but given the occurrence of a fire, the major effort is to reduce the time between occurrence and extinguishment. There are numerous ways to reduce the latter time. Possibilities are early detection through automatic devices; more fire-fighting units; or use of improved technology to extinguish fires. Another possibility, the one considered here, is to reduce the time required to process the alarm and to decide which unit to dispatch to the reported fire.

Dispatching consists of two basic sets of operations: receiving, interpreting, and identifying alarms; and allocating and dispatching fire-fighting units to respond. The first set involves a great deal of noise and activity—counting telegraphic signals, answering telephones, etc.—and is so hectic in busy periods that it readily dominates one's view of the system. Indeed, prior improvement plans—which include new equipment, additional receiving dispatchers, and computer aid for routing tasks—focused heavily on this first set of operations.

At the start of his study, Arthur Swersey of the New York City Rand Institute spent many weeks—night after night—in the Brooklyn Communications Office, the busiest in the City, observing both the unusual and the routine operations of the office. He found that the system had a definite bottleneck when busy, not at the point where alarms were received, but at the point where response decisions are made and carried out. Moreover, his data showed clearly that the time delay in this bottleneck increased significantly with the number of active incidents, not only because there were more incoming alarms to be processed, but also because service of each additional alarm slowed down as there were more active incidents that had to be considered. These and other results were incorporated into a simulation model which accurately depicts the system's performance over a wide range of conditions. The way in which Swersey built and used his model is described in the next section.

The description that follows was prepared as a lecture. The model is presented more formally in Swersey (1972) with considerably more mathematical detail. Because this book is written primarily for nonprofessionals, I have chosen to retain the lecture presentation as being more easily followed.

FIRE DEPARTMENT COMMUNICATIONS: REDUCING DISPATCHING DELAYS

Arthur J. Swersey

I am going to describe the process of building a mathematical model to represent an operating system by outlining a study of the Fire Department's Communication System. I hope that through this description you will get some feeling for the procedure one follows in defining and modeling a problem.

First, what do we mean by a model? A model is a representation of the system. It is an abstraction or simplification. A mathematical model is one which represents the relevant properties of a system in mathematical terms. For example, y is equal to $ax + b$, the equation for a straight line. This mathematical model, although simple, is quite important. In operations research there is a whole class of linear models, called linear programming models, for which all the equations that make up the model are in this form. For example, if x is the number of units produced in a production process, then y might be the production cost.

In general, in a model, you try to express the effectiveness of the system as a function of the input. Effectiveness may be measured by, for example, cost or profit or, if the system is cars arriving at the George Washington Bridge, perhaps the waiting time.

How do we derive a solution from a model? How do we find that alternative which maximizes effectiveness? Let me show you how by an example: the Christmas Tree Problem. Anyone who sells Christmas trees on any corner, whether he knows it or not, is faced with the Christmas Tree Problem. I doubt whether any people actually use this type of analysis, but it is applicable, as you will see. The Christmas Tree Problem is the following: associated with every tree that the man sells is the cost c and the selling price s. In addition, there is some demand d for his trees. Now the demand is something he does not know. He has some history from previous years and may be able to estimate the demand, but he does not know exactly what the demand will be. The problem is: How many Christmas trees should he buy? Well, what happens if he buys too many? The day after Christmas he is left with Christmas trees he can't sell. Now, what happens if he buys too few trees? Then, the day before Christmas, or worse, the week before Christmas, he will have run out of trees. Because he can't go back up to Canada to get more, he loses the profit he could have made if he had more trees. Clearly the problem is to determine exactly how many trees he should buy. Ideally, on the last day when people are buying trees, he should sell his last tree. Now, we can write down what his profit is, as a function of how many trees he sells. Let us call the number of trees he buys n, so the profit he makes is equal to

$$n(s - c), \text{ if the demand } (d) \text{ is greater than } n$$

What does this mean? If the demand is greater than the number of trees, then he sells n trees, and he wishes he had more. Now, on the other hand, his profit might be

$$ds - nc, \text{ if the demand } (d) \text{ is less than } n$$

In this case, he is left with $n - d$ trees. Now, as I said, he does not know what his demand is going to be, but he can construct what we call a probability distribution of demand. That is, from past experience he may be able to plot a histogram that shows the number of trees versus the probability of there being a demand for that number (Fig. 19-1a).

Given this information he can write down an expression for his expected profit. It is just the probability that there is a demand for one tree times the profit that he will make if there is a demand for one tree, plus the probability that the demand is two times the profit for two, etc., and we continue in this

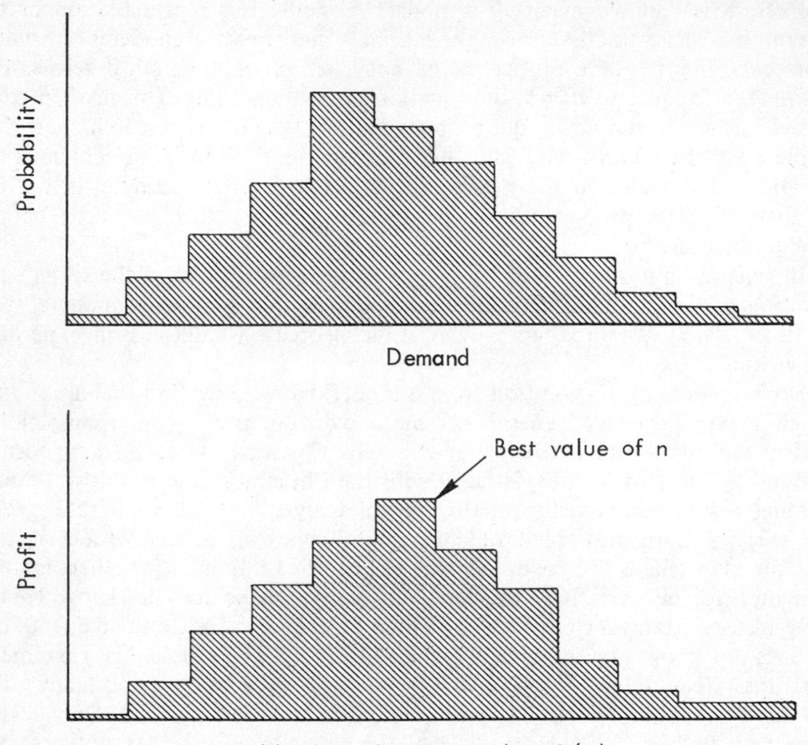

FIG. 19-1(a). Distribution of demand for trees. (b) Relationship between profit and number of trees purchased.

way for every possible value of demand. He wants to maximize his expected profit where the expected profit equation might look something like Fig. 19-1b. How does he find the value of n that maximizes his expected profit? By using analytical techniques (calculus) we can explicitly solve that equation, and find out what this n is. Another way is the iterative method; you pick some n, and then pick another value of n, the one right next to it. If the profit increases, you are going in the right direction. You continue trying new values of n until you get to the maximum value. It is a trial-and-error technique.

In this simple example, the Christmas Tree Problem, you have seen two ways to derive solutions from models. One is analytical and one is a trial-and-error method. Sometimes models are very difficult to solve; we can write down the mathematical equations but we cannot solve them. So we use a third technique called simulation. By simulation we mean imitating the way the system behaves.

For example, let us consider the experiment of flipping a coin. What if we wanted to simulate this experiment to generate a sequence of heads or tails? We do not have to flip the coin. We can spin a roulette wheel and if it stops in the first half, call that a head, and if it stops in the second half, call that a tail. The result is exactly equivalent to flipping a coin. Now let us go further. Divide the wheel into ten equal parts and spin it. Each time we write down the number that results. In this way we generate a sequence of random numbers. We can associate a head with each number between one and five and a tail with each number between six and ten. In simulation, you associate a probability with each event and using random numbers generate a sequence of events that looks very much like a real sequence. We can do this over and over again, and on a computer we can do it very rapidly and easily.

Now I would like to talk about the Communications Office study done for the New York Fire Department. What is the Communications Office? At some point in time a fire occurs, and at some later time an alarm is received at the Communications Office. For each borough there is one Communications Office. If we pulled a fire alarm box to report a fire in Times Square it would ring in a Communications Office in Manhattan. At some time later, fire units would respond to the alarm. We are concerned about the period between receipt of the alarm and the dispatch of equipment.

What is the problem? The dispatchers who work in the Communications Office said the problem was that more men were needed, and the equipment wasn't too good. What did the management of the Fire Department say? Management viewed the problem differently, which is not surprising. They did not think they needed more men but more modern equipment in the short run, and in the longer run a computerized system. What was our view? We became interested in the problem when we realized that a communications office is vulnerable–that if one office breaks down and another one must take over, it might not be able to do so. So we first looked at the problem as vulnerability to overload or disruption.

Notice that there were three distinct views of what the real problem was. We did not yet know what the problem was. We viewed vulnerability as one small problem. The dispatchers' union wrote letters, called newspapers, went on the radio and said they needed more men. But they did not have a model that could relate the output–how long calls have to wait–to how many men are working. So they could say: 10 years ago we had so many men; the number of alarms has increased by so much, and now we have even fewer men. But this really does not prove anything. In the meantime, management was designing some new equipment, and a computer manufacturer was designing a computerized system. This was the situation when we arrived.

There are five or six men who work in the Brooklyn Communications Office. It is very crowded, it is not laid out very well, and it is very old–the equipment was installed around 1900. When someone in Brooklyn pulls an alarm box, it "rings" on a panel and the dispatcher has to count the number of rings in order to identify the number of the alarm box. When it starts ringing, everyone turns around and counts.

On the other hand, someone might call up by telephone to report a fire. Whoever answers the phone asks what is on fire, what is the address of the fire, and writes it down. He looks up the street address in the street index file to determine the number of the nearest alarm box.

At that point, another man retrieves an "alarm assignment card" from a file. There is one such card for every street box, and it lists the engine companies and the hook and ladder companies that respond to that alarm. He then looks at a status board, where each company is represented by a plastic chip. If the chip is on the left side of the board, it means that the company is in its house and available to respond. If the chip is on the right side, it tells him where the company is, because someone has written down the incident the company is now attending. The dispatcher can see whether anyone is available or where they are. He has to decide whether he has any companies to send—because sometimes he runs out of companies—and, second, he has to decide whether this is a new alarm. (Someone else may already have reported this fire, and units are already on their way.) Even if the new box number is different, it still might be on the next corner to a previous alarm. So he must decide whether to send the alarm out or not. If he decides to send it out, it is transmitted by a telegraph key. The number of the box gets tapped out, one tap for each number, and firemen in Brooklyn firehouses listen to the signal. Each company has a list of the boxes that it responds to, and knows whether to go or not.

If some of the assigned companies are riding around, coming from another fire, the dispatcher will have to contact them on the radio and direct them to the incident.

Now the first thing we wanted was to see if a model of the system could represent what goes on. Figure 19-2a shows the sequence of operations following an alarm. There are three stages. In Figure 19-2b, alarms come in on the left side. They are either telephone alarms or street box alarms, and someone answers the phone or "counts the box." We call that the first stage. If dispatchers are busy, talking on the phone of doing something else, then the alarm waits. It forms a queue or a waiting line at the first stage. After the alarm assignment card is retrieved, it goes to the decisionmaker at the second stage. He looks at the status board and decides whether to send out the alarm and, if so, which units to send. Finally, it goes to a third stage, transmission of the alarm.

In the simulation model, we decided to ignore this third stage. In reality that stage is there, but the time it takes to send out the signal on the telegraph key is very small, and we assumed that an alarm is never delayed at stage 3. By eliminating stage 3 we simplify the model but lose very little realism. We can write a simpler computer program. This approach is basic to all model building: to develop a model that approximates reality, but is not so complicated that it is too expensive.

Given this model, we want to measure dispatching time, the time from when the alarm is received until it is transmitted to the companies. The dispatching time is the waiting time at each stage plus the service time. It takes some amount of time to count the boxes, it takes some amount of time to answer the telephone, it takes some amount of time to figure out whom to send, etc. Each of these times is distributed probabilistically—there is a certain probability that the time will be longer than ten seconds, longer than twenty seconds, and so on. The total dispatching time will also have a distribution, which depends on how

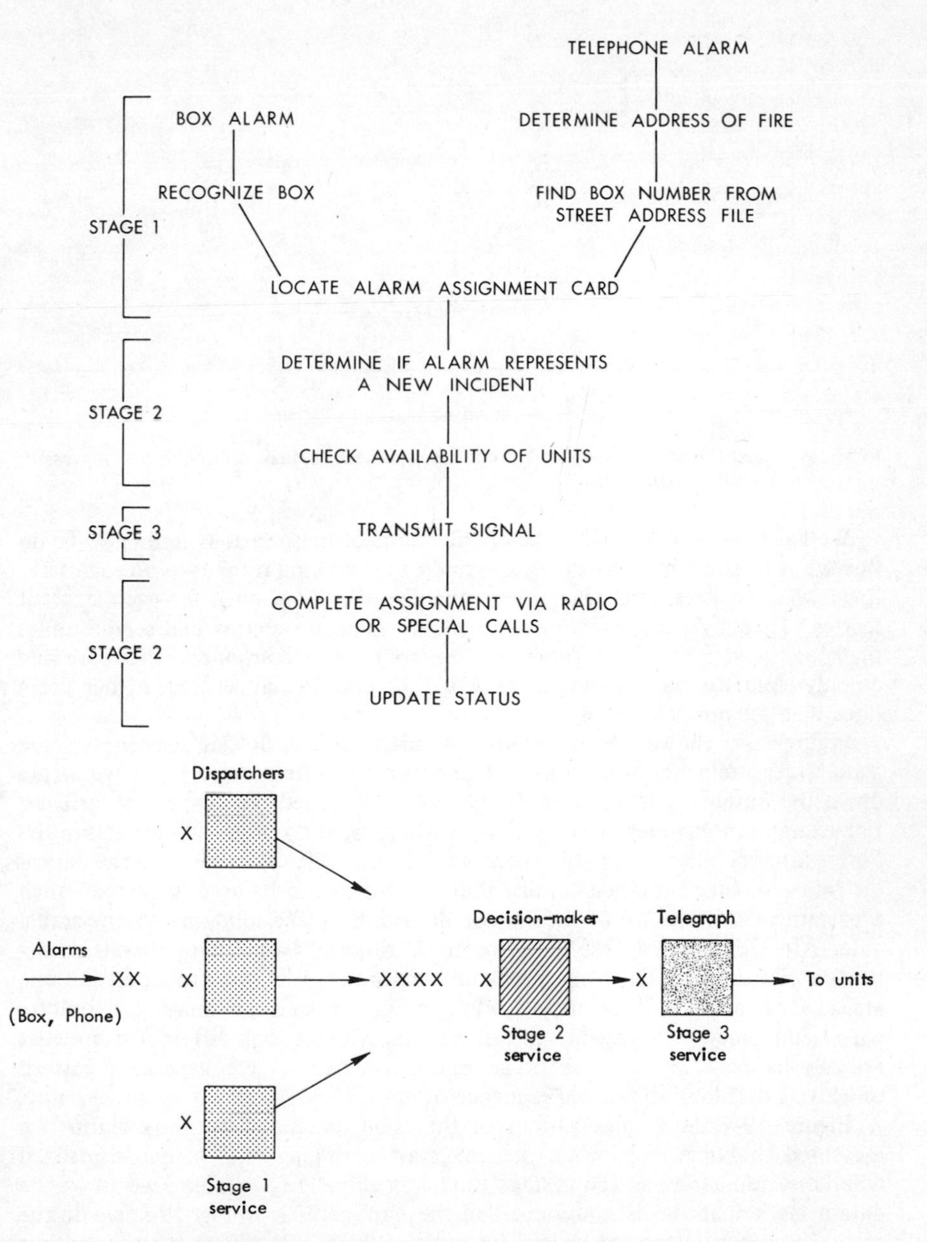

FIG. 19.2(a) Dispatching procedure. (b) A multistage queuing model.

quickly alarms are coming in, whether they are telephone or box alarms, the number of men that are answering the phones or counting boxes at the first stage, and the distributions of service times at each stage.

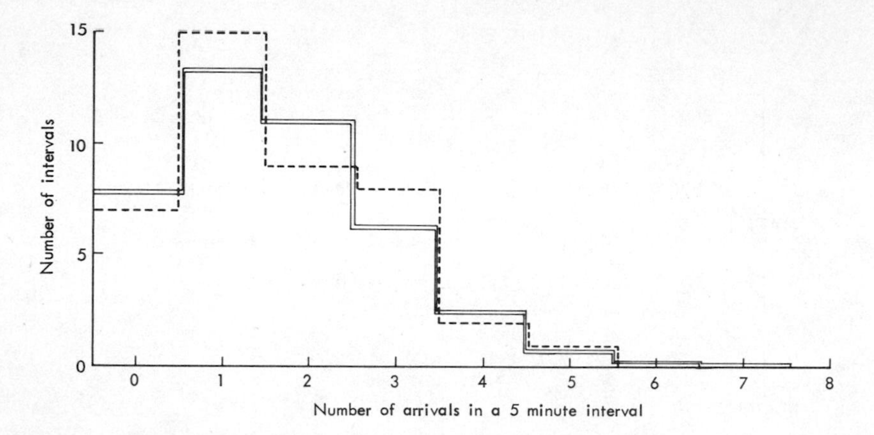

FIG. 19-3. Distribution of arrivals. Dotted line represents actual; unbroken line represents Poisson distribution.

We have to figure out what the distributions of these various times are. To do this we take some measurements and find out how long it takes to do each task. From this we determine what the probability distribution is for each stage of service. Then by using random numbers we generate alarms and service times that look just like a real sequence, but do it on a computer, where we can quickly simulate months or years of actual experience and generate higher alarm rates than are normally received.

Figure 19-3 shows the distribution of arrivals. The dotted lines show some actual measurements. We looked at each five-minute interval, and we wrote down the number of alarm arrivals. The average turned out to be 1.66 arrivals. For example, there were seven instances where we did not get any arrivals in the 5-min interval, there were 15 where we got one, etc. There is a mathematical distribution, called the Poisson distribution, which can be used to describe such a pattern of arrivals. We fit the Poisson distribution (the solid line) to the actual data. After this we can use the Poisson distribution to generate arrivals. Using this distribution, we know the probability of no arrivals, the probability of one arrival, two arrivals, three arrivals, etc., so we can spin our imaginary roulette wheel and generate a sequence of arrivals that will not look different from what actually happens. If we know the average alarm rate, we can generate a pattern of arrivals that looks like a real sequence of alarms.

Figure 19-4 shows observations of the time to recognize a box alarm. We measured the time from when the box started ringing until someone realized what the number was. The average time was about 13 sec. Here, we fit to the data a mathematical distribution called the exponential. In Fig. 19-5, we do the same thing for telephone calls. The average time was about 40 sec, with a minimum of 20 sec.

We then fit an exponential curve to the data for how long it takes to look up the street address (Fig. 19-6), and do the same thing for the time to pull the alarm assignment card (Fig. 19-7), which on the average takes about 10 sec. As

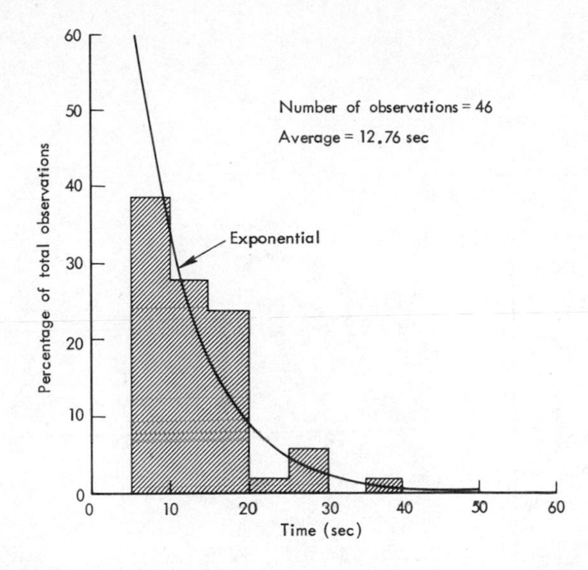

FIG. 19-4. Histogram of the time to recognize an incoming box alarm.

we determine the probabilities of the individual events, we are beginning to reproduce on our computer model what happens in the communications office.

The critical part of the model is the second stage. While observing the system in operation it appeared that the time between when the dispatcher looks at the alarm assignment card and when the signal is sent by the telegraph was usually quite short. This was because the alarm rate during the observation periods was comparatively low. We realized, however, that the service time at the second stage did not end with the transmission of the alarm. The dispatcher had to spend some additional time before he was ready to process another alarm. This time was spent in contacting units on the radio and by updating the status of those units sent to the alarm. Therefore, we divide the service time at the second stage into two parts: the time until initial dispatch of units, and the time from initial dispatch until completion of the assignment and updating of the status board.

It also became apparent that the time to dispatch units increases with the number of incidents in progress. (Actually, it goes up with the number of companies that are busy, but it is simpler to look at the number of incidents in progress than to count the number of busy companies.) Also, the dispatcher must determine whether it's a new alarm. If nothing is happening in the field, he knows each alarm is a new alarm. But if he has five or ten incidents in progress, he has to check each new incident against old ones.

The second part of the service time also increases with the number of incidents in progress. It does so because of the greater probability of having to call units on the radio. The more busy units, the higher the probability that you

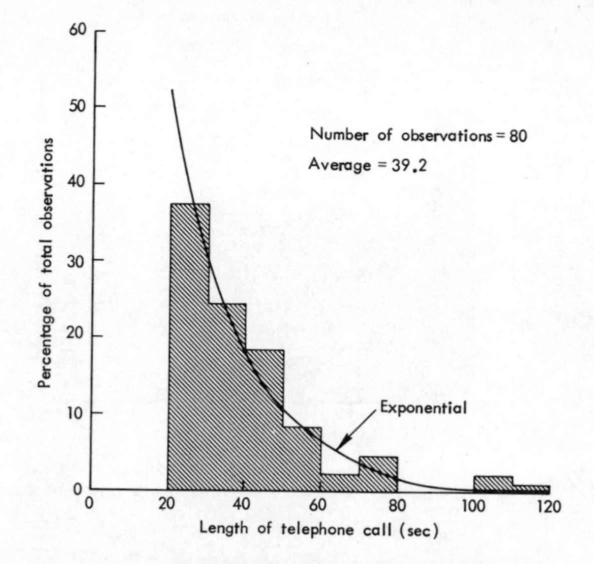

FIG. 19-5. Histogram of the length of telephone calls.

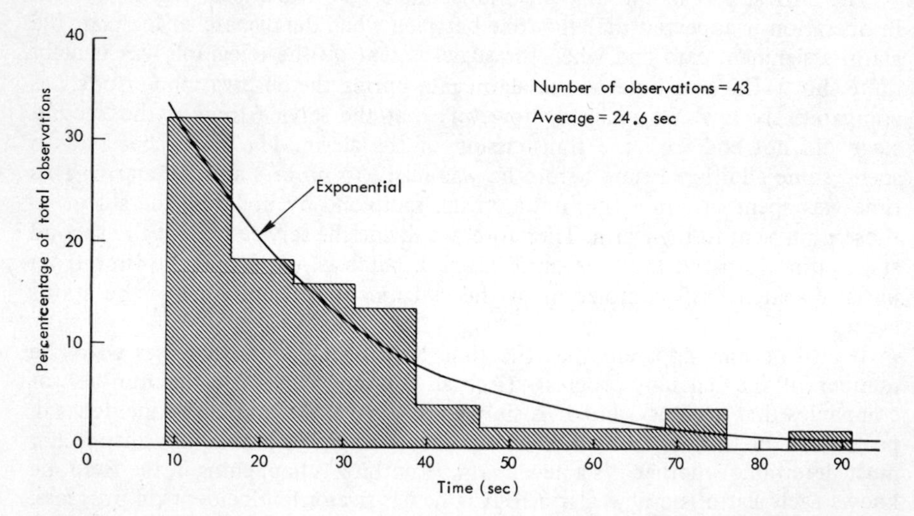

FIG. 19-6. Histogram of the time to find the alarm box number in the street address file.

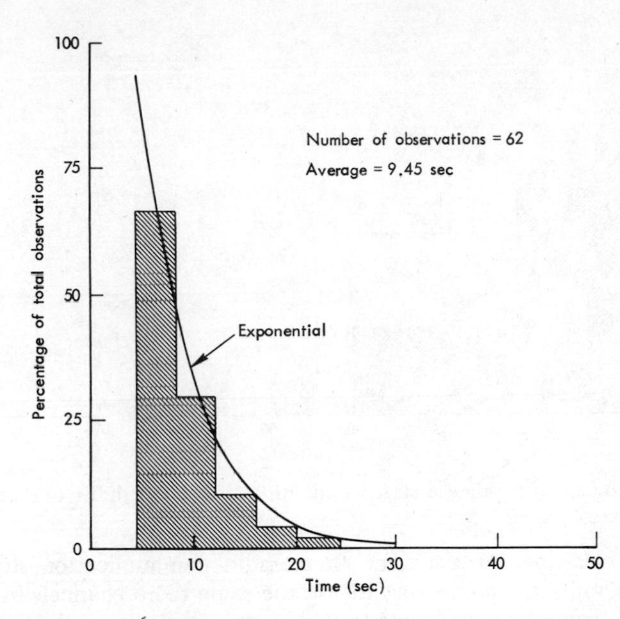

FIG. 19-7. Histogram of the time to locate the alarm assignment card.

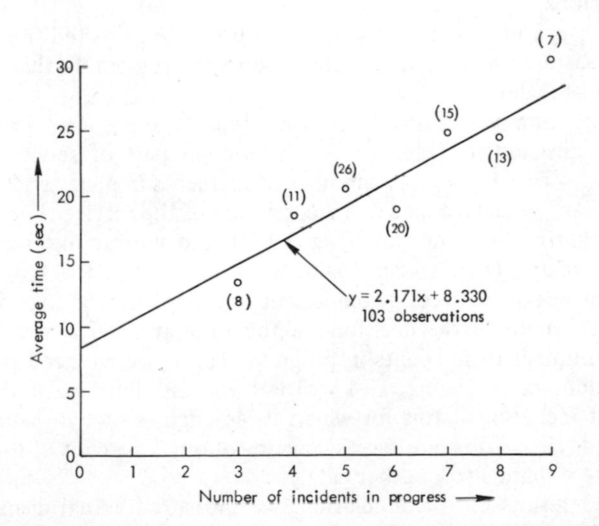

FIG. 19-8. Relationship between time at stage 2 until initial dispatch and number of incidents in progress.

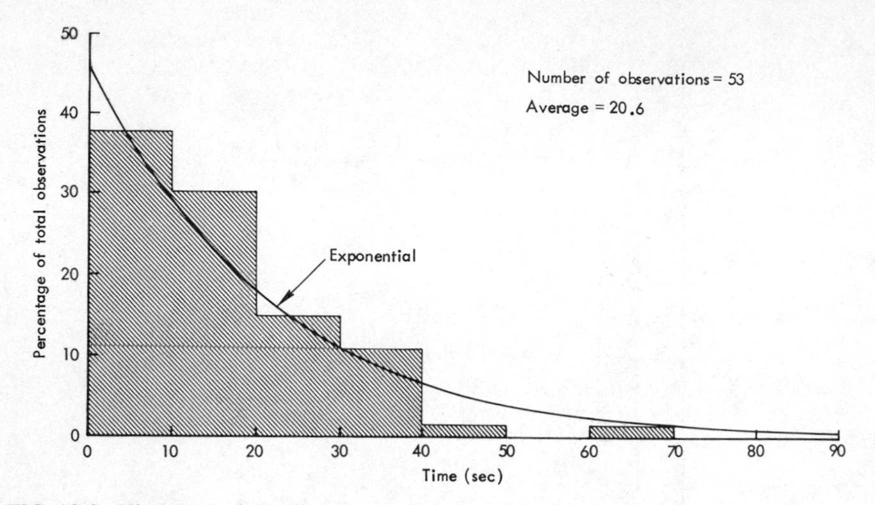

FIG. 19-9. Histogram of the time at stage 2 until initial dispatch with five or six incidents in progress.

will have to call one. At the same time, radio communication slows down because more units are competing to use the same radio channel. In addition, companies returning from incidents to their quarters telephone the Communications Office to report their new status. The dispatcher at the status board does not want to interrupt what he is doing, but he needs this new status information to make decisions.

Figure 19-8 is a plot of the data showing how the expected time until initial dispatch increases with the number of incidents in progress. In this case, we fit a straight line to the data.

Given some number of incidents, the time to make the decision is still exponentially distributed (Fig. 19-9). The second part of service time at the status board also goes up with the number of incidents in progress. To make data collection easier, we determined the second part of the service time at the status board by measuring the total time (Fig. 19-10) and subtracting from it the time until initial dispatch (Fig. 19-8). Again, we assume that, for some number of incidents in progress, the second component of service time is also exponentially distributed. To simulate the decision-making time at stage 2, we need to keep track of the number of incidents in progress. To do so, we need to know how long an incident lasts. Figure 19-11 shows the distribution of the length of incidents, not including alarms for which three engines or two ladders or more will be needed. (Such fires are rare, occurring about 1 percent of the time; their durations were obtained from other data.)

To test the results of the simulation, we measured actual dispatching time (the time from the first ring of the box or telephone until the first tap of the telegraph) over a period of about three hours at the Brooklyn Communications Office. For the test period the alarm rate was 22.4 alarms/hr with 26.8 percent telephone alarms. We ran the simulation at the same rate with the same

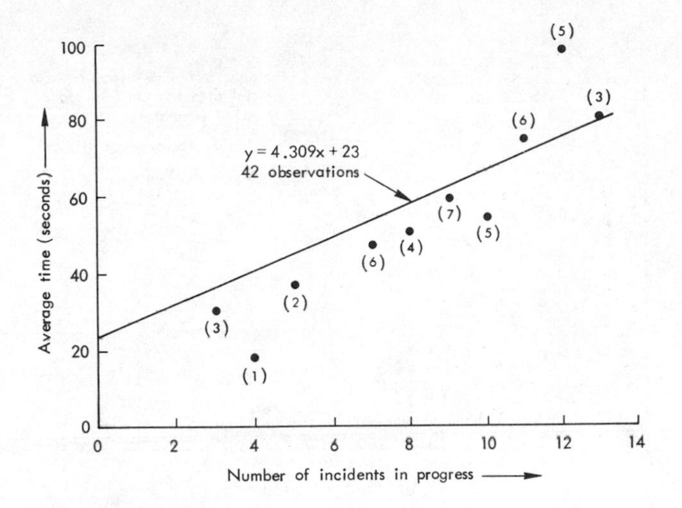

FIG. 19-10. Relationship between the total time at stage 2 and the number of incidents in progress.

percentage of telephone alarms and simulated 5000 alarms. In Fig. 19-12 we compare the results of the simulation with the actual histogram of dispatching time. The simulation predicts an average dispatching of 90.6 secs; the actual average was 87.1. You can see that the curve fits the data quite well. If you look at the left side, between 0 and 40 secs, there is a peak where the simulation underestimated the frequency. But we can understand that since, at the times the Communications Office is not busy, the dispatchers do not follow the procedure sequentially; the two stages often take place simultaneously. However, this "error" is not a serious one because we are concerned mainly with the upper tail of the curve in Fig. 19-12. Given these results, we are satisfied that our model can be used to make predictions about the future.

The results of the simulation are shown in Fig. 19-13. On the x axis we see the arrival of alarms. And on the y axis we see the average dispatching time. At 5 and 10/hr it is less than a minute–very fast. All the companies are in, and the alarm goes out very quickly. As we increase to 20 alarms/hr the curve starts rising. At 25/hr it has risen to almost 2 min; at 30/hr it increases to about 3 min; and at 35/hr it would take about 9 min for the average alarm to be dispatched! What does this mean? At the time of this study, an alarm rate of 20/hr in Brooklyn was considered a busy night, while a rate of 25/hr was considered extremely busy. From Fig. 19-13 we see that, at the time of the study, the alarm was approaching the neck of the curve–the point at which dispatching time begins to increase rapidly and the system begins to break down. That is something the Fire Department did not know until that time the Brooklyn Communications Office had been operating along the flat part of the curve.

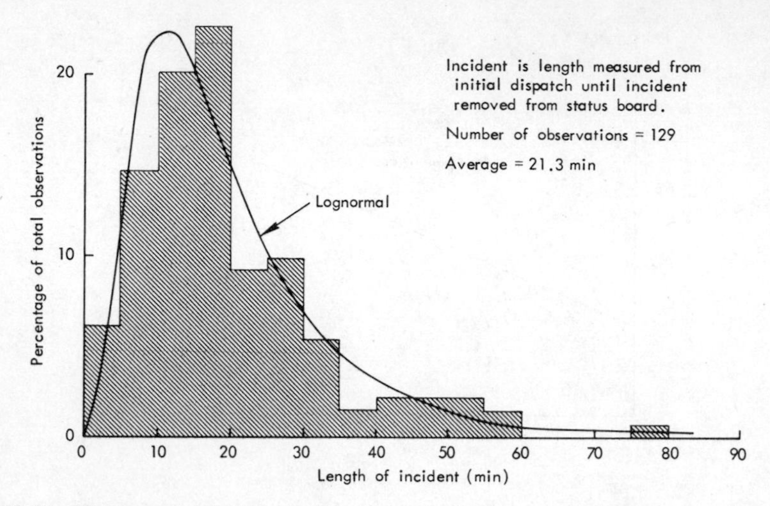

FIG. 19-11. Histogram of the length of incidents, for incidents which require less than three engines and two ladders. Incident length is measured from initial dispatch until incident removed from status board.

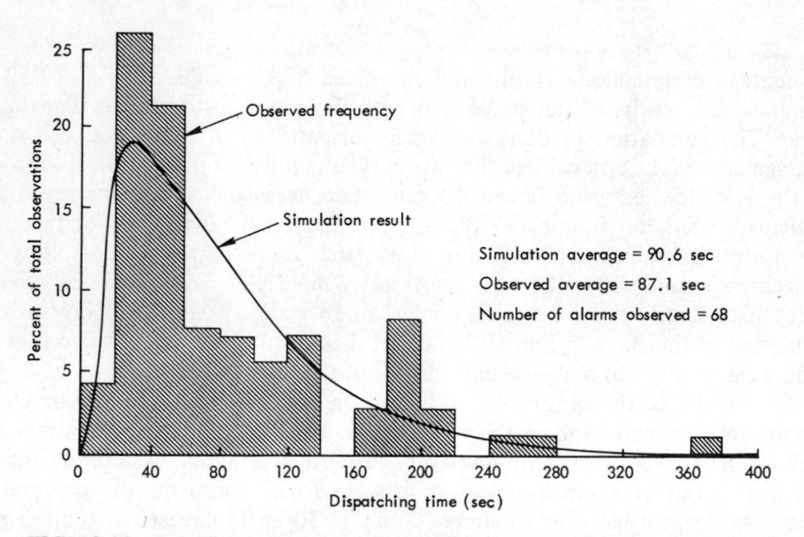

FIG. 19-12. Comparing the simulation predictions to actual dispatching time.

In addition to the average dispatching time, we are interested in the distribution (Table 19-1). At 20 alarms/hr, only about 20 percent take longer than 2 min. At 30 alarms/hr, 49 percent would take longer than 2 min, 16 percent would take longer than 5 min, and nearly 7 percent of the alarms would take more than 8 min.

TABLE 19-1
Percentage of Alarms with Dispatching Time Greater Than *T* Minutes

	Alarm rate			
T	20/hr	25/hr	30/hr	35/hr
2	20.7	28.9	49.1	76.1
3	9.0	14.0	32.7	65.8
4	4.5	7.4	22.8	57.9
5	2.5	4.1	16.3	51.0
8	0	1.0	6.8	34.7

That the problem is at stage 2, the "status board," is clearly illustrated when we look at the components of average dispatching time (Fig. 19-14). At 30 alarms/hr, there is an average delay of only one second. The service time at that stage is only about 35 sec on the average. But we have almost a two minute delay at the status board, and the service time there is well over a minute.

There are several significant observations we draw from the study. One is that the system has limited capacity. The Fire Department did not know this. You could probably send out 180 alarms/hr on the telegraph but the system cannot

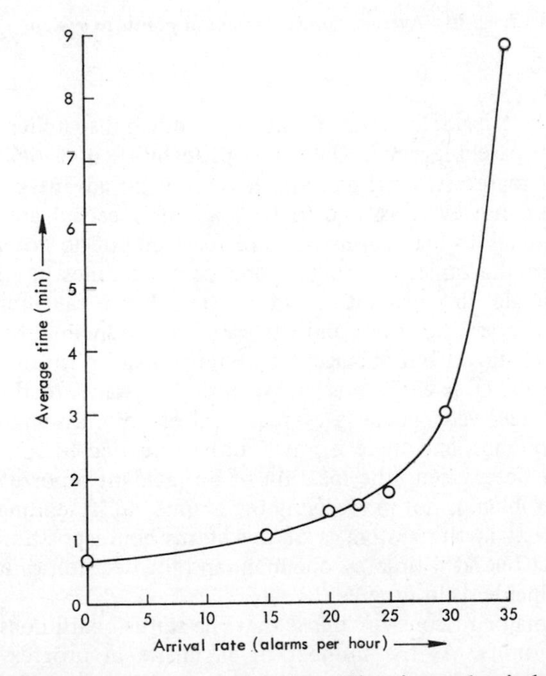

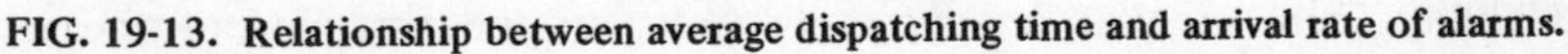
FIG. 19-13. Relationship between average dispatching time and arrival rate of alarms.

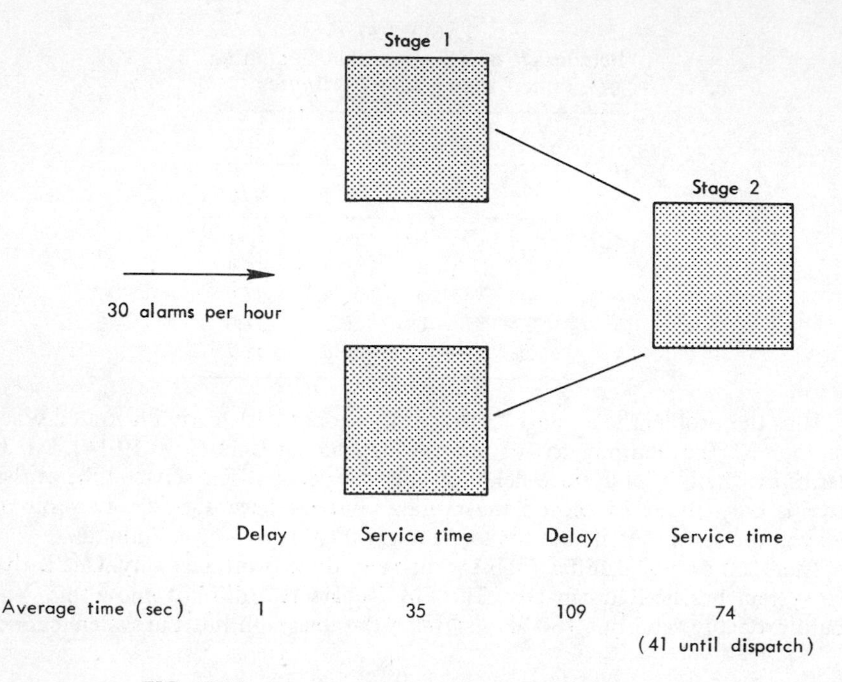

FIG. 19-14. Average duration times at points in system.

get alarms to the telegraph operator that fast. Adding dispatchers at stage 1 does not improve dispatching time. Once the probability of a delay at stage 1 is reduced almost to zero (as in Fig. 19-15), there is no advantage to adding men. The model does, however, assume that all alarms received are transmitted. In reality, there are additional alarms that are received but do not go beyond stage 1. Examples are the same box ringing one or more times in a short period of time, and a single fire generating several telephone calls reporting it. The difference is not great, however, and the conclusion remains the same, that the problem is not at stage 1 but at stage 2, the status board position.

A visitor to a C.O. is easily misled. Attention is drawn to those areas where box alarms are received. There is a great deal of noise associated with ringing boxes and telephones, and there is much activity related to counting boxes and telephones and determining the location of an incident reported by telephone. However, the problem is not in receiving the alarms but in sending them out.

The bottleneck in the system is at the status board position. Decisions are essentially made one at a time by one man, and the decision time increases with the number of incidents in progress.

One of the major reasons for this is that the status board does not reflect the true status of units. As the number of incidents in progress increases, the probability increases that one or more units on the first line of the alarm

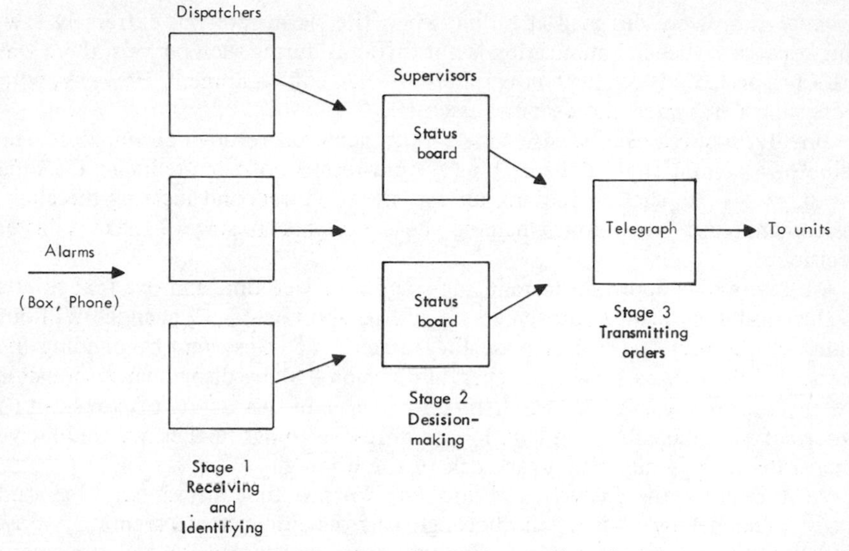

FIG. 19-15. A new dispatching configuration.

assignment card are not available in quarters. In the majority of cases, these units are listed on the status board as being at an incident and, therefore, technically unavailable. However, unless a unit is actually working at an incident, there is a good chance that it is available to respond to the new one. Because of this, the decision-maker may contact the units on the radio, ask: Can you respond to this alarm? and direct them to the new incident if they are available. If this procedure were not followed, either an incomplete response would result, or the decision-maker would search down the alarm assignment card for units that are farther away from the incident. Having identified the cause of the problem, what can we do about it? What are the alternatives for improving the system? One method of reducing dispatching time is to reduce the service time at stage 2. There are several ways to do so. One is to provide more current status information. A substantial part of the decision-maker's time is spent in determining the availability of units that are not in quarters but are on the air. An electronic status reporting system would enable units to transmit this status information rapidly without using the radio channel. An automatic system could also be used by units in quarters to replace the current voice method.

Another approach to reducing stage 2 service time is to update status faster. Status information is received from units in the field and in quarters. Within the Communications Office, this information is relayed to the status board (either verbally or by messenger) or, particularly in the case of radio reports, it is lost. An improvement would result if the dispatcher receiving new status information could transfer it directly onto the status board; for example, by activating switches connected to lights on the status board.

A third alternative is to improve the physical layout. The Communications Offices are generally not laid out to facilitate an ordered flow of information.

Because they were designed at a time when the alarm rate was extremely low, and because verbal communication is not difficult during such periods, there was little reason to consider the configuration of men and equipment. However, with increasing alarm rates, this becomes essential.

Finally, reductions in service time can be achieved through automation. The value of a computer-aided dispatching system results not from reducing the time for stage 1 tasks such as looking up the street address and locating the alarm assignment card but through helping the dispatcher at stage 2 make a faster decision.

An alternative approach to reducing stage 2 service time and one that results in the most significant improvements is adding another stage 2 channel. Without using a computer, we can increase the capacity of the system by dividing the borough into two parts, with a separate decision-making dispatcher responsible for the alarms and units in each half (Fig. 19-15). In this way two alarms can be processed simultaneously. In addition, to improve things further, we could have a separate radio frequency for each half of the borough.

At this point the analysis was finished. We saw that introducing a second stage 2 channel by dividing the borough was the simplest and seemingly most effective solution to the problem. What remained was to develop and implement a new procedure.

We started testing the procedure in February of last year, but the alarm rate was too low to provide an adequate test. By May the alarm rate had increased considerably—in fact, on several evenings it rose to the neck of the curve in Fig. 19-13. Initially, the two-channel concept was tested using two separate status boards. After three evenings of experimentation, we decided to use a single status board, divided, but with all units listed in the center section. The advantages of this arrangement are that it eases communication between the two decision-makers, and that it allows one decision-maker to operate alone during slow periods. Because the two decision-makers are side by side, and because each has access to all the companies, each may easily locate and dispatch units from outside his area, thereby lessening the problem caused by alarms drawing on units from both areas. The final test proved extremely successful. The system ran smoothly for three hours, during which approximately 100 alarms were transmitted. During a particular half-hour period, 27 alarms were sent out, and during an hour period, 43 alarms were sent out, with no appreciable delays. Except for special evenings (for example, on a July 4th) when a reduced response policy is in effect (dispatching fewer units reduces service time at stage 2) these figures are the highest ever achieved by the Brooklyn office.

Even after the successful test there was some resistance to the new procedure. However, within about six weeks the dual status board was introduced in Brooklyn. In addition, the Department has introduced procedures for faster status updating and the equipment needed for it is now being built. This equipment will be introduced in the Bronx Communications Office as well as in Brooklyn. The new Bronx system will also have the capability to split the borough if demand there increases beyond a critical point.

The improvements introduced enable the more effective use of additional manpower. The union had said that more men were needed but, without changing the system, these additional men would have been ineffective.

What does the case show? How does it illustrate the process of systems analysis?

In this study, there were six steps:

1. Identify the problem
2. Develop a tentative model
3. Collect data
4. Improve the model
5. Derive a solution
6. Implement it

After developing a tentative model and collecting data, we attempted to improve it, to make it more realistic. We saw that the service time at the second stage increases with the number of incidents in progress, and we saw how we could model this second stage operation. That is something that took weeks of observation. We then ran the simulation and derived a solution from the model. Using these results and our understanding of the system, we made recommendations that were finally implemented.

The process is a repetitive one. It began with formulation of the problem and it turned out that what we saw as the original problem, the vulnerability of the system, was actually part of a larger problem. We discovered that the system had a limited capacity, and that major improvements could be made. So the process is a continuous one: developing a model, collecting data, improving the model, reformulating the problem, developing a new model, collecting more data, and so on. In addition, we are always observing the system. The model serves as a framework for observation, giving the analyst a structured way of looking at the system. To build a model you must understand what is going on. And trying to build the model helps you to understand the system. Being there and looking at the system is one of the keys to successful systems analysis and was certainly the key to this study.

References

Swersey, Arthur Jay, *Models for reducing fire engine response times.* Doctor's Thesis, Columbia University, 1972.

Swersey, Arthur Jay, *Reducing fire engine dispatching delays,* The New York City–Rand Institute, Santa Monica, California, R-1458-NYC, December 1973.

Chapter 20

PITFALLS AND LIMITATIONS

Policy analysis depends so strongly on judgment and intuition and is so lacking in theoretical foundations that it is pointless for any analyst to expect success merely by following a set of well-defined rules. The problems an analyst can be asked to tackle in the public sector are particularly frustrating. Usually they are urgent and ill-defined. Often they are complicated, and sometimes they change radically during the investigation. The environment, the circumstances under which decisions about public problems are made can compound the difficulties. The very aim of the analysis, which is to assist one or more decision-makers choose a better course of action than they would otherwise have done, tends to introduce all the difficulties and contradictions associated with value concepts, human behavior, and the communication of ideas. These conditions, either inherent in analysis or external to it, are common to any analysis that is not strictly scientific. They combine to create a situation full of pitfalls and difficulties.

The purpose of this chapter is twofold: to alert analysts and their clients to some of the major sources of error that are likely to be found in policy analysis and to the limitations of such analysis. Most of these pitfalls and limitations have been mentioned earlier in this book but they can stand additional emphasis.

A discussion of the pitfalls or hidden traps where the analyst is likely to be led into error should help the analyst avoid them and the user to discover any errors that might result.[1] Errors result from simple mistakes or blunders and from fallacies or failures in logic.

There is little we can say about blunders or how to avoid them. They may be due to ignorance or stupidity, as in the case of the analyst who attributes his results to a computer in an environment where the word computer spells unemployment, but more often they are due to carelessness instead. They sometimes appear as errors in arithmetic or in these days of computers as coding errors. There is little practical advice one can give to discover such errors before it is too late. Careful checking is required but qualitative evaluation of numerical

[1] For an earlier discussion of pitfalls in analysis, see Kahn and Mann (1957). The inspiration for this chapter and many of the ideas in it came from this work. See also Koopmans (1956), Hitch (1956, 1958), Hitch and McKean (1960, pp. 120–25), and Quade (1968).

results also helps. It is unfortunate, but humans just make mistakes and experts sometimes turn out to be wrong.

The analyst must always be on the look-out for errors; not only for his own but also for those of the experts and the professionals on whom he depends for practical knowledge of the situation. Inefficiency, even errors, in routine operations can go undetected for long periods. A case in point taken from World War I is related by Col. Leonard P. Ayres (1940), then Chief Statistical Officer for the U.S. Army:

> When I came into the War Department I did not know much about being in the headquarters of the War Department. I was taken almost the first day into an office where they computed the turnarounds of the troop and cargo ships, and projected ahead the amounts of cargo and numbers of troops which would be landed on the other side month by month–joint Army and Navy affairs. All the computations were working out badly; practically nothing came out right. Week after week they had been making these computations and the actual turnarounds were always a little off. I told them they always would be wrong and that their computations should be changed. Statistics of rates and speeds are never the same as those of quantities and amounts. There is a certain difference when one gets to work with rates and speeds. There are always two ways of doing it and those two ways always tend to get confused and to bring out erroneous returns.
>
> For illustration, suppose your cargo fleet consists of three vessels; one slow vessel which takes 50 days to make a turnaround and two fast vessels which take 25 days each. The average of these three would be 33 and one-third days; that would mean on the average three turnarounds per vessel in a hundred days, and the two fast vessels make four turnarounds each in 100 days and the slow one makes two, and that makes ten turnarounds in all. That means that one answer is nine turnarounds and the other is ten, and both answers are based on perfectly good arithmetic.
>
> The next day an orderly came around and said the Chief of Staff presents his compliments and would like to see the colonel at his convenience. You all know what that meant, and, dropping everything, I hurried to his office and was told he had gone to this secret room of the General Staff. I went there and to my consternation I found that not only the Chief of Staff and his associates, but the Chief of Naval Operations and his associates were seated at the long council table. General March said, "It has been reported that you looked over the data of our turnarounds in the Transportation Section yesterday and said that the computations were erroneous. It is reported that you said that the present methods of computation should be abandoned and that something you called the 'effective average' should be substituted. What is the 'effective average'?" I said, "Sir, in mathematical terms the effective average is the harmonic mean."
>
> The general asked what it was used for, and I answered, "In this connection it should be used where rates and speeds are concerned." Then I found out one of the reasons why he was Chief of Staff. He said, "Define the harmonic mean." I said, "Sir, the harmonic mean is the reciprocal of the mean of the reciprocals of the several variates. The Chief of Staff said, "Precisely so, and it will be computed that way from now on."

Because fallacies represent a false idea or an error in logic, we have more hope for eliminating them than for simple blunders. Why do we have fallacies? One reason, mentioned earlier, is the lack of a well-developed theory. Another is a lack of training and experience. Nevertheless, a certain amount of experience has been accumulated and a few precepts based on that experience plus common sense should help us in avoiding some of the more flagrant fallacies.

The fact that a fallacy occurs in a particular study and is discovered, does not necessarily mean that all of the work is wasted, for it may be corrected. The very fact that someone can point out where an analysis has gone wrong strongly attests to the value of the analytic approach. It is thus a serious mistake not to make any analysis and the judgments on which it depends explicit for, if they are not, we surrender three great advantages that an analytic approach has over its competitors–namely that someone else can examine the work, can redo it with different assumptions, and can modify it or correct it as new information or insight becomes available.

For the purposes of this discussion, I separate the difficulties and pitfalls associated with offering analytic assistance into two categories–those internal to the analysis itself and those concerned with getting it used rather than how it is done. I further subdivide the internal pitfalls into those that are inherent in all analysis that seeks to provide help to someone else–the belief that the analysis can be complete, for instance–and those that we introduce ourselves–the most devastating of which is bias. Suppose we start with the inherent pitfalls.

Pitfalls Inherent in the Analysis

The most serious of these inherent pitfalls are those associated with the presence of uncertainty, with finding suitable ways to handle effectiveness and criteria, and with the necessity for basing whatever help can be offered on an incomplete analysis.

UNCERTAINTY

The major pitfall associated with uncertainty is to neglect it: to reduce the problem to a situation in which uncertainty appears no longer to be present by making a series of assumptions, say based on best guesses, that eliminates everything uncertain and turns the problem into one of "certainty." When uncertainty is present there is no reliable method for predicting a single future in terms of which we can work out a best system or determine an optimal policy. Consequently, in policy analysis, we must consider an appropriate set of futures or contingencies. For any one of these we may be able to determine or designate a preferred course of action, but we ordinarily have no way to determine one for the entire spectrum of possibilities. Therefore, for example, defense planning is rich in the kind of study that tells us what damage could be done to the United States given a particular enemy force structure, but it is poor in the kind of study that tells us how we will actually stand in relation to a potential enemy in years to come.

It is not enough simply to acknowledge that uncertainties exist and to warn the user that some things have been left out of a study because of the lack of information. We must have high confidence that the omissions do not have critical a effect on the final outcome of the study. The user, if not the analyst, has to come to grips with these omitted factors or issues and he needs to know what their effects are likely to be, how likely they are, when he can expect them, and what he might be able to do about them.

Systems analysis can be so striking in its attention to detail and to elaborate calculation that it may create the impression that more of the significant factors have been considered than is actually the case. This impression has enabled systems analysis sometimes to get by with an inadequate treatment of future possibilities on the assumption that uncertainties are best taken care of by desensitizing results and including some well-chosen caveats.

A cherished objective of systems analysis is to produce results that are insensitive to what might actually happen; in other words, to discover a system or policy that will be reasonably effective under any of the future possibilities and therefore be a hedge against whatever happens. We do not deny the desirability of this objective, but even when it can be done (which is probably not very often), this method of dealing with uncertainty may have some considerable disadvantages. For example, if we take insensitivity to be a property of policies that "work for the full range of extremes" and apply it to determine the overall policy, we may find we have overcommitted ourselves. One way to view the situation is as follows. For a given budget level the analysis investigates those alternatives that offer some chance of performing reasonably well over that range of the more important major uncertainties. Very often, none of the alternatives will look good enough—a very important thing for the decision-makers to know for, unless the analysis was defective, it implies that, *at the given budget level,* we cannot "get there from here." To "get there" cannot only be costly in resources but may introduce a rigidity in the policy that makes it unable to cope at all well with other less major contingencies we have not considered.

Another pitfall associated with uncertainty lies in the overconcentration on the statistical type of uncertainty where the probabilities of occurrence of events are more or less calculable. As discussed earlier, uncertainty of this type can be handled in the analysis by Monte Carlo models or other methods. The treatment of such uncertainty, however, is a considerable practical problem and a challenge to the analyst. The pitfall for analysts and model builders lies in accepting this challenge—to the neglect of real or the unforeseeable uncertainties. These latter involve forms of ignorance that cannot be reduced to probabilities and their consequences can be truly devastating. An objective in our policy studies is not to learn how chance can affect a given situation with a specific probability, but to design a system or determine a policy so that any risks caused by fluctuations with known behavior are unimportant.

Because a full investigation of the stochastic variation by means of a complete Monte Carlo calculation may seriously expand the analysis, it is frequently better to carry out a simple expected value investigation, deferring a full treatment of fluctuation phenomena until the qualitative aspects of the problem are fully understood. It may then turn out to be unnecessary to perform these time-consuming calculations, as consideration of the real uncertainties may have shown the effect of any statistical variation to be trivial.

EFFECTIVENESS AND CRITERIA

The pitfalls associated with the selection of appropriate criteria and measures of effectiveness are legion. Ultimate goals tend to be obscure and intangible. The most common pitfall is to substitute something that can be measured, no matter how appropriate. Thus, we find cost used to measure effectiveness (expenditures per pupil, for example, used to measure effectiveness of public school education), response time to measure the effectiveness of ambulance service for health care, and the rise in the price of narcotics to measure effectiveness of ways to reduce flow of narcotics. Other pitfalls associated with goals and criteria lie in the use of ratios, in the neglect of spillovers and in the failure to consider the distributional effects.

One commonly used criterion is the ratio of effectiveness to cost, that is, the ratio of achievement of objective to cost. This seems like a reasonable criterion—provided we can ignore questions of scale. McKean (1964) used an example to point out the difficulties here.

> To bring this point home, let us suppose that we are comparing two dwellings, and we accept floorspace as a suitable measure of what we want. How does the ratio of effectiveness to cost perform as a criterion? Dwelling A has 1500 sq ft and costs $18,000 (a ratio of 1 to 12); B has 2800 sq ft and costs $28,000 (a ratio of 1 to 10). Is B an obvious choice? Clearly we must be concerned with the scale—about the absolute amount of space the house will provide or the absolute amount of money the house will cost. The real question, concealed by the ratio, is the following: Is the extra 1300 sq ft worth an extra $10,000?
>
> Without constraints on the budget or the scale of effectiveness, then, ratios may point to extreme solutions and may not be consistent at all with higher level criteria. Now, suppose that we impose a constraint confining our consideration to a sensible budget range. Indeed, the example above might be considered as a case in point. In that case, the ratio is prevented from carrying us outside the constraint, but its significance is not clear until the range is narrowed to a fixed budget or objective.

For further examples of pitfalls associated with the use of ratios, see Hitch and McKean (1960, pp. 165–173).

In economics, the impacts caused by the actions of one party that lead to gains or costs of others are spillovers. For example, an oil well that forces brine into the underground water supply may reduce the fertility of adjacent farmlands.

The neglect of spillovers can be particularly serious in public policy analysis. They must be taken into consideration in judging the effectiveness of programs. An agricultural policy that leads to improved production can have as spillovers the threat of agricultural surpluses followed by strong pressure for farm support prices and then an accelerated farm to city migration which may mean a vast accumulation of uncontrolled growth, slum conditions, and unrest urban regions. A highway program that makes automobile transportation attractive may lead to the decay of the means of mass transportation, which, in turn, may make the highways increasingly congested with the accompanying air and noise pollution. In fact, one of the great ironies is that all solutions create new problems; the more comprehensive the solution the greater the problems are likely to be.

Sometimes after honest analysis, we are fundamentally uncertain about which of several alternatives is best. The pitfall, then, is to use some specious criterion that is not fundamentally bound up with the problem to decide which of them to indicate as preferred. This can lead to serious error; it is far better to acknowledge the situation and allow the decisionmakers to apply their own criteria without being inhibited by a recommendation from the analysis.

Because objectives, particularly public policy objectives, are likely to be multiple, ill-defined, and sometimes conflicting, measures of their attainment must be approximations, often fairly inadequate approximations. This imperfect coupling between what we want to achieve and what we measure may not be too serious if we are comparing two policies or alternatives that are not radically unlike one another. Two locations for a hospital, for instance. But at higher levels of optimization, gross differences in effectiveness may be obscured by an inadequate approximation to what it is we want to measure.

The inability to determine a perfect match between the measures used to compare effectiveness and the goals we are trying to achieve is a severe limitation on the usefulness of analysis. Suppose we are trying to improve secondary education. If we measure the degree of improvement in terms of skills in mathematics and arithmetic, we are likely to have a different program suggested than if we do our measurement in terms of some other skills.

An additional pitfall lies in putting exclusive dependence on criteria such as "greatest effectiveness for given cost" or "minimum ratio of cost to effectiveness." How the benefits from a policy compare with the costs is important but so also are the questions of who reaps the benefits and who pays the costs. Sometimes they are the same people but often they are not. Unless the decision-makers can supply this information it must come from the analysis. Without knowledge of the distributional effects of the various alternatives it may be impossible to make an acceptable decision, let alone an equitable one.

INCOMPLETENESS

Time, money, and other costs obviously place severe limits on how far any inquiry can be carried. The very fact that time moves on means that a correct choice today may soon be outdated by events and that goals set down at the start may not be final. This is particularly important in public policy analysis, for usually the decision-maker can only wait a very limited time for an answer. The costs of delay may be of more consequence than the benefits of further inquiry because the time at which the decisions can be made successfully may pass rapidly. As we noted earlier, we must often give our assistance before we have completed our inquiries and are fully ready. Hence, there is a good chance we may be wrong on an occasion in which a few more weeks of research would have shown the correct solution. This is a risk public policy analysts must accept, if they expect their work to be used.

Even with no limitations of time and money, however, the analysis would still be incomplete for it cannot treat all the considerations that may be relevant; they may, for instance, be too many in number, too intangible, too difficult to model, or we simply may not have and cannot get access to the data. Moreover, problems do not necessarily come to an end when action is taken to solve them for such actions usually remove only part of the difficulty, resulting, at best, in a partial solution. Analysis of a housing problem can lead one into investigations of industrial location, transportation, fiscal policy, education, and segregation, and these in turn even farther afield. Therefore, it is clear that there will not be a sharp cutoff at some stage of an investigation at which no further work is required, a stage at which we know all there is to know. We may stop because the return for effort is becoming vanishingly small, but there still will remain research that could be done.

Pitfalls Introduced by the Analyst

The most deadly pitfall is bias but there are others that can be almost as devastating; among others, those associated with problem formulation, modeling, neglect of the subjective elements, and failure to reappraise the work.

BIAS

Bias can take many forms, including a deliberate attempt to deceive or to alter the conclusions of a study by selecting false data or assumptions that favor a particular point of view. Bias can be introduced by the decision-maker as well as by the analyst and is probably far more prevalent when policy analysis is not used to help with decisions than when it is. The pitfall we are concerned with is honest but unconscious bias—one form is to allow oneself to be blinded by a preconceived idea as to how the analysis should come out; another is to be so caught up in an organization that one follows its cherished beliefs—the party line—with a mind closed to alternative approaches.

The first of these two forms of bias I called "conclusion bias." Louis Pasteur in the mid-nineteenth century recognized its existence in science, remarking "The greatest derangement of the mind is to believe in something because one wishes it to be so."

Bias of this type shows up clearly when controversial issues about which people feel strongly are being investigated; today (1974), for example, in the effects of segregated or desegrated schools or racial differences in intelligence. Wilson (1973) remarks.

> Because of these considerations, and after having looked at the results of countless social science evaluations of public policy programs, I have formulated two general laws which cover all cases with which I am familiar:
>
> *First Law:* All policy interventions in social problems produce the intended effect—*if* the research is carried out by those implementing the policy or their friends.
> *Second Law:* No policy intervention in social problems produces the intended effect—*if* the research is carried out by independent third parties, especially those skeptical of the policy.

These laws may strike the reader as a bit cynical, but they are not meant to be. Rarely does anyone deliberately fudge the results of a study to conform to preexisting opinions. What is frequently done is to apply very different standards of evidence and method. Studies that conform to the First Law will accept an agency's own data about what it is doing and with what effect, adopt a time frame (long or short) that maximizes the probability of observing the desired effect; and minimize the search for other variables that might account for the effect observed. Studies that conform to the Second Law will gather data independently of the agency; adopt a short time frame that either minimizes the chance for the desired effect to appear or, if it does appear, permits one to argue that the results are "temporary" and probably due to the operation of the "Hawthorne Effect" (i.e., the reaction of the subjects to the fact that they are part of an experiment); and maximize the search for other variables that might explain the effects observed.

People will naturally disagree over whether a given policy evaluation by a social scientist supports either the First Law or the Second Law. Many considerations prevent that argument from being carried on very intelligently—the loyalties and commitments of the scholars involved, the efforts of partisans and polemicists to defend one interpretation absolutely and to reject the other entirely, the defensiveness of whatever government agency is being praised or blamed by the study in question, and the tendency of human affairs to be so complex and ambiguous as to make the possibility of designing and executing a Decisive Experiment all but impossible.

A good illustration in another context of bias introduced by the analyst's strong belief that his views are right and should prevail, appears in testimony before the Senate Armed Services Committee during debate on ballistic missile defense (U.S. Senate, 1970). Here two analysts, using essentially the same data, managed to reach diametrically opposite conclusions by making different assumptions about such considerations as the way the enemy is likely to operate.

Parochialism, a form of bias mentioned earlier in Chapter 8, is particularly strong in close-knit organizations such as those found in the military services or in the "health complex." Often it is recognized (Kent, 1969)

First, decisionmakers are becoming increasingly annoyed that different analysts get quite different answers to seemingly the same problems. We allege that analyses are for the purpose of illumination. Still, at times, the light has a green tinge[*] or a light blue tinge, or a purple tinge. Sometimes the light comes out pure black. Seldom do we produce illumination with pure white brilliance. So the decisionmaker becomes wary—as well he should—of this biased or shaded illumination. There must be something wrong when quantification of some particular problem produces such radically different results. In our blind rush to be worthy advocates we enthusiastically engage in practices which border on perjury. The naive exclaim that the answer appears to have been known ahead of time. The calloused inquire if indeed there is another way.

There is no easy fix. A common suggestion—in the interests of objective analyses—is to establish joint organizations for analysis or have analyses done by people who are "above Service bias." This sounds good, but the theory is better than the practice. We are merely substituting one form of parochialism for another. To be more pointed. The illumination on problems by the Services will predictably reflect their own color. The illumination afforded by JCS studies has a way of coming out black

*Green, Army; deep blue, Navy; light blue, Air Force; purple, Office of Secretary of Defense.

because it goes through all of the filters. Those by OSD come out purple which may or may not be a better (or wiser) color than green, deep blue, or light blue. All too often the analyses are conducted in the context of a pre-conceived position. They become papers for "advocacy" as distinct from papers for "illumination." The quantification is shaped, twisted, and tortured to establish the "validity" of some particular point. But the decision-makers want the facts. They in turn can supply their own distortion.

Analyses by OSD and "think" organizations do not escape this plague. For one reason, their analyses are not so subject to critical review by non-believers as are analyses from the Services. Whatever objectivity is achieved by the Services does not necessarily stem from basic purity but rather from the fear of rebuttal. One could get a single answer to a particular problem by never having more than one analyst work on the problem. But, while the difficulty of getting different answers has been resolved, there is still the nagging concern about parochialism. Such a measure clears up the symptom, but does not cure the disease.

A related pitfall is to expect a man or the organization that created a system or originated a policy to discover its faults or to devise countermeasures or counterarguments. It is almost impossible to get an engineer or an analyst to display much ingenuity in tearing apart a brilliant scheme he may have been working on for months or years. Consider, for example, how the British discovered "window" or, as we called it here, chaff. In 1937, in Great Britain, two sets of scientists were working on ways to detect incoming aircraft; one by radar and one by infrared. Talk about stopping research by infrared because it was vulnerable to countermeasures (whereas radar was not) reached the Scottish physicist, R. V. Jones, the principal investigator working on infrared. He decided he had better do something about it. He stopped his own work and, turning to radar, came up almost immediately with the idea of "window"—an excellent countermeasure. When the radar investigators heard the idea they admitted it might work—but could not bring themselves to order trials. Later, these were ordered by others.

The first step, and the most effective one, in eliminating bias is to recognize that it is likely to exist! The second is to have the assumptions and work looked at by a different set of people with another set of cherished beliefs.

PROBLEM FORMULATION

An analysis must begin with an attempt to define and clarify the problem. A major pitfall is the failure to allocate the total time intelligently so that a sufficient share of it will be spent in deciding what the problem really is. The pitfall is to give in to the tendency to get started without having devoted a lot of thought to the problem. A large share of the effort by the leaders of the project, possibly as much as 50 percent must be invested in thinking about the problem, exploring its proper breadth, trying to discover the objectives of the systems or operations or policies under consideration and searching out good criteria or choice. It is useful to know as much as possible about the background of the problem—where it came from, why it is important, and what decision it is going to assist.

It is also important, if we are going to come up with acceptable results, that the client understand what problem we are trying to solve. The analyst cannot take the attitude that it is not up to him to question the assumptions, but merely to determine what conclusions follow from them. He may have to use the sponsor's assumptions, but he will also determine what happens when other assumptions are used.

The statement, early in the analysis, of the possible conclusions or recommendations is sometimes regarded as a pitfall. This is in itself a mistake. Once we realize that the analysis is an iterative process and that a single cycle of formulation, data collection and model building is unlikely to give the final answer, we should realize that setting forth hypotheses and possible conclusions early in the study is essential to guiding the analysis that follows. The real pitfall lies not in forming a preconceived or early idea about the solution but in being unwilling to abandon such an idea in the face of new evidence. In other words, we have to be flexible. When someone looks at the work and suggests that we might be wrong, we have to at least entertain the possibility that he might be right. A set of tentative conclusions or hypotheses helps to guide the analysis; it tells us what we are looking for while we are looking. In addition, it offers something concrete for others to probe.

Another pitfall is to look at a problem too narrowly. When this is done the solution can produce unexpected results. As an example, consider the "cost plus" system for government contracts. This was conceived as a scheme to ensure that excess profits were not made, particularly during World War II, by paying the contractors a fair percentage profit on their basic costs. The effect, however, was that now the contractor had no incentive to keep his costs down. Moreover, the percentage profit was set rather low and both the contractor and the auditors monitoring his work knew it was too low; so low that instead of bidding for such a contract the company might be better advised to put the shareholder's money into some other business rather than try to operate at the level of profit set. The rate had to be artificially inflated by a generous allowance for costs to keep the contractor in business; the scrutiny of contracts therefore became fairly perfunctory. This system thus may have been a factor contributing to inflation in the R&D business and to the scandals that have risen from time to time in the aerospace industry.

The objectives determine what alternatives are relevant; the decision-makers determine the objectives. The analysts can help with this latter but it is a serious pitfall to assume that analysis can determine public goals as well as aid in fulfilling them. It helps, of course, for the analyst to learn whom the real decisionmakers are, for without that information any investigation of objectives may be a waste of time.

Determining where the real problem lies is important and may not be easy. I remember Herman Kahn's remark that the payoff from designing shelter B to be better than shelter A is likely to be small compared with designing a plan that will get people into either should the need arise.

The analyst must try to consider everything pertinent to the problem including the possible reorientation of the entire activity. A most important qualification of the public policy analyst is his ability to see this fact. It is not always obvious. For example, we have the housing administrator in a large eastern city whose objective was to add to the housing supply in the city. When analysts began to work on his problem, they discovered that he did not really realize that this could be done by decreasing the rate at which dwellings were withdrawn from the housing supply as well as by increasing the rate at which dwellings were built.

MODELING

Overconcentration on the model and model building can be a serious problem. It is easy for an analyst to become more interested in the model than in the problem itself. Technical people with specific training, knowledge, and capability like to use their talents to the utmost. It is easy for them to focus their attention on the mechanics of the computation or on the technical relationships in the model, attempting to make these more representative of the situation. At the same time, they may neglect the important questions that should be raised in the study. They may thus find out a great deal about inferences that can be drawn from model, but very little about the question they are trying to answer.

A great pitfall of quantitative analysis and modeling is to quantify and model what we can, *not* what is relevant, neglecting the difficult–like Kaplan's drunk looking for his key under the street light even though he had dropped it in the dark around the corner. It is almost standard operating procedure for policy analysts to strip away large parts of the problem with simplifying assumptions. The pitfall is to carry this to such extremes that the remainder bears little resemblance to the reality with which the policy-maker must deal.

Excessive attention to detail is another pitfall. Although there are dangers in oversimplifying the model, it generally pays to be simple. Complicated formulas, or relationships so involved that it is impractical to reduce them to a single expression, are likely to convey no meaning at all, while a simple relationship may be understood. A major error may invalidate the more complicated expression and yet, in the general complexity of the formulation, pass unnoticed. In uncomplicated expressions, serious error is apt to be obvious long before the computation is complete, because the relationships may be simple enough to reveal whether or not the behavior of the model is going to be reasonably in accord with intuition. The most convincing analysis is one that a nontechnician can think through.

After working with the model for some time, the analyst may tend to mentally identify the model with reality and so come to accept oversimplified conclusions that apply to the model but not to the real world. One such error is to accept as useful output from the model the results of computations that are merely incidental to the question the model was designed to answer. For example, a simulation designed to yield some information about housing demands may also be made to yield some information about the demand for location and number of public schools. Such information must be regarded with suspicion unless the model is specifically designed to yield that information also.

It is a serious pitfall to attempt to set up a model that treats every aspect of a complex problem simultaneously. What can happen is that the analyst finds himself criticized because the model he has selected has left out various facets of the situation being investigated. He is vulnerable to these criticisms if he doesn't realize the importance of the point made earlier about models: the question being asked, as well as the process being represented, must determine the model. Without attention to the question, he has no rule for guidance as to what to accept or reject; he has no real goals in view and no way to decide what is important and relevant. He can answer criticism only by making the model bigger and more complicated. This may not stop the criticism for something must always be left out but it may make the model very difficult to handle and dilute the results that he is trying to obtain. The size of the model is then determined not by what is relevant but by the capacity of his computing machine.

Let us illustrate this dependence of the model on the question by considering a hypothetical study concerning flying saucer systems taken from [Specht, 1964].

> This study is about flying saucers made on Mars and dispatched on pioneer reconnaissance flights to the United States.
>
> While the saucer is being built, it may represent to the costing expert only a pair of numbers—its serial number in the production of this model and the number of man-hours (or Martian *augenblicks*) required to produce it. These numbers are the essential ones in determining the learning curve with which future costs may be estimated.
>
> After production the saucer is shipped to the depot by boat—canal boat, naturally. Here the saucer can be replaced by a different set of numbers—linear dimensions and weight, together with the freight classification (3B in the case of saucers). The machinery of the model, in this case, consists merely of the set of tables that give freight rates in terms of weight, cubage, classification, and route.
>
> After the saucer has been launched and is in free flight in the gravitational fields of Mars, Earth, and Sun, not to mention Phobos, Deimos, and Luna, a discussion of trajectory necessitates a different model. The saucer can now be idealized as a point-mass having position and velocity. Any practical man could object that we are being quite unrealistic; that we are neglecting size, shape, material; that the saucer has a span of 100 ft, is colored bright red, and carries a crew of three. But these things have little effect on the answers to the questions we ask of this model. The interrelations between the factors of the model—that is to say, the inner machinery or structure of the model—may be given by a computing schedule expressing Newtonian gravitation. Into this schedule we enter positions and velocities and from it we calculate future path.
>
> The saucer now enters the earth's atmosphere and becomes an object of interest to the aerodynamicist rather than to the astronomer. Where our last model was that of a point-mass we now have to deal with shape and drag coefficient as well as velocity.
>
> If the Air Control and Warning network of the Air Defense Command picks up the saucer on its Early Warning radar, then the saucer is merely a radar echoing area as determined by material, size, form, and aspect.
>
> If the saucer proves to be hostile and an interceptor makes a firing pass, then a

different model comes into play. For vulnerability calculations we are interested in the two-dimensional profile, fuel storage, and other vulnerable components.

And so on. The same part of the real world may be modeled in many ways. Factors relevant to one model may be completely immaterial to another.

One approach to designing a model is to attempt to reduce the real situation to a logical flow diagram. The dangers of this approach are that the model may tend to be too detailed and components of the real process will be included which contribute nothing to the question to be answered. For this reason, it is advisable to design the model around the questions to be answered, rather than in imitation of the real world.

The people for whom analysis is undertaken might do well to become aware of another aspect of big simulation models. Almost invariably the time required to set up the computing program for a large simulation is underestimated. Instead of taking the few months that might be estimated, it tends to take a year or longer. Ideas change fairly rapidly in the field of public policy, and questions that people think are important soon become unimportant. A large model man always runs into the danger of having spent most of his time looking for ways to make the computing machine better approximate some relatively trivial aspect of the real world rather than setting out to study the problem he is supposed to solve. He may thus learn a lot about programming and very little about the question. For these reasons, as we said earlier, we should turn to the mathematician as much for his ingenuity in escaping mass computations as for his skill in organizing them. Indeed, the role of the big model in analysis has been more to verify and work out the details and to prepare for presentation ideas that are fairly well understood than to discover a solution. There is a good chance that ideas originated and the analyst became convinced of their validity through intuition aided by very rough and elementary calculations.

Capturing reality is an elusive business. Mere bigness in a model is not the road. Acquiring and organizing the data to support an all-inclusive model can cost a great deal in time, money, and attention, and much must still be left out. The capability of the analyst to understand and manipulate his model decreases rapidly as analytic size and the mass of data increase. It is easy to reach a point of vanishing returns.

Another dangerous pitfall lies in attempting to force a complex problem into an analytically tractable framework in order to have ease of computation. Now compromises must always be made in model building but this is not the direction in which to go. It is almost always better to sacrifice workability in order to represent the process being modeled more adequately even though we must depend on judgment more.

It is a serious pitfall for the analyst to concentrate so completely on the purely objective scientific and quantitative aspects of his analysis that he neglects the subjective elements or fails to handle them with understanding. Quantification is desirable, but it can easily be overdone; if we insist on a completely quantitative treatment we may have to simplify the problem so drastically that it loses all realism. Judgment and intuition are an integral part of analysis and cannot be avoided.

The first author of a book on systems analysis—at least the first book I know about that is clearly on that topic—understood the role of mathematical calculations well—A. M. Wellington (1887) wrote

> The mathematical form of discussion has been intentionally avoided . . . chiefly, because mathematical methods of solution are not only inexpedient, but positively dangerous for the class of problems considered. When the difficulty of a problem lies only in finding out what follows from certain fixed premises, mathematical methods furnish invaluable wings for overcoming intermediate obstructions; but whenever the chief difficulty of a problem lies in the multiplicity and dubiousness of the premises themselves, and in reconciling them with each other, there is no safe course but to remain continuously on the solid ground of concrete fact.

In spite of these objections, I must not imply that large computer models have no role in policy analyses. The prediction of something like the noise level contours around a proposed new airport may require such a model. Working out how a change in the tax laws will distribute the burden throughout the nation may require such a model. But data must be available and the underlying phenomena must be well understood if models of this type are to give valid results.

FAILURE TO REAPPRAISE THE WORK

Clients sometimes feel that one of the worse characteristics of the analysts they engage is their desire to make drastic changes in a study after the work is almost done. As they see it, the result is a great deal of wasted effort and deadlines are not going to be met.

It is, of course, quite true that making a major change in a study at a late stage means that much of the early work cannot be used, and even that, because a change may involve a great deal of additional work, the time of completion may be delayed well past the deadline. For these reasons, some analysts when they are one-half, two-thirds, or three-fourths through a study, hesitate to pause to evaluate what they have done thus far. A periodic reappraisal is essential, however, because as the study progresses the analyst broadens his understanding of its scope and purpose and of the relationships involved. If stock-taking then results in junking a major portion of the work, the results clearly indicate that a reappraisal was especially necessary. To head off this problem, careful attention to problem formulation and early stock-taking are helpful.

MISCELLANEOUS REMARKS ON ANALYST-ASSOCIATED PITFALLS

We learn by experience but not necessarily what we should—thus the cat that jumps on a hot stove learns never to jump on a cold stove.

Alternative explanations and alternative means abound and the analyst cannot seize on the first explanation or alternative merely because it seems so logical. Teaching experience, for instance, is positively and highly correlated with pupil excellence. The usual explanation is that experienced teachers make better pupils—but it may be the other way around. Schools with the better pupils get the experienced teachers since those teachers have the experience to get transferred there.

Evidence is seldom categorical. Expert witnesses have seen "flying saucers" that turned out to be such things as weather balloons, Venus, aircraft, mirages, ice halos, and so forth. Because so much data turn out to be wrong, the analyst must do constant checking. A possible way to avoid some of the pitfalls here is to include lay experts on the analyst's team who are informed at least in all the important fields with which the study is concerned.

Whereas the analyst will strive for efficiency and consistency himself, he must be careful in how far he pushes the client in those directions. Striving for efficiency in one area may sacrifice it overall. The Russians have a proverb for this—the better is the enemy of the good.[2] Instituting incentives, things like extra payments for early achievement, for instance, can result in frantic efforts that neglect everything not constrained. Striving for consistency can escalate conflict. Organizations can tolerate inconsistency—for instance, the federal budget recently has included funds both to discourage tobacco smoking and to promote tobacco growing—but possibly not conflict.

People, including the analysts among them, tend to ignore certain costs—imparing incentives and generating dissatisfaction in the organization they are trying to help are among them. Many of the difficulties we discuss next are aggravated because analytic assistance is sought by an individual but the actions it calls for must be taken by an organization, often one of which the decision-maker is not a part.

External Pitfalls

By external pitfalls I mean factors or actions that handicap the analysis but do not originate within the analysis itself. These involve such things as communication with the client, the myth that there is a decision-maker with authority, and the belief that in a particular area there is a universal framework or model that can answer all questions and thus substitute for the decision-maker.

COMMUNICATION WITH THE CLIENT

Communication of ideas is always difficult and a stage full of pitfalls is the one at which the transition from an analytic study to decision-making occurs.

There are good reasons why communication with a client may be difficult even though we put a lot of attention on briefings and reports—we sometimes speak almost in different languages. Moreover, the client is likely to have strong

[2] Luchshee—vrag khoroshego.

views about his own problem including, if not a preferred solution, at least a preferred type—a fire department likes to do its job by putting out fires, not by preventing them through some social action program. We are thus often trying to tell the client something he does not want to hear. He may then simply not listen.

For this form of communication failure, I have no words of good advice. Stafford Beer (1966, pp. 59–60) calls it "thought block."

For those of you who think blind resistance to change of opinion is not realistic, consider the horse cavalry (Katzenbach, 1958). By the end of the nineteenth century the introduction of automatic and rapid-firing weapons, the machine gun and the French 75, for example, had convinced military analysts that the traditional cavalry was obsolete. The lessons of the Boer War (1898–1901) and the Russo-Japanese War (1902–1904) reinforced this. By 1904 the horse cavalry was recognized as an anachronism in the eyes of all except the cavalryman himself. The cavalry made some changes: the U.S. Cavalry began to practice charges with 45 in hand instead of a saber and the British and other European armies gave up the lance. Yet by 1907 the lance had come back and it stayed for more than 30 years. In fact, by the start of World War I, almost all military men had forgotten that the cavalry was obsolete. The introduction in World War I of the truck, the tank, the trench, and barbed wire, followed by aircraft and poison gas a little later soon made it clear all over again that horse cavalry was obsolete. Yet it did not die.

One of the most remarkable instances in the survival of the cavalry occurred during the summer maneuvers at Fort Benning, Georgia, in 1936. The cavalry rode and the infantry was trucked to the area. The infantry arrived first and had ample time to get into position. They waited, well camouflaged and with some excitement until the cavalry passed the forward concealed units. Suddenly, the infantry rose shouting from entrenched positions waving bedsheets. What the infantry expected followed; horses were thrown into a panic, men and horses were injured—and the infantry commander may have even been reprimanded for making a practical joke out of serious maneuvers.

Before the somewhat skeptical Congressmen of the Military Affairs Subcommittee of the Appropriations Committee, the Chief of the U.S. Cavalry expounded the virtues of the horse in the spring of 1939 and again in March 1940. The reasons he advanced illustrate some common pitfalls (Iklé, 1971).

(1) Other important military powers are retaining the horse, therefore it must be worth retaining.

In the spring of 1939, the Congressmen were told by the Chief of Cavalry that "both Poland and Russia are maintaining great masses of horse cavalry and even France on the western front is maintaining a larger percentage of horse cavalry than we are." Comes March 1940, with part of the evidence destroyed, the conclusions of the argument nonetheless remain unshaken: "What happened to the Polish Cavalry, far from being a proof that the days of horse cavalry are over, supports the lessons from Spain. . . . had Poland's [horse] cavalry possessed modern armament. . ."

(2) The man on his horse has greater "fluidity and flexibility" than competing systems.

"Mechanized cavalry is going to be chained largely to the roads. . . ," the Chief of Cavalry explained to Congress in 1940, "it has not the flexibility of horse cavalry for each man on his horse, armed with his rifle and pistol, is an independent fighting unit."

(3) All alternatives to the horse are vulnerable to countermeasures.

According to the Chief's testimony, ". . . there is no combat car or tank built or to be built for which there has not been or cannot be manufactured a multitude of antitank weapons at less cost than of the vehicles and which will shoot holes through them in the same manner that the rifle or machine gun may shoot through a man or a horse."

(4) Forms of warfare for which the horse cavalry is useless are the wrong kind of warfare; it is the purpose of doctrine to prohibit these kinds of warfare.

"It is readily admitted," the Chief of Cavalry testified, "that [horse] cavalry has little or no value in trench warfare. But wherever there is movement the use of cavalry in large groups is vital. The whole idea of modern doctrine, organization, and training is to restore movement to the battlefield. If this attempt is successful, it will largely eliminate those obstacles, particularly trenches and wire, which render cavalry, as such, useless in stabilized situations."

(5) The American "horse industry" is far ahead of those of other nations, hence we must take advantage of our natural asset.

"We are particularly fortunate," the Chief concluded his presentation in March 1940 before the Subcommittee, "in having great resources both in horses and motors. There are more than 10,000,000 horses in this country . . . Therefore we should not mistakenly ape the example of European nations . . . whose resources in animals and forage are such as to further restrict their use of the horse. Ours is an American problem."

In spite of these things, the U.S. horse cavalry survived until 1951 and the last mule was decommissioned in 1956. As far as I know the last serious effort to revive the cavalry came in 1956. The Belgian General Staff suggested that for the kind of dispersed war that low-yield atomic weapons necessarily create, the horse, which in Europe could be independent of depots, should be introduced into the NATO weapons system.

Analysts are also a source of communication failure. They must avoid the pitfall of appearing to be arrogant "whiz kids" making "God-like" pronouncements or behaving as if they were decisionmakers. Although only very few analysts ever claim infallibility of judgment, many more sometimes act as if they would like to.

THE MYTH OF THE DECISION-MAKER

Another pitfall associated with implementation is the "myth" that there is a unique decision-maker. Analyses are ordinarily designed and carried out, although perhaps not always deliberately, as if they were to assist a solitary decision-maker who had full authority over acceptance and implementation. This may sometimes be the case but it is *not* the usual situation, even in the military, and almost never when broad social issues are involved. Even when there is a single decision-maker, his staff at a minimum supplies the details of any policy that is set. In view of this, as one Air Force officer put it to me—don't consider your task to be one of telling the Chief of Staff what bomber to select, view it as helping the general's staff to solve an organizational problem in which selecting the next bomber is the driving element. If the analyst expects to see his recommendations implemented, steps in the right directions are to give considerable attention to "political" or organizational feasibility, as well as to technical matters, and to the problems of the various secondary interest groups as discussed in the chapter on Implementation (Chapter 17).

Influencing organizational behavior can be quite different from influencing the behavior of an individual and, since we understand so little about it, can constitute a pitfall for policy analysis. Government organizations do not go about satisfying their wants in the same way as individuals. Although individuals may learn it quickly, changing the doctrine or policy of an organization is a slow process. The significant difficulties in getting agreement from all subunits necessitate an approach that is a deliberate one. Organizations have great capabilities for resisting change.[3]

SUBSTITUTING A MODEL FOR THE DECISION-MAKER

The failure to realize the importance of the question in the design of the model leads to another pitfall: the belief (by decision-makers if not by analysts) that there are "universal" models—one model, say, that can answer all questions about a given activity and which therefore can be used to evaluate without supplemental judgment a full range of alternatives. There are, of course, models which allow the user to experiment with a wide choice of parameters and assumptions. A number of large-scale urban simulations have this property. It has, for example, been proposed a number of times (even to the extent of writing a study contract) that a general computer model for strategic air war be set up to supply weapons designers with a systematic evaluation of their design concepts and to enable the Department of Defense to evaluate the worth of contractors. Attempts are being made to do the same sort of thing in the area of urban models.

One argument for a "standard" model notes that "the choice of assumptions, the forecase of the future, and the methods of analysis have a marked influence on the performance and physical characteristics of the system set forth as preferred or optimal"; therefore, a uniform framework would mean that "the results obtained by the various contractors would be comparable since the effects due to variation in the assumptions they might have chosen to form their models would have been eliminated." This may indeed be the case, but will the end result be desirable? A rigidly specified framework may mitigate one sort of undesirable bias—by making it difficult for an analysis to be used to rationalize

---

[3] For an interesting discussion of this matter, see Agency Strategies for Dealing with PPB, particularly the section on Some Techniques to Make PPB work in Your Favor or Not at All in Hovey (1968).

conclusions already otherwise derived–but only at the severe risk of introducing other biases.

A fundamental objection is that a uniform framework necessarily conceals or removes by assumption many extremely important uncertainties, thus tending to lead to solutions that disregard the value of hedging against those uncertainties. Another is that even if efforts were made to keep the model "up to date," this would turn out to be impossible, for the analyst must be able to modify his model in the terminal stages of his study to accommodate information acquired during the early phases. Indeed, in a problem involving something like educational reform, there are so many factors of shifting importance, and such radical changes in objectives and tactics are likely, that most models are obsolete long before the recommendations from the study can become accepted policy. Finally, if a model or a mathematical formula were used to indicate which proposal to select, the proposers' emphasis would soon focus on how to make his design look good in terms of this analytic definition, not on how to make it look good against reality–a much harder problem.

The Honorable C. J. Hitch (1964), while Assistant Secretary of Defense, made this last point with an analogy.

> Another kind of problem that might be encountered with an analytically based contract would be "rule beating." An analogy can be found in the case of some of the handicapped rules drawn up by yachting organizations. The intent of these rules is to allow the owners of often greatly dissimilar sailing yachts, basically designed for cruising, to compete against each other on an equitable basis. The rules are generally empirical in nature, and take into account such factors as the dimensions of the hull, the amount of sail area, and so on, resulting in a handicap for each yacht which reflects its theoretical speed. The rule is expressed in terms of a formula which may be rather complex. So long as the competitors are all sailing relatively conventional yachts, the goal of generally equitable competition can be achieved.
>
> However, once such a rule is established, the serious competitor has a considerable incentive to study it very carefully when he is considering a new yacht, or even a new rig for his old yacht. In such an environment, from time to time, there have appeared some fairly unconventional yachts, designed not in the usual way, but in a way specifically tailored to beat the rule. From a practical point of view, these yachts are freaks; nobody would have designed such a thing or wanted to own one save for the existence of the rule. They tend to be undesirable in most ways save that of winning races through the establishment of an unusually favorable handicap. Whenever such freaks start to win most of the races, of course, there is a strong tendency for the rules committee to plug the previously unsuspected loophole in the rule.
>
> To the extent that such rules have loopholes, emphasis is shifted away from beating other yachts towards beating the rule itself. By the same token, setting up a weapons system contract on the basis I have described would not really mean that the contractor will, by definition, be motivated to develop and produce the best possible system. Rather, he will be motivated to develop and produce the system which best meets our analytical definition of the best possible system. To the extent that our definition is incomplete, or subject to unsuspected loopholes, the product may tend to diverge from what we really have in mind. Thus, this sort of contract would be subject to "gaming" on the part of the contractor–either deliberate or unconscious. He may be able to develop a system which meets the necessarily artificial time-cost-effectiveness model beautifully, but which is, in fact, a rather poor weapon system.

Out of context, these pitfalls I have mentioned seem so obvious that one wonders how they could have led analysts into error; one has only to examine actual analyses to find out that they are still occurring. By discussing them, our hope is not that we will make them vanish completely for that is probably impossible, but that we will at least make them become much more obvious.

References

Ayres, Colonel Leonard P., The uses of statistics in war. Army Industrial College, AIC 195, March 4, 1940, pp. 8–9.

Beer, Stafford, *Decision and control.* Wiley, London, 1966.

Hitch, C. J., Cost considerations and systems effectiveness. Address presented at the SAE-ASME–AIAA Aerospace Reliability and Maintainability Conference, Washington, D.C., June 30, 1964.

Hitch, C. J. & McKean, R. N., *The economics of defense in a nuclear age.* Harvard Univ. Press, Cambridge, Massachusetts, 1960, particularly pp. 120–225.

Hitch, C. J., *Professor Koopmans on fallacies: A comment,* The Rand Corporation, Santa Monica, California, P-870, May 1956 (This paper was also published in *Operations Research* on the pages immediately following the paper by Koopmans).

Hitch, C. J., *Economics and military operations research.* The Rand Corporation, Santa Monica, California, P-1250, January 1958.

Hovey, Harold A., *The Planning–Programming–Budgeting Approach to government decision-making.* Praeger, New York, 1968.

Iklé, Fred C., unpublished paper, The Rand Corporation, Santa Monica, California, 1971.

U.S. Senate, Subcommittee on National Security and International Operations, "Defense Analysis: Two Examples" in *Planning–Programming–Budgeting,* U.S. Senate, 91st Congress, 2nd Session, Washington, D.C., 1970, p. 659.

Kahn, H. & Mann, I., *Ten common pitfalls.* The Rand Corporation, Santa Monica, California, RM-1937, July 1957.

Katzenbach, Jr. & Edward, L., The horse cavalry in the 20th century: A study in policy response. *Public policy,* Graduate School of Public Administration, Harvard, 1958.

Kent, Major General Glenn A., USAF, Keynote Speech, 24th Meeting of the Military Operations Research Society, 1969.

Koopmans, Bernard O., Fallacies in operations research. *Operations research,* August 1956, 4, (4), 422–426.

McKean, R. N., Criteria. In E. S. Quade (Ed.), *Analysis for Military Decisions,* North Holland, Amsterdam, 1964, Chapter 5.

Quade, E. S., Pitfalls and limitations. In E. S. Quade and W. I. Boucher (Eds.), *Systems analysis and policy planning applications in defense.* American Elsevier, New York, 1968, Chapter 19.

Specht, Robert D., Why and how of model building. In E. S. Quade (Ed.), *Analysis for Military Decisions,* North Holland, Amsterdam, 1964.

Wellington, A. M., *Economic theory of the location of railroads.* Wiley, New York, 1887.

Wilson, James Q., On Pettigrew and Armor: An afterword. *The public interest,* Winter 1973, (30), 132–135.

INDEX

Selected List of Rand Books

Bagdikian, Ben H. *The information machines: Their impact on men and the media.* New York: Harper and Row, 1971.

Brodie, Bernard. *Stategy in the missile age.* Princeton, New Jersey: Princeton University Press, 1959.

Bruno, James E. (Ed.) *Emerging issues in education: Policy implications for the schools.* Lexington, Massachusetts: Heath, 1972.

Canby, Steven L. *Military manpower procurement: A policy analysis.* Lexington, Massachusetts: Heath, 1972.

Cohen, Bernard & Chaiken, Jan M. *Police background characteristics and performance.* Lexington, Massachusetts: Heath, 1973.

Coleman, James S. & Karweit, Nancy L. *Information systems and performance measures in schools.* Englewood Cliffs, New Jersey: Educational Technology Publications, 1972.

Cooper, Charles A. & Alexander, Sidney S. *Economic development and population growth in the middle east.* New York: American Elsevier, 1972.

Dalkey, Norman (Ed.) *Studies in the quality of life: Delphi and decision-making.* Lexington, Massachusetts: Heath, 1972.

DeSalvo, Joseph S. (Ed.) *Perspectives on regional transportation planning.* Lexington, Massachusetts: Heath, 1973.

Downs, Anthony. *Inside bureaucracy.* Boston, Massachusetts: Little, Brown, 1967.

Fisher, Gene H. *Cost considerations in systems analysis.* New York: American Elsevier, 1971.

Haggart, Sue A. (Ed.) *Program budgeting for school district planning.* Englewood Cliffs, New Jersey: Educational Technology Publications, 1972.

Hirshleifer, Jack, DeHaven, James C. & Milliman, Jerome W. *Water supply: Economics, technology, and policy.* Chicago, Illinois: The University of Chicago, 1960.

Hitch, Charles J. & McKean, Roland. *The economics of defense in the nuclear age.* Cambridge, Massachusetts: Harvard Univ. Press, 1960.

Kershaw, Joseph A. & McKean, Roland N. *Teacher shortages and salary schedules.* New York: McGraw-Hill, 1962.

Levien, Roger E. (Ed.) *The emerging technology: Instructional uses of the computer in higher education.* New York: McGraw-Hill, 1972.

Lubell, Harold. *Middle east oil crises and Western Europe's energy supplies.* Baltimore, Maryland: The Johns Hopkins Press, 1963.

McCall, John J. *Income mobility, racial discrimination, and economic growth.* Lexington, Massachusetts: Heath, 1973.

Meyer, John R., Wohl, Martin & Kain, John F. *The urban transportation problem.* Cambridge, Massachusetts: Harvard Univ. Press, 1965.

Nelson, Richard R., T. Paul Schultz, and Robert L. Slighton. *Structural Change in a developing economy: Colombia's problems and prospects.* Princeton, New Jersey: Princeton University Press, 1971.

Nelson, Richard R. Peck, Merton J. & Kalachek, Edward D. *Technology, economic growth and public policy.* Washington, D.C.: The Brookings Institution, 1967.

Newhouse, Joseph P. & Alexander, Arthur J. *An economic analysis of public library services.* Lexington, Massachusetts: Heath, 1972.

Novick, David (Ed.) *Program budgeting: Program analysis and the federal budget.* Cambridge, Massachusetts: Harvard Univ. Press, 1965.

Novick, David (Ed.) *Current practice in program budgeting (PPBS): Analysis and case studies covering government and business.* New York: Crane, Russak, 1973.

Park, Rolla Edward. *The role of analysis in regulatory decisionmaking.* Lexington, Massachusetts: Heath, 1973.

Pascal, Anthony. *Thinking about cities: New perspectives on urban problems.* Belmont, California: Dickenson, 1970.

Quade, Edward S. (Ed.) *Analysis for military decisions.* Chicago, Illinois: Rand McNally and Company, and Amsterdam, The Netherlands: North-Holland Publ. Co., 1964.

Quade, Edward S. & Boucher, Wayne I. *Systems analysis and policy planning: Applications in defense.* New York: American Elsevier, 1968.

Wolf, Charles, Jr. *Foreign aid: Theory and practice in Southeran Asia.* Princeton, New Jersey: Princeton University Press, 1960.